The Behavioral Ecology of the Komodo Monitor

Ora 19W, dominant male of Loho Liang, Komodo.

The Behavioral Ecology of the Komodo Monitor

Walter Auffenberg

A University of Florida Book
University Presses of Florida
Gainesville

University Presses of Florida is the central agency for scholarly publishing of the State of Florida's university system. Its offices are located at 15 NW 15th Street, Gainesville, FL 32603. Works published by University Presses of Florida are evaluated and selected for publication by a faculty editorial committee of any one of Florida's nine public universities: Florida A&M (Tallahassee), Florida Atlantic University (Boca Raton), Florida International University (Miami), Florida State University (Tallahassee), University of Central Florida (Orlando), University of Florida (Gainesville), University of North Florida (Jacksonville), University of South Florida (Tampa), University of West Florida (Pensacola).

Library of Congress Cataloging in Publication Data

Auffenberg, Walter.
The behavioral ecology of the Komodo monitor.

"A University of Florida book."
Bibliography: p.
Includes index.
1. Varanus komodoensis. 2. Reptiles—Indonesia—Komodo Island. I. Title.
QL666.L29A93 597.95 80–26683
ISBN 0–8130–0621–X

Printed in U.S.A.

Acknowledgments

The support that this study received during the course of two years cannot be adequately acknowledged in a few words. Almost the entire program was financed through the generosity of the New York Zoological Society, specifically the Charlotte Ordway Fonderas Wildlife Protection Fund. Dr. F. Wayne King, formerly curator of conservation of the society and currently director of the Florida State Museum, conceived the study, helped establish the base camp, and served in the important position of liaison officer between the field crews and the society. The University of Florida provided me with a Faculty Development Grant in 1969–70; the Florida State Museum granted me a year's leave of absence and the use of considerable equipment during the field work.

The completion of the study would have been impossible without the official assistance and cooperation of Messrs. Hasan Basjarudin Nasution, Directorate of Forestry, and Walman Sinaga, Department of Conservation, Bogor. Without their help it would have been impossible to study the monitor on any of the islands. Dr. Otto Soemarwoto, National Academy of Biology, Dr. S. Somadikarta, Museum Zoologicum Bogoriense, and especially Miss Sjamsiah Achmad, Indonesian Institute of Sciences, were instrumental in facilitating our research program. Brigadier Djendral Soekertijo, Bali, was particularly helpful in making all the officials in his district aware of our presence and of the purpose and importance of our study. Admiral Abdul Kadir, Jakarta, shipped food to Komodo via an Indonesian naval vessel when our usual supply boat could not serve us.

The late Mr. Hilmi Oesman, Surabaja Zoological Garden, was most helpful in providing much information based on his numerous trips to Komodo, as well as on his observations of lizards kept in the Surabaja Zoo. Mr. Benjamin Galstaun, Jakarta Zoological Garden, and Drs. Endung Dasuki and J. Busone, Faculty of Veterinary Medicine, both associated with the Jogjakarta Zoological and Botanical Gardens, provided pertinent data on Komodo lizards maintained in the Jogja and Jakarta zoos. Messrs. Boedhi Setiawan and Umar Shirmohammad, both from Surabaja, were exceedingly helpful in arranging our monthly supply shipments. Pamong Abdul Hamid (Kampung Djarek), Kepala Niso Habat (Kampung Nggoer), and particularly Kepala Mohammad Tiaep (Kampung Komodo) were extremely gracious and cooperative during our studies in their respective administrative districts.

The Catholic missions of western Mangarrai, Flores, deserve considerable thanks for their hospitality during our studies on Flores, especially Reverend Franz Schaarf, S.V.D., Procurator Missi Ruteng. The Reverend J. A. J. Verheijen, S.V.D., was helpful in matters regarding local dialects and word meanings. Reverend Erwin Schmutz, S.V.D., Missi Nunang, Flores, deserves special notice and thanks. Without his assistance, advice, and data on the oras of his parish, our 1971 studies on Flores would certainly have been much less successful than they were.

Dr. Harry Hoogstraal and the staff of the U.S. Naval Medical Research Unit No. 3, Cairo, Egypt, identified tick samples obtained from the lizards and their prey species. Dr. Robert Woodruff, Division of Plant Industries, Florida Department of Agriculture, Gainesville, identified beetles associated with carrion on Komodo; Dr. Fred G. Thompson, Florida State Museum, identified mollusks eaten by oras; Dr. Sam R. Telford, currently with the Division of Malaria and Other Parasitic Diseases, World Health Organization (formerly of FSM), studied all the parasitological material; and Dr. S. David Webb, Florida State Museum, identified some of the mammalian remains.

The personnel associated with several zoological parks deserve considerable thanks for their attention to my many inquiries: Marvin L. Jones, San Diego, California, for his encouragement and particularly for the use of his voluminous notes on the individual histories of Komodo monitors exhibited in the zoos of the world; Dr. E. M. Lang, Zoologischer Garten (Basel); Mrs. J. Schultz, J. H. Staedeli, and Dr. J. Bacon, San Diego Zoological Garden; E. Imaizumi, Ueno Zoological Gardens (Tokyo); and particularly T. H. Reed and Mrs. Jaren Horsely, National Zoological Park (Washington, D.C.), for their continued interest and cooperation in many different ways. Mr. J. DePrato's notes on captive oras in the same zoo were exceedingly interesting and informative.

The staff of the Florida State Museum deserves thanks for much assistance: the illustrations I have not done myself are the work of either Mary Glen Rogers, Barbara Irvine, or Margaret "Peg" Estey; Howard Converse was an able field assistant in 1972; Sylvia Scudder, Richard Franz, and the late Ruby Kuykendall were extremely helpful in the innumerable library problems associated with the project. Portions of the manuscript were criticized by my colleagues, J. C. Dickinson, Jr., S. R. Humphrey, J. B. Iverson, and S. D. Webb, whom I thank for calling attention to at least some of my errors. R. Zweifel (American Museum of Natural History, N.Y., N.Y.), H. Marx (Field Museum of Natural History,

Chicago, Ill.), and G. Zug and the late J. Peters (National Museum of Natural History, Washington, D.C.) graciously made preserved material available for study. I wish particularly to acknowledge the work of Rhoda J. Bryant. She has not only retyped the manuscript several times, but has been exceedingly helpful in an editorial capacity, correcting my inconsistencies, questioning my wording, and advising me in countless circumstances. She deserves much credit.

As a graduate student in 1969–70, Ir. Putra Sastrawan spent the entire year with us on Komodo studying the diel activity cycle of the monitors and was responsible for collecting much of the climatic data. In 1971 he was very helpful in the Flores study, and in 1972 he returned to Komodo with me to complete the work. His dedication and thoroughness during all three expeditions are greatly appreciated. Hendryce P. Matur, a student from Universitas 17 Augustus 1945, Java, obtained much of the data on burrow ecology.

These acknowledgments would be incomplete without mention of my wife, Elinor, and my son, Kurt. Without Elinor's constant encouragement and her activities as cook, nurse, and companion, the field work would probably not have been concluded. She cheerfully joined into all aspects of the project and remained steadfast during the long months of isolation on an island where conditions were often far from ideal. Kurt contributed to the study in many ways, but I especially thank him for his help as an eager and dependable field assistant, as well as an effective liaison in our dealings with the local villagers.

Contents

1

Introduction

This publication reports the findings of a 13-month field study of the ecology and behavior of *Varanus komodoensis*, commonly known as the Komodo dragon or "ora." (Ora is the proper local name in the Mangarrai dialect. See Appendix 2 for other regional names for this species and for the water monitor, *V. salvator.*) Work was conducted on the islands of Komodo and Padar (July 1969–June 1970; November 1972) and Flores (June–July 1971), all located in the Lesser Sunda Islands group, Republic of Indonesia (Figs. 1-1, 1-2). If this study is successful it should contain considerable basis for thought, whether the reader has a purely scholarly interest in the Komodo monitor as one of the world's largest living reptiles, an applied interest in its maintenance, conservation, and/ or management, or is mainly concerned with the local ecology and the place of the Komodo monitor in it.

In some cases circumstances have made possible the accumulation of more data than could reasonably be published—notably in the fields of geology and ecology. More details than usual are included because such information is unavailable elsewhere and they are basic to a thorough understanding of the ecological factors to which the Komodo monitor has become adjusted. Such information is also required in the more practical aspects of conservation, management, and maintenance in zoological gardens. It is hoped this volume will provide those data needed by future workers faced with decisions regarding the ora. Some data, such as those relating to the physical microecology of burrows, that can be extrapolated from general ecological studies are nevertheless included here because of their practical implications in zoo maintenance programs.

No extensive field studies have been published on any of the highly predatory species of the family Varanidae, though a number of short contributions report relevant information for certain members of the genus *Varanus*. The most important contributions regarding the behavior of *Varanus komodoensis* in captivity

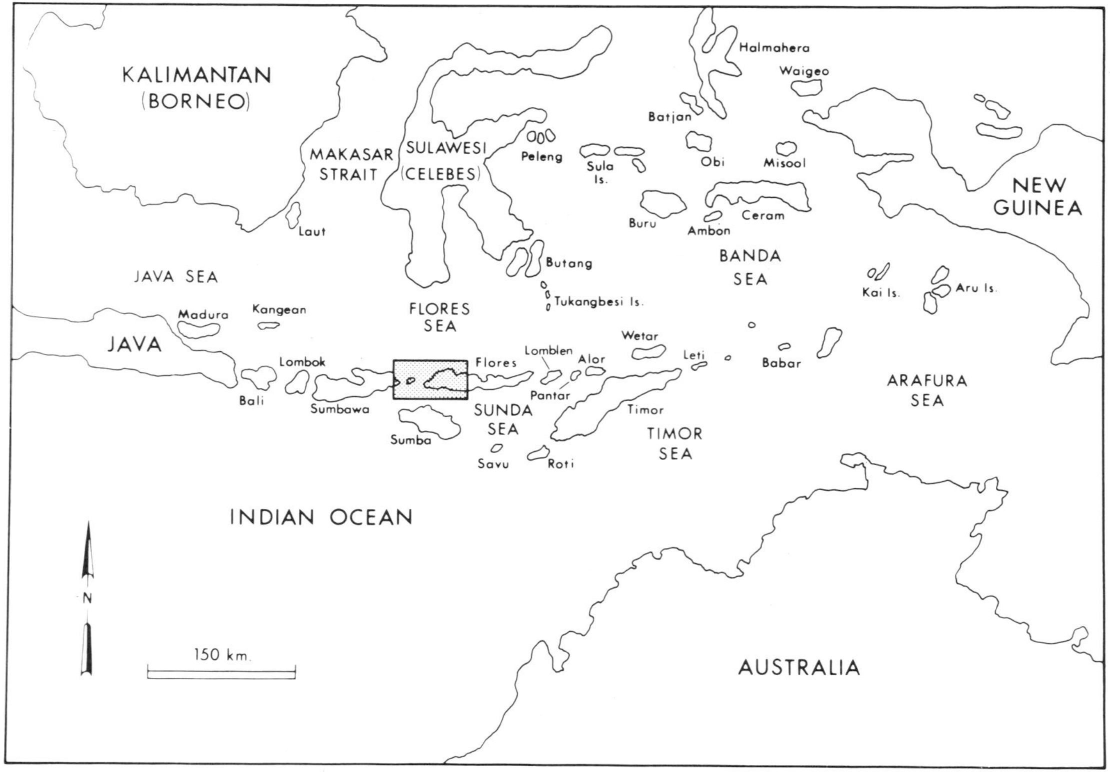

Fig. 1–1. General area of Komodo (located in approximate center of rectangle).

are found in Lederer (1942). The observations of Burden (1928), de Jong (1929, 1937), and Hoogerwerf (1954a) on the behavior of free-ranging individuals and the ecology of the area in which they are found are particularly important for their thoroughness. Appendix 1 lists pertinent information on all major expeditions to study, capture, or photograph the Komodo monitor through 1975.

While some sections of this work are well supplied with pertinent or related information, the lack of certain important data is most distressing. In addition to the paucity of physiological information, data pertaining to demographic and reproductive factors are scanty, largely due to the generally low population densities, the large activity ranges of individual animals, and the visual obscurity provided them by the vegetation at the level in which these monitors spend most of their time. Laboratory work on this thoroughly protected species is nearly impossible due to present restrictive, albeit necessary, legislative, ethical, and practical policies. Therefore, the proofs required to support some of my hypotheses must be sought in future work. However, I must caution against too much emphasis on means, medians, and modes in certain behavioral studies. Some behavioral patterns are biologically important regardless of how few times they occur. That a monitor lizard is capable of occasionally killing an animal as large as an adult water buffalo or that it may attack humans is highly significant, even if it occurs only rarely. That such behaviors are reasonably common in only certain individual Komodo monitors becomes even more important and illuminating. These behaviors not only help define the limits of the behavioral repertory but represent a potentially important part of the raw material upon which selection and the ultimate evolution of behavior patterns may depend.

A perusal of the recent literature in behavioral herpetology suggests that our present notions of what a lizard is and what it does are based mainly on information on two closely related families (Agamidae and Iguanidae). The tacit assumption by some workers is that the behavioral patterns of other lizard groups are probably very similar. Recent studies of turtles and crocodilians suggest that reptilian behavioral patterns are much more variable than previously believed; this work suggests the same for lizards. While the ora possesses a significant number of traits common to all terrestrial reptiles, it also possesses many that are unique. Although ora behavior does not represent a gross departure from presently recognized behavioral patterns in lizards, it suggests that monitors in general possess a life style different in certain important essentials from the Agamidae and Iguanidae.

Study Areas

Ora populations were studied in two major areas, western Flores and Komodo.

Flores.—Primary emphasis was placed on the study of the monitors in a 38 km^2 area in extreme western Flores, approximately 24 km south-southeast of

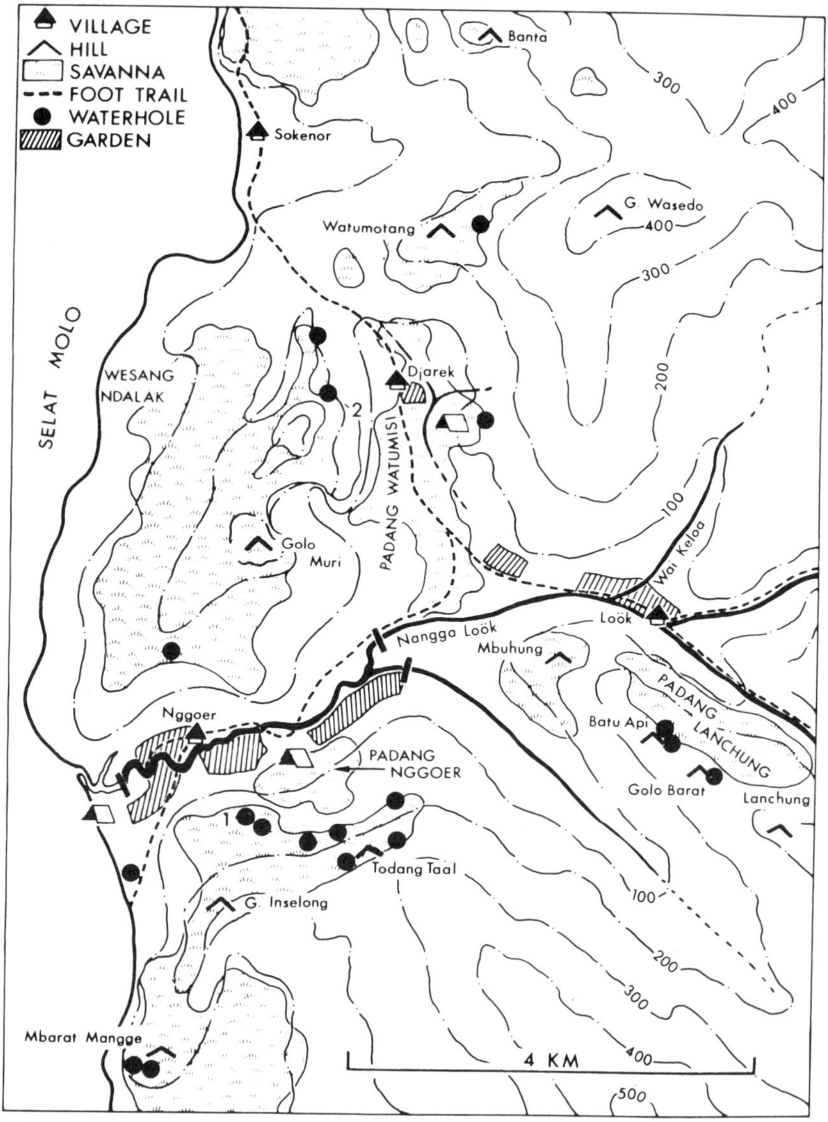

Fig. 1–2. Study area in Mangarrai District, extreme western Flores. Tents show camp sites during the study. Numbers 1, 2 show location of *Varanus komodoensis* nest sites.

Labuanbadjo, including the villages of Djarek, Nggoer, and Loök (misspelled as Looh on many maps) in the valley formed by the only permanent stream in entire region—Nanga Loök (Fig. 1-2). The valley is about 1 km wide, with rather steep walls 200–400 m high. In addition to the river gallery forests, the study area included open beach, several savanna associations, mangrove forests, steppe-covered hills, monsoon forests, and extensive bamboo thickets, as well as both cultivated and abandoned gardens.

Komodo.—Though trips were made to many parts of the island of Komodo, and to nearby islands on which the monitor is found, the area in which the major part of the work was concentrated is located on the northern part of Telok Slawi, a bay on the southeast coast of Komodo (Figs. 1-3, 1-4). The large valley containing intermittent streams emptying into this part of the bay is called Loho Liang. It was in the northeast part of this valley that the major part of the work was conducted (Fig. 1-5).

The area of concentration was chosen for its habitat diversity and comprised approximately 8 km^2. Included were several intermittent streambeds (Fig. 1-6), large patches of open grassland, two different savanna associations, and a fairly large mountain separating Loho Liang from a smaller, drier valley to the east, Loho Kalo (Fig. 1-7). At the western end near the coast it included a small mangrove swamp. Inland it extended nearly to the highest ridge of the island. Also included was an old overgrown garden (*ladang*) and several natural savannas (*padang*).

Padar.—Two short trips were made from Komodo to the island of Padar (Fig. 4-2). Most work was concentrated on the west side, though almost the entire island was searched for droppings.

Methods

Traps.—The lizards in the Komodo study area were captured by means of three types of traps, all based on a funnel design. The largest and most elaborate was constructed of saplings cut near the site, driven vertically into the ground, and lashed with lontar palm cordage and rope (Fig. 1-8). By trial and error it was found that to be most effective the fence should lean inward slightly, be at least 1.5 m high, and have spaces between the poles of less than 5 cm. The stockade was roughly heart-shaped, approximately 10 m wide. At the lower apex a small box with a drop door was built into the wall. A ramp of strong planks was constructed at the wide end. Immediately inside of the ramp was a large tripod, from which bait was periodically suspended—usually a dead domestic goat, wild hog, or feral dog, sometimes fish heads and entrails. Two portable funnel traps, baited with rotten flesh, were used for small specimens. The funnel design was used on all the traps, because it had the potential of capturing many animals without being reset. If already marked, the monitor was released by means of the

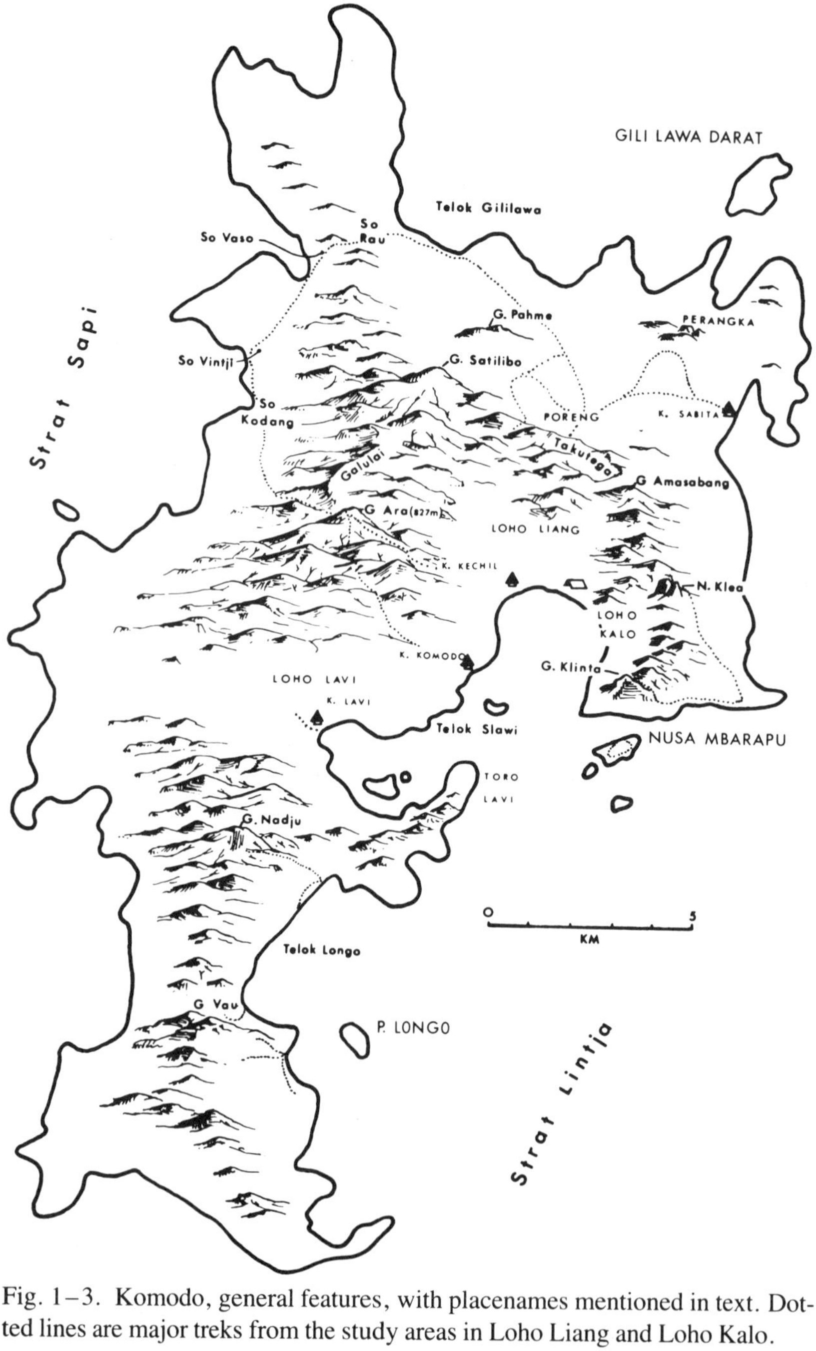

Fig. 1–3. Komodo, general features, with placenames mentioned in text. Dotted lines are major treks from the study areas in Loho Liang and Loho Kalo.

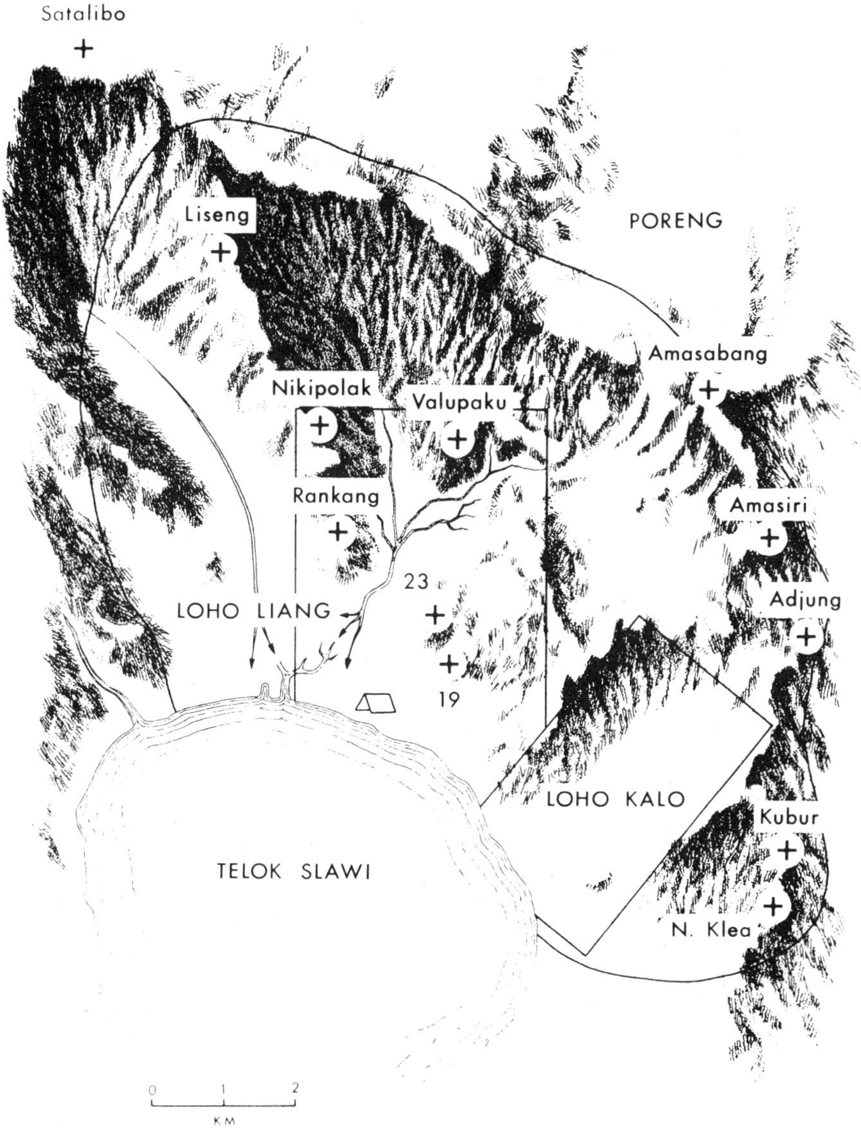

Fig. 1–4. Major study area on Komodo enclosed by boundary line. Rectangles relate to figs. 1–5 and 1–6. Crosses represent mountain peaks. Numbers 19 and 23 are two small hills frequently mentioned in the text. The tent shows the location of the base camp.

box door. If unmarked it was restrained, weighed, measured, tagged; the number of ectoparasites were counted; and scars, sex, and other characteristics were noted. Occasionally other features such as skin, buccal, and cloacal temperatures were recorded.

Though as many as five individuals were found in the trap at one time, one or

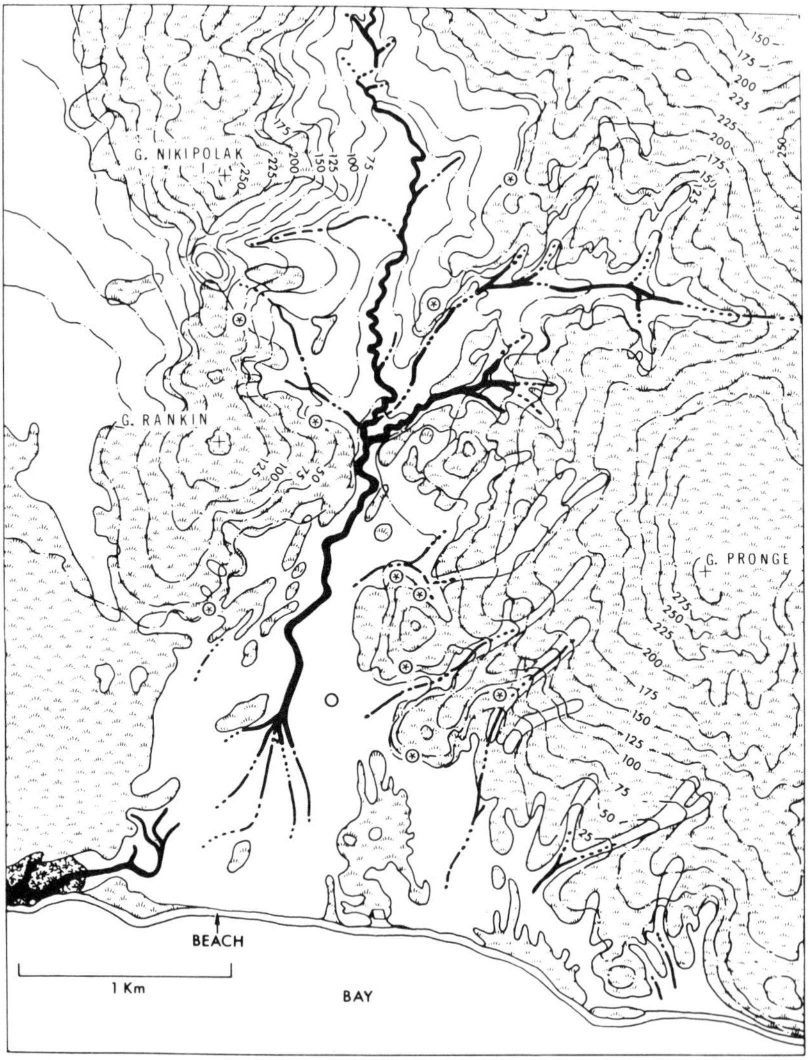

Fig. 1–5. Loho Liang, Komodo, study area. Contours in meters. Stars in circles are traditional resting sites (see text). Land areas not shown as savanna are largely monsoon forest.

two were most common. Lizards smaller than 1.5 m were often simply grasped by the tail and neck and quickly lifted off the ground. Covering the eyes quieted them before tying or taping the legs. The mouth was simply held shut. Larger individuals were first lassoed in the pen, often with two ropes pulled in opposite directions. It required three men to manipulate lizards 1.5 m or larger. When the

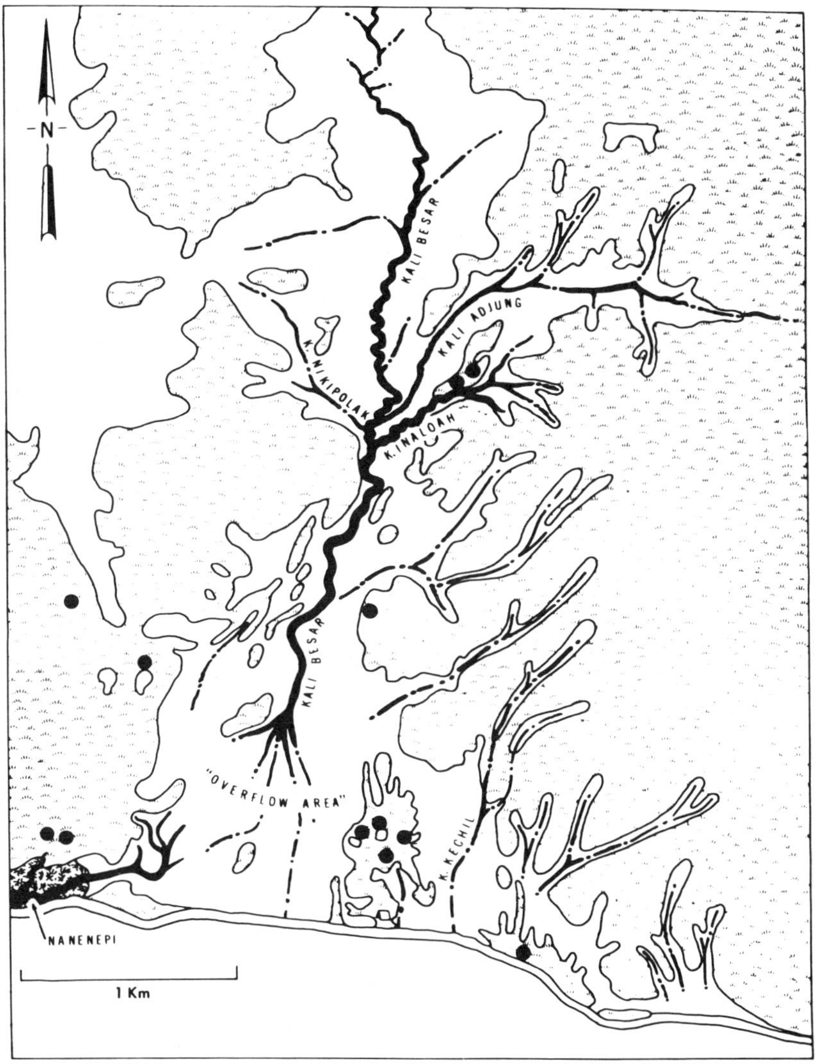

Fig. 1–6. Loho Liang, Komodo, study area, showing streams (all intermittent), temporary water holes (solid circles), and the mangrove swamp called Nanenepi.

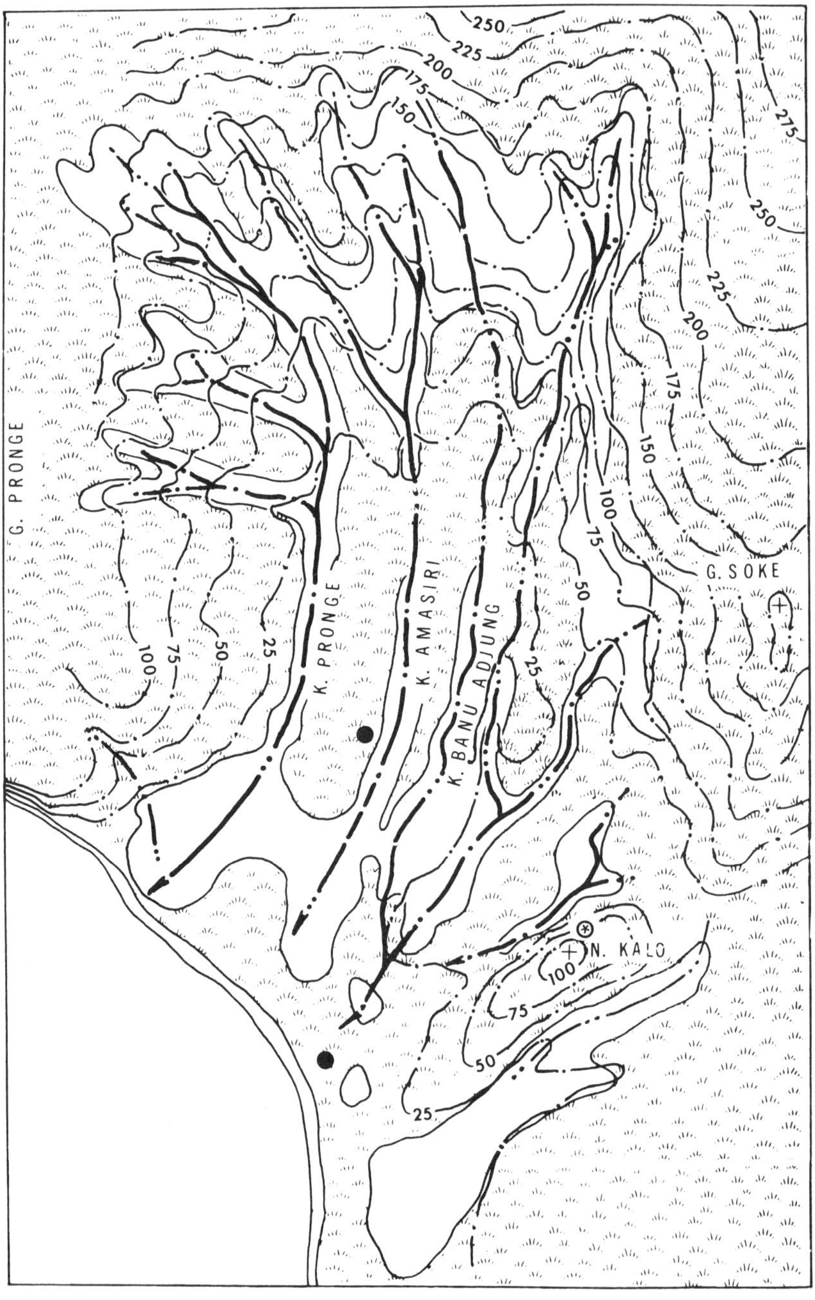

Fig. 1–7. Loho Kalo, Komodo, study area. Symbols as in figs. 1–5 and 1–6.

lassoes were passed over the neck, pulled taut, and tied, the body was lifted off the ground by the tail. One person clamped the mouth shut and the eyes were covered by another's hands; the third worker tied the front and hind legs and fashioned a rope muzzle for the snout and a body harness for suspending the animal. Figure 1-9 shows the locations of major trapping areas in Loho Liang.

Fig. 1–8. Activity pen (*above*) and trap style (*below*) used in Loho Liang study area.

Tags.—Weighing, measuring, and tagging were usually carried out immediately next to the trap. Each monitor was marked individually by means of a two-piece cattle ear tag (Nasco Roto-tag) driven through a hole punched in the lateral body fold with a large awl. Each tag had a different number. All tags were of white plastic; replacements were yellow. Tag loss during the first year was 4

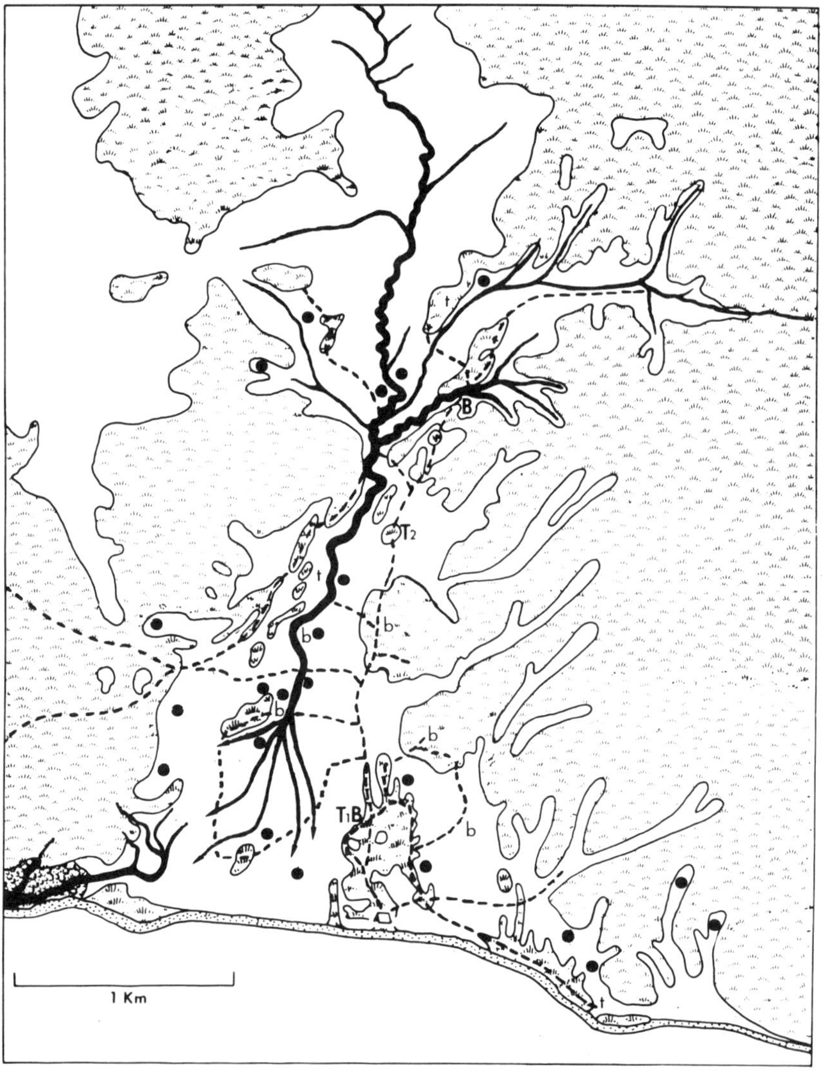

Fig. 1–9. Loho Liang study area, showing major (T or B) and minor (t or b) trapping and baiting areas. Dashed lines are trails cut and used during the study. Solid circles show location of megapode bird nests (see text).

percent, usually because it had been placed too near the outer edge of the skin fold. In addition, the tag number of each monitor was marked in large numerals with white paint (thinned, outdoor, synthetic enamel was the longest lasting) on the side of the body, allowing for distant identification. Individuals were repainted if they had shed their skin or the paint had worn off (which was rare).

Individual characters.—Total weight was obtained with a large beam balance suspended from a nearby tree. Several dimensions were recorded to the nearest millimeter. These and the several scale characters tabulated are provided in Auffenberg (1980).

Observation blinds, platforms, and towers.—Three major types of blinds were used from which to observe the free-ranging lizards: a platform type approximately 6 m high in a tree at the base of hill 19, an observation tower about 2.5 m high near the base camp, and two different surface blinds—permanent and portable. The permanent blind (or "observation blind") (Fig. 1-10A) was constructed of an A-frame covered on one broad side with coarsely woven bamboo badek. The ground was covered with palm matting. The ends of the A-frame were normally left open, though matting was sometimes draped on them when particularly nervous or aggressive lizards were in the immediate area. Holes large enough to receive the lenses of our cameras, and through which we could observe the lizards, were provided in the badek surface. Bait, usually goat, was suspended from a large, trimmed *Tamarindus* tree in front of the blind, and the latter could be moved closer or farther away from the tree as desired. A rope supporting the goat was passed through a small pulley hung in the tree and then drawn to the observation blind; thus the bait could be raised or lowered from within the blind. On one side of the bait site small inconspicuous sticks were driven into the ground at 1 m intervals. In addition, a board 2 m long, marked in distinct blue and white 1-ft bands for use in estimating lizard size, was placed horizontally behind the bait so that it could be easily seen from the blind. Unmarked individuals were sprayed with white paint in distinctive patterns and then sketched, so that they were later recognizable. The paint was applied by means of a trombone-type insecticide sprayer in which the nozzle had been modified to release a fine stream.

The portable blind (Fig. 1-10D) was constructed of materials that were easily carried. A framework of nonflexible plastic 32 mm tubing fashioned into a light A-frame could be quickly erected and held in place by several nylon cords. The cover of coarsely woven nylon cloth was provided with several holes for observation and photography. This blind was most often used in four established baiting sites some distance from the base camp and near the edges of the study area: Vai Nggulung, Loho Kalo, Kali Inaloah, and Nanenepi. From behind the blind we were able to observe and photograph oras feeding and interacting. Again, lizards were marked by paint spraying as described above, except that a different color was used at each site.

The viewing angle in the tree platform blind (Fig. 1-10B) at the base of hill 19 was not satisfactory for detailed observation; furthermore, the blind was uncomfortable to be in for long periods of time.

A bamboo observation tower was constructed along one of the frequently used ora trails (Fig. 1-10 C). Though larger and more comfortable than the tree platform, the tower was infrequently used because its viewing angle made it unsuitable for many behavioral studies.

Tracking.—Two primary tracking techniques were used, spoor and radio telemetry. Spoor tracking in 1969–70 was facilitated by the fact that several months prior to our arrival the eruption of Gunung Io (160 km E, on Flores) generated a heavy ash cloud that drifted over Komodo. Tons of ash were deposited in a layer 1–6 cm deep over the entire island. Its texture was so fine that highly accurate impressions of scalation, including the subdigital series, could

Fig. 1–10. Types of blinds used during the study.

be seen. Footprint lengths could be determined as accurately as if the living animal was being measured. Furthermore, details of some movements associated with locomotion, such as lateral undulation of the tail tip as it was dragged along the surface, were easily observed and measured. While all but the smallest oras could be trailed, only the spoor of the largest size class could be assigned to individual specimens—always by a combination of hind foot lengths and individual peculiarities of tail tip movement.

Similar tracking was also possible in the sand of the creek beds and on the beach, as well as on some of the friable loamy soils. However, critical individual identification was less often possible. Smaller individuals could be similarly trailed but were rarely identifiable if not seen making the trail. We obtained 618 tracking segments by this means.

An additional series was gathered by tracking specifically identified individuals for one or more days. Two investigators usually worked cooperatively on the tracking segment, keeping in touch with one another by means of 27 Mhz transceivers. Spoor were picked up and the individual identified and followed until nightfall, when its evening shelter was noted. Telemetric tracking was resumed before sunrise the next day and continued until the trail was eventually lost. Tracking sequences lasted from 1 to 6 days and our best results were with the larger monitors. We obtained our most reliable data with this technique, because our close pursuit not only detailed where the individuals went but also often indicated what they had been doing along most of the route. A total of 21 daily segments was obtained by this means.

Radio telemetry was also successfully employed to track oras and obtain data on deep body temperatures. Appropriately modified 3-channel 27 Mhz transceivers were used in conjunction with hip-harness mounted transmitters. The circuit was based on that used by Stebbins and Barwick (1968) on free-ranging *V. varius*, but it was modified so that tone and oscillation frequency were adjustable. This allowed placing several transmitters on one frequency, so that a total of nine parameters could be tracked on the 3-channel receivers. The circuit and harness arrangement is shown in Appendix 3. Three adults were each monitored for 10 successive days, providing a reasonably complete picture of general movement within the foraging area.

Weather and climate.—Wind direction and intensity were measured in the *Zizyphus* savanna immediately behind the beach. In the shade nearby we obtained all data on barometric pressure, soil temperature 8 cm deep, surface soil temperature, air temperatures 10 cm and 1 m above the surface, and relative humidity. Nearby, in full sun we obtained data on precipitation (nonstandard gauge) and on temperatures 8 cm below the surface, at the surface, 10 cm above the surface, 1 m above the surface, and black bulb 1 m above the surface. All readings were taken with an Atkins telethermometer and matched appropriate probes. Relative humidity was measured with an Atkins pistol psychrometer probe and checked occasionally for continuing accuracy with a Barach sling psychrometer. All remaining macro- and microenvironmental parameters were measured with the same equipment.

Tree ring analyses were also conducted on *Tamarindus indicus*, a species in which a "sensitive" ring series (Fritts, 1965; Stokes and Smiley, 1968) is found in appropriate habitats. No core borer was available, so data were obtained from cut stumps of known age, smoothed off to allow transfer of the ring series to paper strips. Stumps in dense stands were avoided because competition among closely growing trees may modify or change otherwise normal ring patterns (Stokes and Smiley, 1968).

Vegetation.—Vegetative cover diminishes light intensity or obstructs vision, both of which are important to the ora in prey detection and to the prey in successful avoidance. For purposes of this study, it was assumed that both the light intensity beneath the cover being measured and the obstruction of vision are functions of cover (de Vos and Mosby, 1963). Two methods were used to evaluate cover density: measuring vision obstruction caused by plant cover at the predator's level by the method Wight (1938) proposed, and measuring, by photoelectric devices (see Allard, 1947), the influence of cover on light exclusion at the predator's level. The latter was found to be of little or no significance in the behavioral ecology of the ora.

Though relationships among dominant plants and animals in plant communities have received considerable attention in many parts of the world, this is not true in Indonesia. In fact, few plant communities have even been defined, much less studied intensively. Therefore, one of the primary goals of the work was to characterize the plant communities of which the ora is an integral part. Plant identifications were provided by the Bogor Herbarium staff. In the field our investigation emphasized density of the dominant types (5 percent or more in each sample) per unit area and importance of specific plant types as either shelter or food to the ora or its prey species.

Recent research has shown that the shape of the sample plot is of considerable importance (Oosting, 1958). A circular plot, rather than a rectangular one, was used for studying all but the trees. Included areas were 0.01 acres for grass plots and 0.25 acres for shrub plots. A belt transect system of 1.25 acres for forest plots was used.

Trail system.—We cut and maintained approximately 13 km of trails in the Komodo study area (Fig. 1–9). These were used to study monitor and prey species movements and for the construction of transects useful in determining both plant and animal densities. The trail system included as many major plant associations as was practical.

A second, special trail system, used as a disturbance strip, was constructed along the periphery of the savanna located at the base camp. A narrow strip approximately 1 m wide by 1000 m long was hoed and cleaned of all surface vegetation and debris and then covered with beach sand. All animal tracks were periodically tabulated and then swept from the trail, providing information on the number and kinds of animals moving through the savanna on a diel and seasonal basis.

Mapping.—The entire study region was mapped during the early part of our work on Komodo, with particular emphasis placed on the area within 2 km of the base camp. General island characteristics were taken from available Dutch, Indonesian, and United States sources, particularly recent satellite photos (Loc. Nos. 121,066 and 122,066).

In almost all instances the common local usage has been followed for names and spellings. This may differ from island to island because of dialect and language variations or because of usage changes with a political basis (e.g., Dutch to Indonesian, Mangarrai or Bima languages to Indonesian, etc.). A glossary of placenames is provided in Appendix 10 for convenience in defining terms and locating places.

Baiting.—The locations of the five primary baiting stations are shown in Figure 1–11; each one was selected to answer basic questions about monitor movement and differential density. The baiting station at Nanenepi was located on the edge of a mangrove habitat; three stations (Nggulung, Inaloah, and Base Camp) were on or near a savanna–monsoon forest ecotone, and the remaining station (Kalo) was in an open savanna. In addition, Kalo was separated from the others by a mountain mass extending from near the middle of the island to the edge of the sea. Nanenepi and Vai Nggulung were located along the same drainage system, and the Base Camp and Inaloah stations were along another parallel system.

A blind was constructed at each site, where carrion was used in a regular, alternating pattern. Hung high in a tree, the bait was allowed to putrify for one day before any observations were begun on the monitors. Notes on behavior and aggregation were made continuously during the next 3 days from 0800 to 1700 hr (local time), after which the site was temporarily abandoned and a new carrion bait taken to one of the other stations. The Base Camp site was thus used for 123 days and Kalo, Inaloah, Nanenepi, and Nggulung for 24 days each. Baits were usually goats, but feral dogs, wild pigs, deer, and waste meat (entrails, fish heads, chicken feet and heads) from our camp and the village were also used.

Hole utilization.—Certain burrows in the study area were periodically checked throughout the year during our daily survey of the trail system. Burrow use was often easily established on the basis of tracks or by the simple expedient of placing small twigs and sticks in the opening.

Activity pen.—Immediately adjacent to the Base Camp and within the savanna, we constructed an activity pen approximately 15 m long and 8 m wide. The sides were comprised of corrugated metal buried in the ground 0.5 m and extended to a height of 1.5 m above the surface. A tower approximately 4 m high was constructed adjacent to the pen in such a way that an observer in the tower could see the entire pen area. Telethermometer probes were put in a number of places throughout the enclosure. This arrangement was used to study the diel activity patterns of oras, especially in reference to temperature changes during the day and night (Fig. 1–8).

Unsuccessful techniques.—Color-coded, self-affixing collars (see Romanov, 1956, and Taber and Cowan, 1963) were set at five major game trails near the

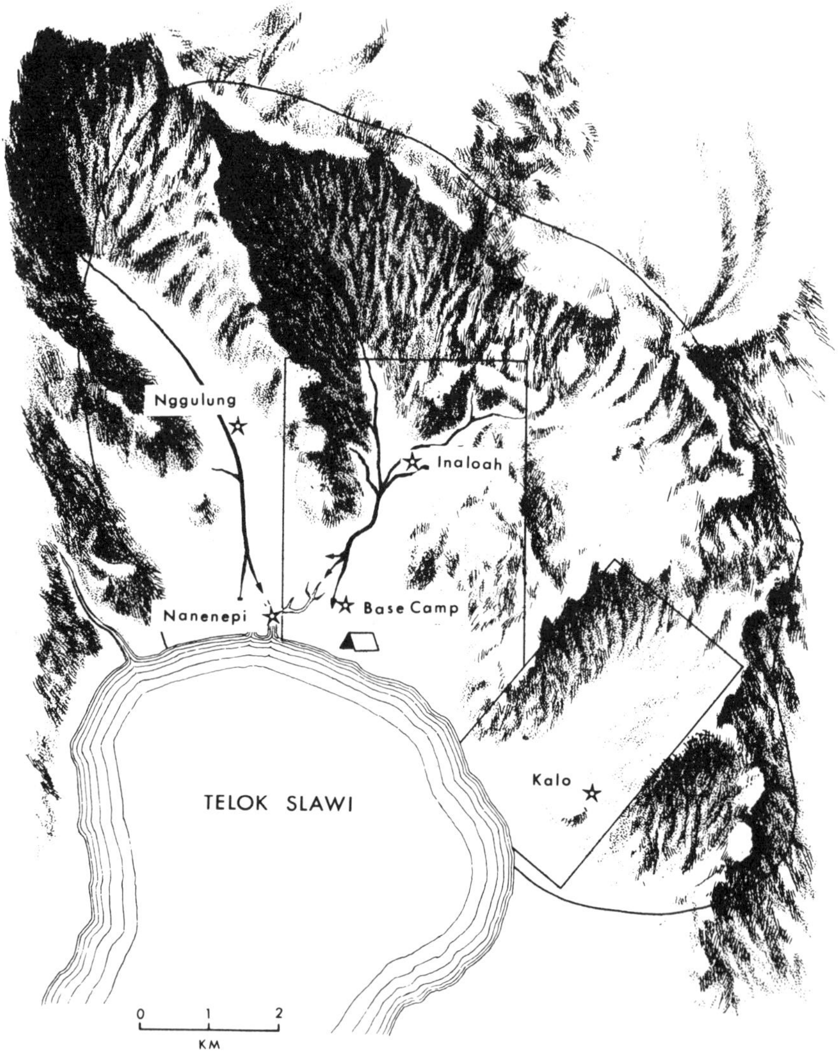

Fig. 1–11. Location of primary baiting stations established in the Komodo study area.

base of hill 19, along which oras were known to travel. These were intended as adjuncts to movement data obtained by other means. However, since the snares had to be set at a center height of approximately 15 cm, hogs and deer knocked them down regularly. The technique was abandoned after a 2-week trial.

Partly in an effort to save bait and partly to test the effectiveness of certain olfacients, various parts of macerating animals (including oras) were offered in 5 gal. tins hung from trees in six different trials. Only the tins containing macerating stomach contents of wild pigs attracted any oras, so the tests were stopped.

We also studied the effect of commercial animal scents on ora movements. Both North American doe and buck musk (National Scent Co.) were liberally sprinkled on a "scent bag" and dragged in patterns that would have altered the normal diel movements of the large oras had they followed the trails. We found no evidence of the scents being attractants, and they may have even had an opposite effect, perhaps because the scents were from a nonnative species of deer. Broughton (1936) stated that linseed oil was used to attract lizards to traps. Unfortunately, we did not check this in the field. The entire matter of scent attractants warrants further investigation, for despite these negative results I am convinced that animal scents will be found useful in studying oras.

Several types of predator calls were utilized under conditions similar to those in which they might be useful for attracting mammalian predators. The calls were Herter's Professional Predator Call, stages one (Rabbit Alarm Call) and two (Rodent Alarm Call). As far as is known, oras did not respond to the calls, though feral dogs came on several occasions. On September 3, 1969, both types of game calls were used within sight of a large hunting ora without any apparent effect (however, see the section on Hearing, Chap. 6).

Morphology

Size and Mass

Length.—Much of the notoriety attached to the Komodo monitor is related to the fact that it is the world's largest living lizard. As might be expected, this circumstance has led to confusion, exaggeration, and misinformation. The longest specimen recorded is a male, reputed to be 10 ft 2½ in. (3.13 m) from snout to tail tip.[1] It was maintained in the St. Louis Zoo for only a short time before it died. To my knowledge, no part of the specimen is available for verification. A specimen confiscated from an animal dealer and purported to have been 3.10 m in total length is now represented by only a skull in the Jakarta Zoo. However, a comparison of its skull measurements with those of specimens of known length prove that total length was only about 2.8 m.

When discussing animal attributes with nontrained observers, it is often difficult to separate fact from fancy. Local tales are often based on myth as much as on fact. However, most of the people in the Sunda Islands are remarkably observant and honest. Piazzini (1960) was incorrect when he referred to "wildly exaggerated dimensions claimed for their dragons by the terrified fisherfolk, who babbled of monsters 30 ft long and more than a ton of weight." Collins (1956) and Hoogerwerf (1954a) did not hear such statements nor did we. Villagers are quite explicit regarding the length of large oras. Collins stated, "When I put down two sticks 23 ft apart and said that some people claimed to have seen lizards this long, they (the villagers) laughed. They put the sticks about 10 ft apart and said, 'that's as long as they grow.' " Almost all the people we asked also agreed that the longest oras they have seen were approximately 3 m. Pfeffer

Note: Only those morphologic parameters of possible significance in the behavior of this species are discussed.

1. Heuvelmans (1965) stated, "...there are persistent rumors from New Guinea of a living monster 20 ft long, which is by no means improbable." He must be referring to *Varanus salvadori,* known to attain a length of approximately 3 m. A probably exaggerated newspaper article (1974) stated that a specimen of this species 14 ft in length had been killed in West Irian, New Guinea.

(1959) reported a similar experience. On Flores villagers pointed out that while occasional specimens reached lengths of 3 m several decades ago, there were none that length in western Flores at present. In only one instance did we hear of a larger specimen. In Lenteng, Flores, villagers told us that a particularly large ora inhabits the area near Tandjung Tongkir, on the southwest coast of Rintja. They stated it was impossible to dry fish anywhere near the point without an ora, approximately 3.5 m in length (11.5 ft), making its appearance. In my opinion native Sundans do not exaggerate ora sizes as often as do Europeans and Americans—particularly through newspaper sensationalism.

When Governor van Steyn van Hensbroek landed on Komodo in 1910 to obtain the first specimens ever sent to a biologist, he was immediately confronted with exaggerations by the Dutch pearl fishermen Aldegon and Kock. They related tales of fantastic encounters and asserted that oras were malignant and attained maximum lengths of between 5 and 7 m (16–23 ft). Van Hensbroek repeated the size maxima, but he never obtained specimens longer than approximately 3 m. Later a diver who had worked with Aldegon and Kock told Collins (1956) that their party never really saw any 7-m lizards. De Rooij (1915), following van Hensbroek's original remarks, perpetuated the misinformation by stating that Komodo monitors attain a length of 7 m. However, Staring (1926) reported that a Sgt. Becker (who was evidently with the Dutch Colonial Forces and had apparently caught many Komodo monitors) had never shot or seen one over 3.60 m. Von Mecklenburg, the explorer-naturalist, was also said to have shot a specimen approximately 3.60 m long, though Sachs (1927) reported that the skin, given to the Berlin Zoological Museum, was only 2.40 m long. Sachs was incorrect when he stated that the largest specimen taken by the Burden expedition was 3.65 m; the longest obtained by that expedition was 2.76 m (9 ft 2 in.) (Burden, 1928).

In 1934 the Surabaja Zoo expedition, represented by Schoenmakers and Hompes, in cooperation with the American archaeologist L. Griswold and ethnologist William Harkness, traveled to Komodo to study the island and its fauna. Conflicting stories regarding expedition details appeared in several newspaper articles. Griswold is reported to have stated he saw oras 5 m in length. Other articles, published at approximately the same time, stated that a large, 5-m ora was to be shipped to the Artis Zoo, and in passing reported that von Mecklenburg once shot a specimen 6 m long! In 1936, Lady Broughton said, "not only did we see small ones, but also the 12 ft giants." In other places she stated, "I spent days watching at close range dragons of all sizes up to about 12 ft in length," and "they sometimes weigh 250 lb and attain a length of 12 ft." It is this report that is often used as the authority for similar statements in encyclopedias both here and abroad.

The Swedish film producer P. Fejos, after a visit to the west slope of Komodo in 1937, wrote to the Swedish consul in Jakarta (9 July 1937), "I have measured the distance between the two points on the grass, against which I have seen the animal, and found it fully 15 ft. I can vouch for the accuracy of this statement.... The following day the same, and another even larger beast was seen . . . 20–21 ft long and was shot" (though no specimens were ever produced). Hoogerwerf (1954a) and Pfeffer (1959) misinterpreted Collins' (1956) sarcasm when he referred to the exaggerations of these earlier expeditions. Heuvelmans (1958) referred to an unsubstantiated report by an airman who stated Komodo oras attain lengths of at least 12 ft. Wendt (1959) extended to 13 ft the length of the dragons reported by a presumed downed airman. Pfeffer (1965) stated that R. P. Jacobus Gerrades, a missionary who lived on Flores for approximately 18 years, said that he knew of an approximately 4.50-m ora (about 14.8 ft) from that island.

Even within the last decade several authors have referred to exceptionally large specimens, undoubtedly exaggerations. Thus, Attenborough (1957) stated, "he was enormous. From the tip of his narrow head to the end of his long-keeled tail he measured a full 12 ft" (3.65 m). Pfeffer (1965) correctly pointed out that the photograph of this "monster," published by Attenborough, is not of an extremely large specimen and the "cave" from which it was said to be caught is not large enough for a 12 ft specimen.

It is, of course, possible (even probable) that in decades past oras may have attained greater lengths than they do now. However, it is probably significant that the relatively few expeditions that have studied the ora extensively in its natural habitat and that were accompanied by trained biologists have never seen specimens over 3 m (about 9 ft 10 in.) long. Almost every report of exaggerated length was made by an untrained individual. It is, of course, conceivable that in remote areas, particularly on Rintja and the east slope of Komodo, specimens over 3 m might still be found. Oesman (1970) was correct when he stated, "it is difficult to judge accurately the size of a Komodo lizard in the field, in trying to do so myself I was mostly proved wrong as we subsequently caught the animal. That a large lizard appears larger, and a small one smaller, than it actually is, probably is caused by the fact that the large ones (over 5 ft long) are plump-bodied, while the small ones are slender." Piazzini (1960) made a similar observation when he referred to statements by two other members of the expedition who had just seen a large monitor feeding on a carcass: one person said the monitor was at least 12 ft from head to tail; another increased this estimate to 14 ft. However, by means of picks driven in the ground they were able to take more accurate measurements of the individual when it appeared again. Piazzini stated, "in this case we had been deceived by the creature's air of power and strength; he was actually not much over 10 ft long." I believe that though the length correction was moving in the right direction, Piazzini's final estimate was still too long. In fact, Pfeffer (1959), who accompanied Piazzini on the same expedition, stated that they saw no oras longer than 2.75 m (about 9 ft). Pfeffer believed that estimates of the total size were extremely difficult in nature, largely because of the emotional circumstances associated with seeing one of these animals in the wild.

The largest specimen presently available is located in the Senckenberg Museum (Frankfurt, Germany). Its capture was arranged in western Flores by the Assistant Resident, H. R. Rookmaaker. It is a male, with an estimated measured length of 3.04 m; approximately 25 cm is missing from the tail (about 9 ft 9½ in. total length). Of 27 oras captured during the 1926 Burden expedition, none was longer than 3 m. The total lengths of the 15 largest specimens to have ever reached zoos vary from 1.77 m (5 ft 10 in.) to 3.13 m (10 ft 2½ in.), with a mean of 2.33 m (7 ft 10 in.). Of the 10 largest specimens to reach zoos, the average length is 2.56 m (8 ft 5 in.).

The mean total length of 50 specimens we measured from Loho Liang, Komodo, was 1.70 m (5 ft 7 in.). The *estimated* total length of 60 additional specimens from the same area on Komodo produced an average of 1.83 m (6 ft). The difference is due to the difficulties in making the estimate, rather than to a real difference in length. Average snout-vent length (of the measured individuals) is 74.5 cm.

The populations at Kali Inaloah, Loho Kalo, Base Camp, Vai Nggulung, and the combined Djarek–Nggoer localities in western Flores have essentially the same mean body length (body length is in this case taken rather than total length, since tails are often incomplete) (Fig. 2-1). However, for 11 marked individuals in the population at Nanenepi (located in a *Lumnitzea racemosa* mangrove swamp), the average body length was approximately 10 cm longer than the other populations (Fig. 2-1). Smaller oras are probably not found at Nanenepi because all small prey species, such as rats, mice, lizards, snakes, and insects, are absent, though wild boar is common.

Dunn (1927b) was one of the first to suggest that males attain a larger size than females, stating that he knew of no females longer than 2 m. Most animal dealers still insist the females are never longer than this. However, Mertens (1942) and Lederer (1942) stated that the Senckenberg female was 2.22 m. The largest female (2.36 m) we measured on Komodo was only slightly smaller than the largest male (2.52 m). (Both specimens had incomplete tails. Measurements were made at the beginning of the study; both individuals grew several centimeters during the year.) Still, males do tend to become somewhat longer than females. The averages of the 10 largest specimens of each sex measured in Loho Liang, Komodo, were males 2.26 m, females 1.76 m.

The hatchlings of *V. komodoensis* (total length 253–555 mm, $\bar{X} = 304$ mm, $N = 37$) are considerably longer than those of any of the other large varanid species (total length 190–278 mm).

Weight.—Because length is often exaggerated, stated weight is also frequently incorrect. Most published field weights are estimates, often grossly inaccurate. Weights given for captive animals are correct, but the specimens are usually overweight. During this study field weights of 50 captured specimens were obtained by suspending the animals from a beam balance.

Total length is correlated with weight in Figure 2-2. The heaviest field weights obtained were 54 kg; however, other specimens of the same length may weigh as little as 35 kg. Specimens at both extremes were healthy individuals, without noticeable distension due to food in the gut. The average weight of large speci-

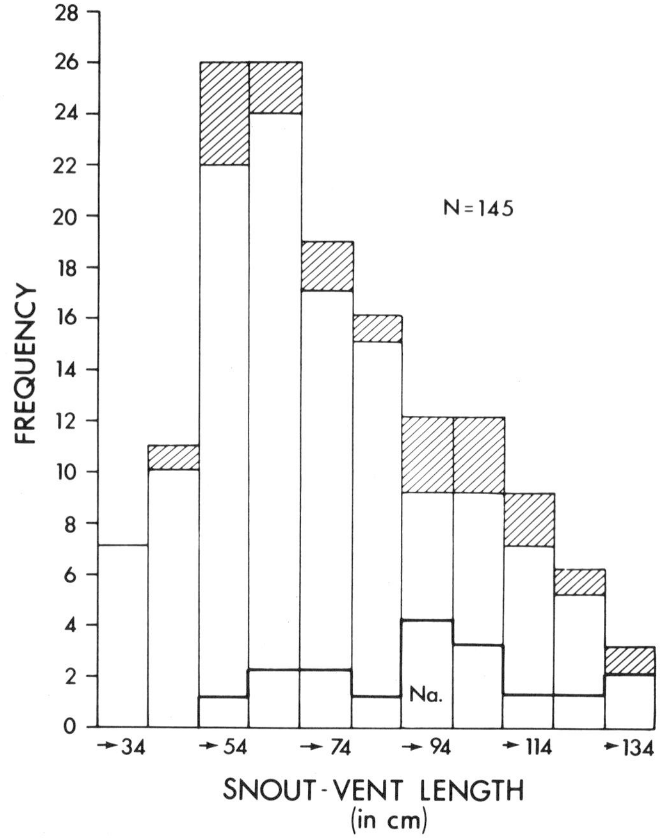

Fig. 2–1. Comparative lengths of specimens from Nggoer, Flores (shaded), and Loho Liang, Komodo (clear). Smaller histogram delineated by darker line represents the Nanenepi, Komodo, sample (see text).

mens, 225–260 cm total length (N = 12), is about 47 kg ± 7–12 kg. Weight variation of these same individuals is ± 35 percent of the average; average weight is only 8.6 kg (N = 50). Hatchling ora weights average 80.3 g (N = 29).

As pointed out elsewhere, feeding oras eat prodigious amounts of food—up to 80 percent of the total weight of the empty ora. Thus, weights both in captivity (because of obesity) and in the field (because of amounts eaten) must be cautiously interpreted. The common statement that wild, 9-ft long oras weigh as much as 250 lb (Broughton, 1936) is incorrect, being much too high. However, for 10-ft individuals maximum hunting weight could be as high as 250 kg (550 lb) and as low as 102 kg (223 lb). Provided the extinct Pleistocene giant monitor (*Megalania priscus*) was of the same general stature and length/weight ratio, an adult of 4.5 m would have weighed about 2200 kg.

The relationship of mass to thermal physiology is discussed in Chapter 5.

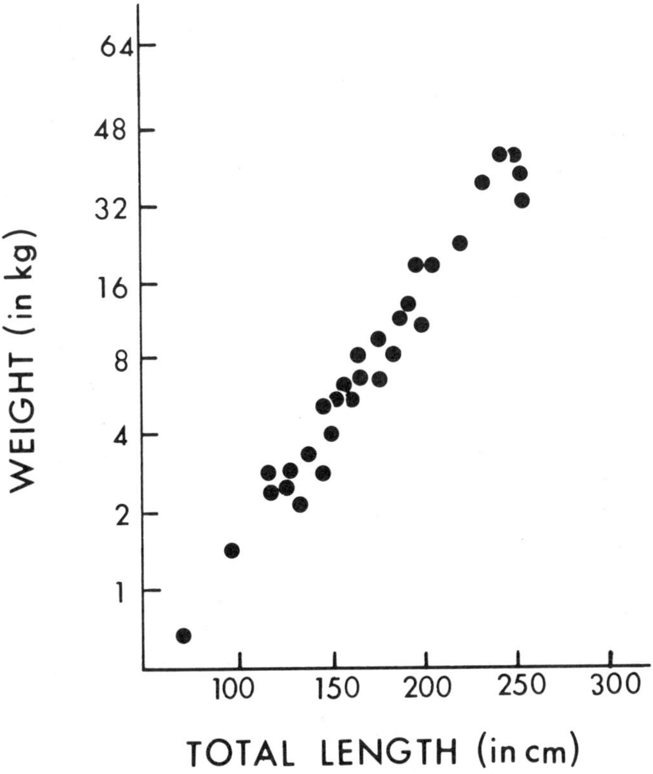

Fig. 2–2. Length-weight relationship in *V. komodoensis*, Loho Liang, Komodo.

Form

A number of workers have correctly stated that, in general, the body of large Komodo monitors is proportionately more massive than that of smaller specimens (Fig. 7-6) and also tends to be flatter. Sexual differences in body form suggested by Pfeffér (1959) are due to size and have no sexual basis. Though I was told by animal dealers that females have a proportionately narrower head than males (also reported by Lederer, 1942), this was not found to be true (Auffenberg, 1980). The hind legs are stout and relatively short in adults, longer in juveniles. All the toes have very sharp recurved claws which are relatively short when compared to those of other varanids. Though Loveridge (1946) stated that *V. komodoensis* has a tail only slightly longer than the head and body, Oesman (1970) correctly pointed out that the proportionate tail length varies directly with total length. The tail in adults is proportionately much shorter than in the young (Figs. 2-3, 7-6; Auffenberg, 1980).

Teeth

The effectiveness of both attack and feeding in oras is partly due to dental specializations. The teeth are highly adapted for cutting flesh (Burden, 1928); they are serrated, compressed laterally, curved posteriorly, and have a sharp tip and a broad base (Pl. 1). All are strongly pleurodont and their number is remarkably constant (7 premaxillary, 13 maxillary, and 12 mandibular). The replacement rhythm is of a rapid, regular, wave-like type (Edmund, 1969), and most of the teeth show signs of active replacement. Though many *Varanus* species have one or two replacement teeth at each position, *V. komodoensis* may have as many as four or five (Edmund, 1960), suggesting a rapid turnover rate, as postulated by Lonnberg (1928). There is almost no ontogenetic change in tooth shape, and heterodonty is exhibited only in the more rounded and inwardly curved teeth of the premaxillary series when compared to the compressed, nonlingually curved teeth of the maxillary series.

Mertens (1942) was careful to point out that the number of teeth preserved in well-macerated skulls is not the total number and, indeed, may be only half the functional teeth in life. Old teeth are loosened by osteoclastic activity, and the new teeth are not ankylosed until their broad bases are almost completely formed (Edmund, 1969). The adult female specimen maintained by Lederer (1942) shed only single teeth. His careful records show that during a 5-year period 17 teeth were found in the enclosure holding the specimen. Though Lederer felt that this rate was nearly correct, it is much slower than Edmund's (1969) figure of approximately 25 tooth spaces per year for *Iguana iguana*. Though tooth replacement in crocodilians becomes irregular and slower with increasing age (Ed-

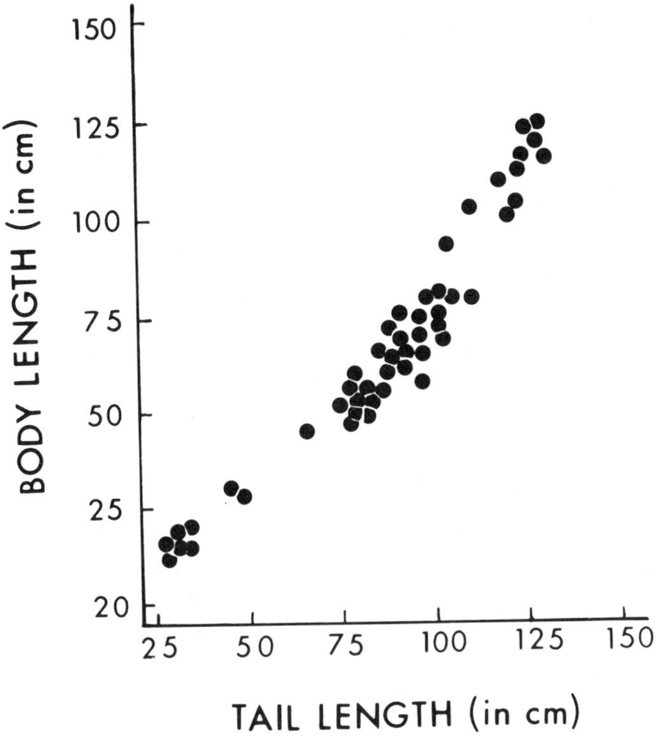

Fig. 2–3. Tail-body length relationship in *V. komodoensis,* Loho Liang, Komodo.

mund, 1960), there is no evidence indicating that this occurs in snakes or lizards (Edmund, 1969). In *V. bengalensis* the average functional life of a tooth is 3.1 months (total time tooth visible in radiographic studies, 6.8 months) (Edmund, 1960). This agrees well with similar studies on *Heloderma suspectum* (Bogert and del Campo, 1956), which showed each tooth was replaced approximately four times a year. With a total of 35 teeth in the jaws, 140 new teeth are produced annually. In *I. iguana* replacement is approximately five times per year (100 total teeth $\times$ 5 = 500 annually) (Edmund, 1969). These studies all serve to indicate that Lederer's (1942) estimate of tooth replacement rate in the *V. komodoensis* was much too low. Presuming replacement rate in the ora is similar to that in *V. bengalensis*, the total number of teeth replaced annually should be 200–250.

In 13 skulls of *V. komodoensis* examined, including those of some of the largest specimens ever taken, the longest teeth are approximately 2 cm from their tip to the basal tooth foramen (Table 2-1). Pfeffer's statement (1965) that

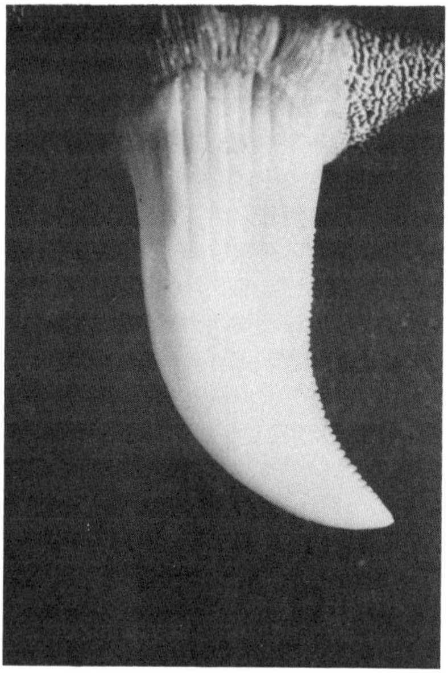

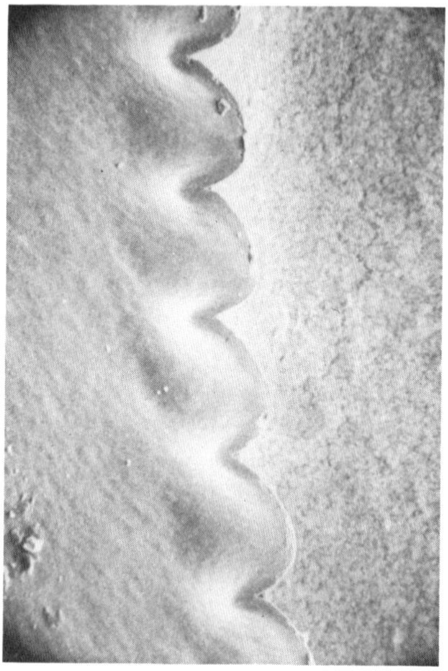

Pl. 1. *Left,* maxillary tooth of subadult (UF 29000). *Right,* serrations along posterior edge of same tooth (100 ×).

Table 2–1. $\overline{X}$ number of posterior serrations and $\overline{X}$ ratio of tooth length (from tip to basal tooth foramen) per condibasilar length of skull at selected upper jaw tooth positions

Skull length classes (mm)	Posterior serrations on tooth number				Ratio tooth length for tooth number			
	4	8	12	16	4	8	12	16
100–119 (N = 3)	19	32	27	16	0.50/1	0.91/32	0.71/27	0.41/16
120–149 (N = 12)	28	39	36	16	0.52	0.90	0.76	0.34
150–199 (N = 8)	23	45	38	28	0.53	1.03	0.87	0.49
200–249 (N = 11)	30	55	51	55	0.55	0.93	0.87	0.51

the teeth of the Komodo monitor are 3–4 cm in length is apparently not based on any data.

The anterior tooth edge does not usually have serrations, though they may occur just below the tip. However, from near the tip to the gum line the entire posterior edge is serrated in the maxillary series (Pl. 4). The total number of serrations varies from 14 to 55. Generally, premaxillary teeth do not have serrations. While there is no significant change in tooth proportion from small to large oras, the number of posterior serrations is positively correlated with tooth and/or ora size (Table 2-1). In lateral view, the entire tooth profile is convex (Fig. 2-4), important in slicing flesh (see Chap. 9). Other specializations include the position of the cutting edges on both the distal and the mesial edges of the tooth (Fig. 2-5), particularly in the longest teeth of the maxillary series. The arrangement of the edges forces the flesh of the prey apart by an action similar to that of a plow, presumably so that the teeth that follow can make an easy entry into the wound. The tips of the upper jaw teeth are slightly lateral to those of the lower jaw.

Scalation

Miscellaneous scale characters are provided in Auffenberg (1980), and the possible importance of scale organs in courtship is discussed in Chapter 6. What remains is our discovery of an important sexual dimorphic scale character. In

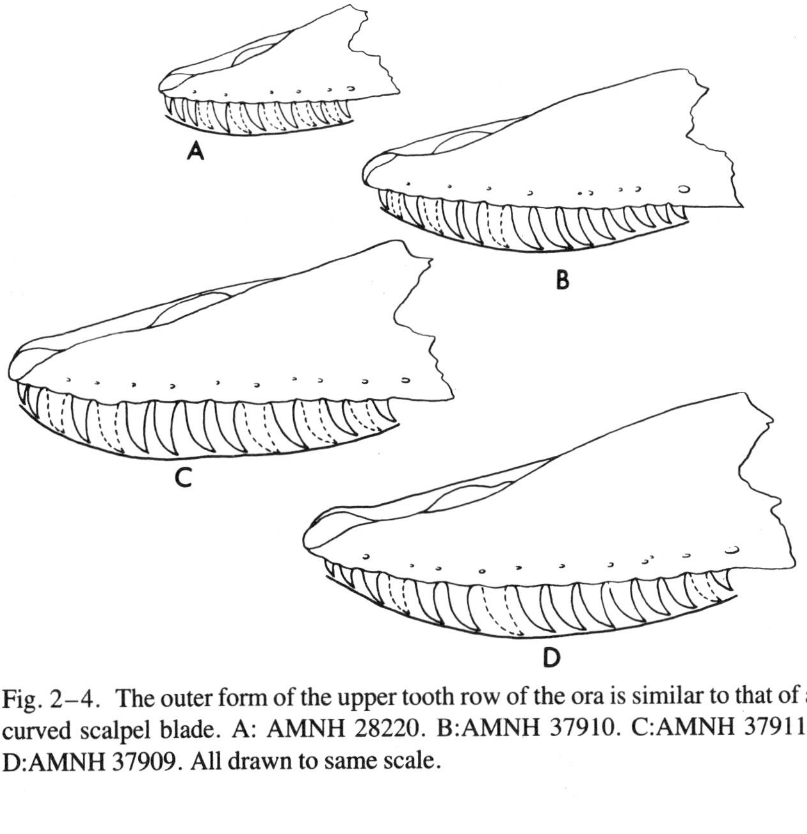

Fig. 2–4. The outer form of the upper tooth row of the ora is similar to that of a curved scalpel blade. A: AMNH 28220. B:AMNH 37910. C:AMNH 37911. D:AMNH 37909. All drawn to same scale.

males of all sizes there are two small rosettes of scales located in the second to third scale rows anterior to the vent, each on either side of the midline (Fig. 2–6). The rosettes usually enclose a small depression in a centrally located scale. In juveniles the pit is often filled with a yellowish-brown, waxy secretion plug, similar to that occurring in femoral pores of other lizard families, particularly during the breeding season (see Cole, 1966, for a review). The number and disposition of the surrounding scale are quite variable. In young males the scale rosette is at approximately the same elevation as the surrounding ventral scales. However, in large males the rosette is slightly raised. Female specimens lack the circlets. Microscopic studies on the skin are presently in progress, and until these are completed there is little that can be said regarding the microanatomy or function of these scales (see Cole, 1966).

Skeleton

Duméril and Bibron (1835) were the first to mention that some species of *Varanus* possessed small osteoderms in the skin. Dunn (1927b) was the first to emphasize their presence in *V. komodoensis*. He correctly stated that at least on the head there is sometimes more than one osteoderm under each scute; they sometimes form a network of small, almost boomerang-shaped bones on the dorsal surface of the skull. Fejevary (1927) used the presence of these bones as

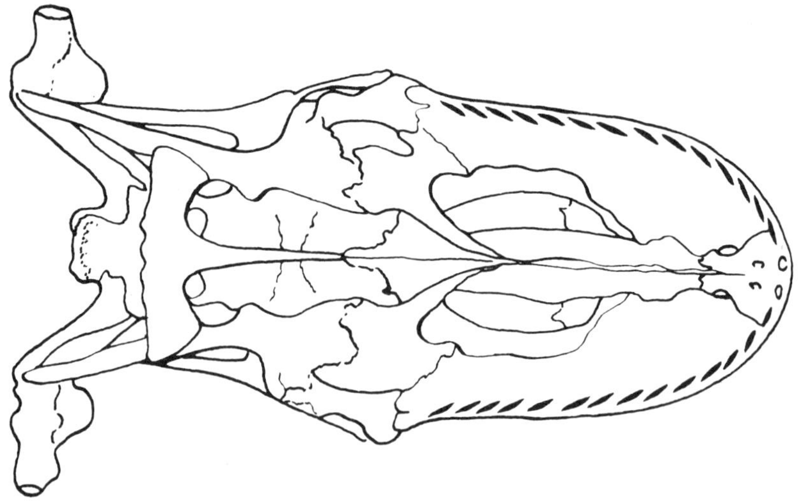

Fig. 2–5. The cutting edges of the teeth (crown view) are aligned at an angle, forcing each cut section lingually as the tooth row is pulled posteriorly. This reduces binding of the flesh between the teeth.

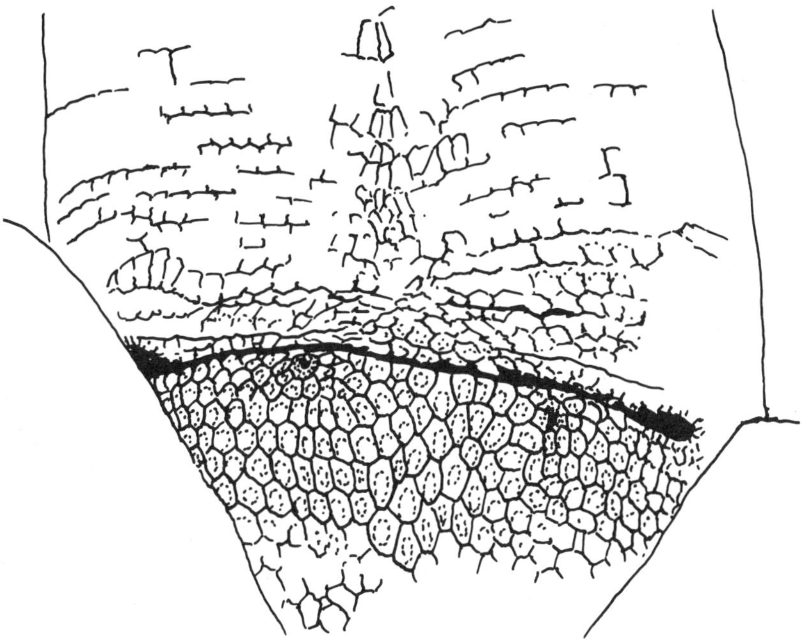

Fig. 2–6. Cloacal area of adult male *V. komodoensis* (anterior direction downward), showing the small-scale rosettes just cranial to the vent that are characteristic of males in this species.

the basis of his new genus *Placovaranus*. However, Mertens (1942) replaced *komodoensis* in the genus *Varanus* on the basis that the genotype, *V. varius*, also possesses them. The bones are also found in *V. giganteus*, *V. gouldi*, and *V. salvator*, though not in the young of any of these species. Smith (1935) stated that they are located under the skin on the front side of the legs, at the tail base, on the neck, and on the belly. He did not find them in *V. griseus*, *V. flavescens*, *V. dumerili*, *V. rudicollis*, or *V. bengalensis*.

The vertebrae and limb bones of *V. komodoensis* have been discussed in several papers (Fejevary, 1918, 1927; Dunn, 1927b; Mertens, 1942; Auffenberg and Franz, MS). Smith (1935) reported that this species possesses a well-developed *os penis*, although I found no indication of such a structure in the skeletons I prepared and/or examined. Giltay (1931) published a short paper following examination of two skeletons of *V. komodoensis*. He reported traces of an ancient tail fracture in one. Other caudal vertebrae in this same specimen were malformed; a few were fused in a way that suggested a congenital condition rather than an injury. He regarded the several types of skeletal anomalies, particularly of the tail, as an indication of the "degenerative evolution" and specialization of the species, believing that this species is subjected to relatively little natural selection (with which I disagree). Mertens (1946b) described bone anomalies in a captive.

The earliest description of the peculiar skull of adult *V. komodoensis* was provided by de Jong and Brongersma (1927a). The most complete descriptions are by Lonnberg (1928) and Mertens (1942, 1946b), who pointed out that the skull of *Varanus* is mesokinetic, with the parietals fused to the supraoccipitals. There is no movement between the frontals and parietals. They showed that the maxillary bones are capable of considerable transverse movement, widening the distance between the two outer edges of the mouth. Their studies indicated that such movement was not possible in the skull of *V. salvator* and only barely possible in *V. niloticus*. The details of skull movements and their importance in feeding are discussed in Chapter 9.

Musculature

In comparing the musculature of *V. salvator* with that of *V. komodoensis*, Giltay (1931) found considerable differences in insertion of the large caudofemoral muscle complexes. In *V. komodoensis*, the insertion of this muscle mass on the internal trochanter of the femur was considerably and proportionately larger than in *V. salvator*, probably (he believed) due to the way in which the tail is carried in the two species. When walking, *V. komodoensis* carries its tail much higher than does *V. salvator*.

After examining the hearts of several species of *Varanus*, Vorstman (1933) concluded that a more or less pronounced separation of the dorsal ventricle into a right and left heart, in addition to the spongy trabecula, is of general occurrence in the genus. However, the vertical separation of the dorsal part of the ventricle showed wide variation, and he was not of the opinion that *V. komodoensis* occupied a unique position in this regard among the Varanidae. However, the vertical ventricular septum is more pronounced in *V. komodoensis* than in the other eight species examined.

Meinertz (1952) has reported the results of his very complete studies of the heart and associated circulatory structures on *V. komodoensis*. He showed that, in general, the heart of the ora is very similar to that of *Python reticulatus*. In comparison to other lizards (though no specific examples were given) and *P. reticulatus*, however, there is less possibility of a mixture of oxygenated with unoxygenated blood in the heart of the Komodo monitor.

Miscellaneous Anatomy

Van Dijk (1935) published results of his study of the liver and the pancreas, comparing them with other reptile genera. This work, together with that of Beddard (1888), showed that *V. salvator* and *V. komodoensis* are very similar in their hepatic duct anatomy. The cystic and hepatic ducts leading from the liver and gall bladder to the intestine form a network, unlike the single bile ducts in other *Varanus* species. The gall bladder of the ora is apparently extraordinarily large for the animal's size (van Dijk, 1935), a fact related, I presume, to the prodigious meals irregularly eaten (Chap. 9).

The tongue of *V. komodoensis* (and *V. griseus*) is reported to be proportionately short compared to that of other species (Mertens, 1942).

Color

A detailed description of the color of three different sized individuals shows a significant ontogenetic change from the multihued, speckled young to the nearly uniform, clay-colored adults (Auffenberg, 1980).

I did not find any obvious sexual color dimorphism in spite of the comments to the contrary by Javanese animal dealers. However, the flanks of the adult females have more red than those of most males, and the yellowish-green spots on the snout are more common in males.[2] However, these spots may be a characteristic of the largest Komodo monitors, and for this reason are most common in males.

Relation of Form and Color to Life History

Ontogenetic changes in form are clearly related to major life history phases (Pl. 2). The slenderness of juveniles, their dexterity, and their light weight obviously account for the ease with which they scramble in the trees. The adults are much too stocky and clumsy to move about in the trees. The long tail of the juveniles, often used for balance and as a prop in the trees (Fig. 10–4), becomes short and heavy in the adult and is therefore a more effective defensive weapon. The heavy, strong front legs of the adult become important in climbing about the mountainous terrain, in holding down parts of the prey carcass as pieces are torn off with the teeth, and in assisting in tearing by pushing the body away from the carcass while the teeth are clamped onto the flesh. The change in proportionate posterior head width is partly the result of greater muscular development in the adults, undoubtedly related to carrion feeding, and partly the result of a wide parallel-sided snout. The latter is important in slicing large pieces of flesh from the carcass and in swallowing them entire.

The juvenile color pattern is clearly advantageous in the rich vegetational fabric that clothes the island during the monsoon, when the oras are hatched. The

2. Pfeffer (1959) reported that females on Flores had more yellow head spots than did males; but I doubt that his sex determinations were correct, for they were based on characters I found to be inapplicable for such means.

Pl. 2 *Top,* hatchling ora (83 cm in length), showing proportions and major features of color pattern. *Bottom,* adult ora (2.4 m in length), showing loss of bright color pattern in later life.

highly complex pattern obviously falls into the category of disruptive, cryptic patterning in color. Henne am Rhyne (1903) came to the same conclusions for baby *V. salvator*. The adults become uniformly clay-colored and thus achieve a cryptic color and pattern when ambush techniques in obtaining prey become important.

The bright yellow tongue color does not change ontogenetically but is species specific. Thus, within the subgenus *Varanus*, *V. giganteus* and *V. salvator* possess a blue tongue, *V. varius* and *V. komodoensis* a light yellow tongue, and *V. dumerili* and *V. grayi* a pink tongue.

3

Relationships in Time and Space

Distribution

Horizontal distribution.—Varanus komodoensis has the smallest range of any of the large carnivores in the world. At present the species is restricted to a few islands in the middle of the Lesser Sunda Islands, Republic of Indonesia (Fig. 3–1). In this area it was previously known from Komodo, Padar, Rintja, and Flores. Several reliable Komodo and Flores fishermen, well acquainted with the ora, state that it also occurs on the islands of Oewada Sami, off Rintja, and Gili Mota, off the southwestern tip of Flores. Though there is presently no proof that the oras are permanent residents, the habitat is correct and appropriate prey species are abundant (Table 3–1). Individuals appear irregularly on Nusa Mbarapu, off the southeastern coast of Komodo (Chap. 7).

In spite of Piazzini's (1960) statement that oras are not residents of Padar, they occur over the entire island—from near the highest pinnacle to, and including, the mangrove swamps along its shores. Oras are found over almost all of Komodo and Rintja, although only occasionally on the highest peaks, and in some lowland areas density is very low. The west and north slopes of the northern half of Komodo have seldom been investigated by expeditions. We assessed the ora habitat and general density on that side of the island for 3 days, concluding that oras are very abundant in certain valleys on the west slopes of Gunung Ara and Gunung Satilibo.

When the several expeditions are grouped, it can be seen that Padar has been completely covered by foot, though the more interesting island of Rintja is poorly known. Most expeditions that have stopped there visited the north coast, usually Loho Buaja, but Hoogerwerf and the earlier de Jong expeditions covered much more ground. The combined Indonesian–Soviet expedition in the early 1960s explored more of the interior than did any other group. We visited Padar on several occasions and found oras wherever we went, though densities were low.

Regarding Flores, most authors simply state that the ora is restricted to the western end. Only de Jong (1937) attempted to detail the limits of distribution on this island. He stated that the animals were found only in the wide valleys of western Flores, specifically from the Loök-Nggoer area northward along the coast to near Labuanbadjo. He believed the eastern limits were near Labuanbadjo in the north, the Beliling Mountains centrally, and Tjerumba in the south. Though de Jong said the ora was unknown east of Labuanbadjo, Rookmaaker (in a letter to de Jong) suggested the eastern boundary was located near Pota, Mangarrai Province. Pfeffer (1959) followed de Jong but emphasized the irregular nature of the distribution (however, he was not correct in asserting that the range is determined chiefly by the distribution of deer).

The ora's eastward range along the south coast was unknown, and although Assistant Civil Administrator Rookmaaker told de Jong the ora was found in an isolated (?) colony at Ngondo (at the northeastern edge of Ende District), de Jong (1937) doubted that the range was that great. Captain Petrus Sina, then chief officer of our supply ship, *Stella Maris*, and a native of Ende on the south coast of Flores, told us he once saw a very large monitor, "identical to those we were studying on Komodo" about "100 km west of Ende" (or near Aimere—could he have meant *east*, or near Ngondo?), far east of previous records (other than Rookmaaker's unconfirmed report to de Jong). Shortly thereafter, Rev. E. Schmutz, Nunang Mission, Mangarrai District, Flores, told me the ora extended eastward along the south coast as far as Nangelele, and provided several records of sightings between this village and Nggoer at the western tip of the island.

It was partly to verify the eastern limits of the ora on Flores that we visited the island in 1971. Two teams were organized: F. Thompson to establish the eastern boundaries and the other, including myself, to study the density, ecology, and vertical limits of the ora on Flores. We traveled nearly the length of the island on the Maumere-Ruteng road, which parallels much of the south coast. En route we asked both villagers and missionaries about monitors in nearby regions. Descriptions of water monitors (= *weti* in most places) were common; no one said oras lived nearby or at the closest coast. Near Aimere local people were familiar with the ora, but they said it occurred only westward in Mangarrai District. Our expedition established that only occasional large (no small) specimens are seen east of the river (Vai Djamal) at Nangelele (= Tiwunonto on some maps), though both young and adult individuals are common west of it. There are no oras on Nusa Mules Potjo Kepi (a small island southeast of Nangelele) or at Bara (8 km north of Nangelele). Thus, in this area the lizards are found only on the xeric southern and southeastern slopes of Potjo Dedeng, not east of the river on the more mesic side of the valley or on the western slopes of Gunung Ntodo.

Our studies also showed that oras have a continuous, though narrow, distribution along the southwest coast. In the Nanga Loök valley they occur inland to between Loök and Tjereng, and (according to the independent observations of

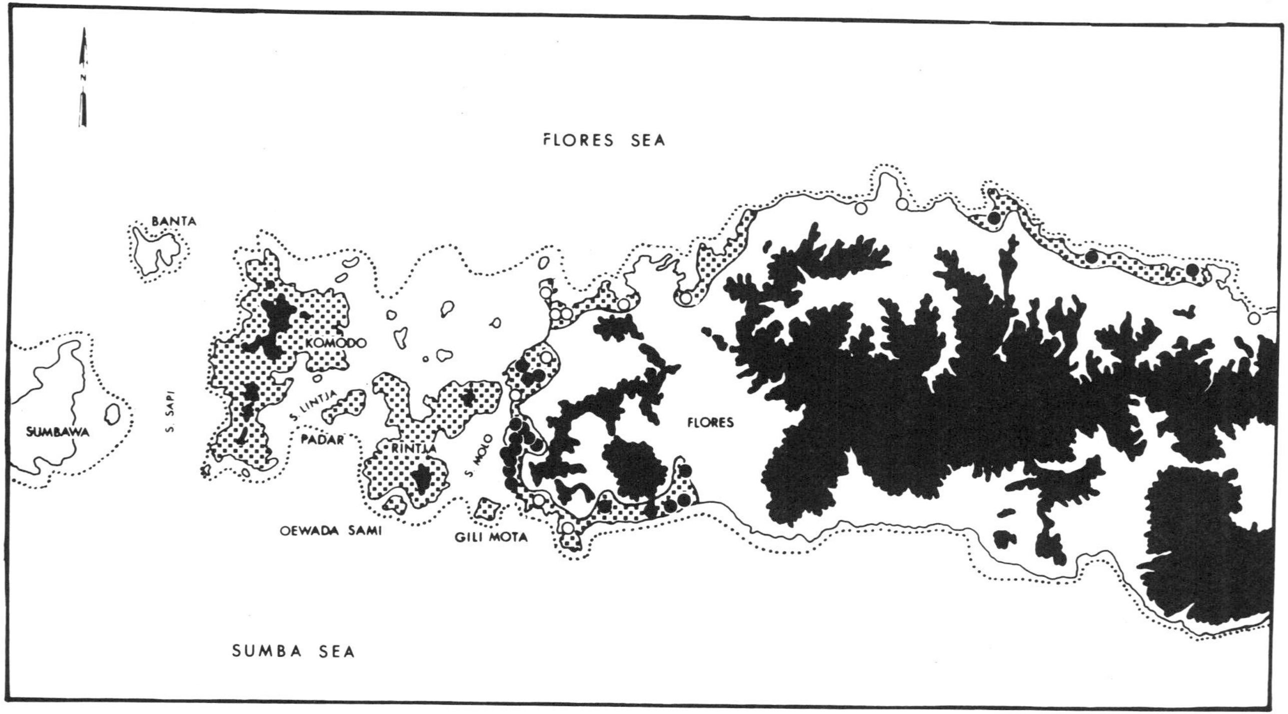

Fig. 3–1. Known range of *Varanus komodoensis*. Stippled areas indicate medium to high density areas. On Flores solid circles represent our records, hollow circles are literature or villagers' reports. Circles outside stippled areas represent extinct populations or reports of transients or both.

both F. Thompson and E. Schmutz) they are absent from the upper end of the valley at the bottom of the east slope of the Beliling Hills. The range becomes very narrow where the eastern flanks of Gunung Wasedoo come close to the coast, but north of this mountain the range becomes as broad east-west as it is in Nanga Loök valley. It was here that de Jong did his work on two previous expeditions to Flores and where the new nature reserve, Wai Wuul, has been established. However, farther north there is a hiatus in the range in the low, largely irrigated lands south of Labuanbadjo. Oras occur on the peninsula north of Labuanbadjo, though densities are low. They are not found on Pulau Seraja Besar or other small islets nearby.

On the north coast of Flores the range is both narrow and discontinuous, being broken by expanses of poor ora habitat, agricultural sections, and hills extending nearly to the sea. This finding is based on the following data.

The ora is found in small populations in savannas near Boleng and on the north flanks of Gunung Namppar near the village of Tereng. East of Tereng it does not extend beyond the west slopes of Gunung Lamas, where the vegetation becomes much more mesic. The manager of a coconut plantation, who is familiar with oras, told me he has seen one individual at Bari, and several villagers at both Labuanbadjo and Reo mentioned having seen oras at Robeck. At the coastal village of Kedindi (the harbor area of Reo) I was told that several years ago a 2-m ora entered the village and attacked a dog, whereupon the villagers killed the ora with spears. This was the only time anyone had ever seen an ora in the area, though water monitors are very common.

Several missionaries in Ruteng and officials as distant as Labuanbadjo told me that oras are common near Dampek and Pota. A friend of the Kedindi coconut plantation manager shot a 3-m ora at Pota for its skin, and our expedition saw a large ora, several burrows, and some droppings near the same place. We also learned that oras are abundant from Dampek east to Nanga Loök (not the same as Nanga Loök mentioned above) at the coastal village of Buntal, near the boundary of Mangarrai and Riung districts. However, in this region the distribution is a narrow band only 1–2 km wide.

There is every indication that oras found between Tereng (and perhaps farther west) and Dampek are stragglers. All are large adults; no young have been seen in this area. However, the population in the Dampek-Pota area is well established and possibly a small part of what was formerly a much more extensive range along the north coast of Flores.

That the northern coastal population does not extend much farther west than Dampek is clearly shown by the statements of villagers in that area that the ora occurs no farther than a kilometer or two west of the village but has a long range along the coast to the east. There is a decided change to more mesic forested conditions west of Dampek, which continue to Reo.

From Pota, our expedition went east to Nanga Loök at Buntal, where we located a burrow and other signs proving that the animals actually ranged nearby. There we were told that oras were common locally, extending up the valley about 5 km, and eastward from Buntal and Kambe (another nearby village) only an additional 5 km more, presumably near the northern foot of Wolo Tanah Peta. Missionaries from Riung, who had resided in the area for many years, told me

Table 3–1. Distribution of oras and other large vertebrates in the Komodo area

Island	Size km² (mi²)	Kerabau	Deer	Boar	Goat	Monkey	Horse	Ora
Flores	12,032.0 (4,700)	+	+	+	+	+	+	+
Komodo	393.4 (153.8)	+	+	+	+	–	–	+
Rintja	278.0 (108.6)	+	+	+	+	+	+	+
Padar	13.5 (5.3)	–	+	–	+	–	–	+
Gili Mota	11.3 (4.0)	–	+	+	–	–	–	+
Oewada Sami	9.6 (3.8)	–	+	+	–	+	–	+
Pulau Longo	4.5 (1.8)	–	–	–	+	–	–	–
Seraja	3.6 (1.4)	–	+	–	+	+	–	–
Siaba	3.5 (1.4)	–	–	–	+	–	–	–
Sabajor	2.6 (1.0)	–	+	–	+	–	–	–
Nusa Vadju	1.8 (0.7)	–	–	–	+	–	–	–
Papagaran	1.5 (0.6)	–	–	–	–	–	–	–
Tatawa	1.1 (0.4)	–	–	–	+	–	–	–
Pulau Misa	1.1 (0.4)	–	–	–	+	–	–	–
Mengjatan	1.0 (0.4)	–	–	–	+	+	–	–
Gili Lawa Darat	0.9 (0.3)	–	–	–	+	–	–	–
Gili Lawa Laut	0.9 (0.3)	–	–	–	+	–	–	–
Langkoe	0.8 (0.3)	–	+	–	+	–	–	–
Nusa Mbarapu	0.6 (0.2)	–	–	–	+	–	–	a
Nusa Rondong	0.5 (0.2)	–	–	–	+	–	–	–
Boegies	0.4 (0.2)	–	–	–	+	–	–	–
Logo	0.3 (0.1)	–	–	–	+	–	–	–
Sabalan	0.3 (0.1)	–	–	–	+	–	–	–
Gili Bodo	0.3 (0.1)	–	–	–	+	–	–	–
Padar Kechil	0.3 (0.1)	–	–	–	+	–	–	–
Nusa Pinda	0.3 (0.1)	–	–	–	+	–	–	–
Sumbawa	10,624.0 (4,150)	+	+	+	+	+	+	–
Sangeang	20.0 (7.8)	–	+	+	+	–	–	–
Banta	3.0 (1.2)	–	–	–	–	–	–	–
Kelapa	0.5 (0.2)	–	+	+	–	–	–	–

a. Periodic; see Chapter 2.

that large oras were rarely found in the vicinity. These may also be stragglers. We have no records that the animals have ever been seen eastward of this place. Missionaries and villagers from Maumere and Larantuka in extreme eastern Flores told me repeatedly that oras do not occur there, though water monitors are found in appropriate habitats.

Thus, it is clear that the Flores ora population can be divided into at least two major units: extreme western Flores, from Labuanbadjo to Nangelele and a north coast population from near Dampek to just west of Riung. Stragglers are occasionally found in the intermediate area. In view of the relictual features of the Flores distribution, other isolated populations remain a possibility. As an example, Rookmaaker's report of oras northeast of Ende bears investigation, though our travel nearby suggests that the habitat is probably too mesic. When viewed from the sea, the habitat along the north coast almost due north of the western edge of Teluk Ende seems very favorable but is quite remote from the closest populations to the west. Perhaps the most interesting possibility for further work along this line is suggested by the report of a military attaché from Sumbawa Besar (presumably familiar with the water monitor) to van Heurn (1932) that oras are also found along the south coast of Sumbawa.

Ecology is a major factor in this distributional pattern but not the only one. Human population density is quite high on the north coast of Flores, and in at least peripheral sectors of the ora's range monitors are often killed. Human interference may be the most important factor in reduction of past ora populations in Flores. However, slash and burn agriculture on at least the north coast of Flores (with shorter fallow cycles being employed as human populations increase) leads to *padang* (secondary grassland) formation in areas that were formerly mesic forests with a more closed canopy. The open padang habitat, though not optimum, allows at least large oras to pass out of the ordinarily rather restricted species range. However, viable populations have apparently not been established in padang regions, perhaps because of the infrequent visits by wandering adults, because prey densities are low or, most probably, because these padangs are closely associated with higher human populations.

As far as we know, the smallest area permanently occupied by an ora population is Oewada Sami, which is 9.6 km^2. Nusa Mbarapu, the smaller island southeast of Komodo, is only 0.6 km^2 and is apparently too small to maintain a permanent population, though it occasionally harbors temporary individuals. The largest block of undisturbed ora habitat is on Komodo, though Rintja and Flores ranges are of similar size. The Rintja and Komodo oras are well protected, and population densities should remain reasonably high. However, the Flores populations are ultimately bound to shrink, and I do not think the small reserve of Wai Wuul will be large enough to save a significant part of even the local populations. A narrow reserve along the north coast near Pota is needed.

The following list includes all the localities at which oras or their spoor have been seen at least once by reliable observers. All populations of reasonably heavy density are italicized. In general, the Flores localities are arranged from east to west along the south coast, north along the west coast, and eastward along the north coast. Elevations are approximate.

Flores

Mangarrai Province

Badjo District: nr. Siru (or Siroe), 250 m; Nangelele (or Tiwunonto), 10 m; Djoreng, 5 m; Wai Pogong (7 km W Nangelele), 3 m; between Sessok (or Ndaang) and south coast, 300 m; Nisar, 150 m; Angkor, 250 m; Gunung Sorkaka, 100 m.

Mata-wae District: Wai Mbeku, 100 m; Wai Peropoi, 100 m; Wai Kanupur, 100 m; Wai Nupa, 100 m; *Wenepa*, 30 m; *Nggoer* (several localities), 100–250 m; Randong, 500 m; Termeda, 900 m; Kandong, 750 m; *Djarek*, 100 m; *Padang Banta* (nr. Lenteng), 300 m.

Mburak District: Warsavi, 40 m; *Gunung Mboering* (nr. Berloka), 0–150 m; *Mburak*, 75 m; *Wai Wuul*, 30 m; *Gunung Manis*, 0–400 m (?); 1 km S Naneganae, 5 m; Kenari, 70 m; Loh Mbekoe, 100 m.

Nggorang District: nr. Tandjung Waturamba, 100 m; Binongko, 100 m; Teluk Rangka, 100 m.

Boleng District: nr. Boleng, 10 m; nr. Tereng, 20 m.

Berit District: nr. Bari, 16 m.

Ruis District: nr. Robek, 10 m.

Reo District: Kedindi, 30 m.

Lamba-leda District: 1 km W Dampek, 100 m; Tandjung Kurungbadjo, 100 m.

Pota District: *Pota*, 100 m.

Rambong District: *Buntal*, 100 m; *5 km S Buntal*, 150 m; about *5 km E Buntal*, 100 m.

Riung Province

Riung District: W Riung (?).

Komodo: generally distributed over entire island.

Nusa Mbarapu (off Komodo): oras occur here irregularly.

Padar: generally distributed over entire island.

Rintja: generally distributed over entire island.

Oewada Sami (off Rintja): over entire island.

Gili Mota (off western Flores): over entire island (?).

On Komodo the highest mountain is called Gunung Ara, a name some islanders claim is derived from the word ora. It is interesting that on Flores the name ara is used in two areas within the range of the ora: about 4 km inland from

Pota is a village called Kampung Ara, and Wai Ara is the name of a creek on the eastern edge of the ora's range on the south coast.

Vertical distribution.—Most *Varanus* species are found at low elevations, but some occur as high as 2000 m (Witte, 1941; Mertens, 1942). Though the region the ora inhabits is very mountainous, *V. komodoensis* normally occurs below 700 m. On Komodo it is rarely found above 500–600 m (de Jong, 1929, and our own observations). On Flores a large ora has been seen often at Randong, Mangarrai Province (500 m). There are reports of individuals at higher elevations, but these animals are rare transients (a large adult has been seen as high as 900 m, above Termeda, and another was killed between 800 and 900 m in 1967 at Kandong, along Wai Sano).

Generally, population densities are highest at low elevations in extensive flatlands surrounded by low, grass-covered hills. With the exception of Flores, most alluvial valleys are rather small, and oras are found throughout them. However, oras are absent from parts of some of the large valleys on Flores. Although one might expect to find oras at Ort Naga (near Nisar), they extend up the valley only as far as Angkor (1 hr downriver from Naga). The same is true at Wai Wake, where the species extends only to Siru (2 hours coastward from Wai Wake). Both Angkor and Siru are 10 km north of the south coast.

On Flores, there is a medium to high ora density over a total area of approximately 400 km^2 (Table 3-1). This is about the same size as the entire island of Komodo (393 km^2), where densities are generally quite high. Rintja has about 278 km^2, and to this must be added the areas of Padar (13.5 km^2), Gili Mota (11.3 km^2), and Oewada Sami (9.6 km^2). Thus, the total range for resident oras is approximately 957 km^2. Transient adult individuals on Flores extend this area to at least 1950 km^2. The smallest area in which they are definitely resident is Gili Mota (11.3 km^2). Though they are reported from Oewada Sami (9.6 km^2, off Rintja), these *may* be transient individuals, as they certainly are on Nusa Mbarapu (0.6 km^2, off Komodo). In general, ora densities are not only lower on smaller islands, such as Padar (Chap. 7), but individuals tend to be smaller (Chap. 2). While common prey of large adult oras (deer and boar) are absent or low in density on many small islets, this is apparently not a major factor, because oras inhabiting these places (temporarily or permanently) feed largely on rats, birds, and sea turtle eggs (Chap. 10).

Barriers to dispersal.—Mesic to hydric habitats function as partial barriers to dispersal. However, as shown above, isolated adults sometimes wander far from their usual fire-adapted, xeric habitat. This is true both vertically and horizontally. In doing so, they may also move across open stretches of fresh or saline water.

That oras swim reasonably well was reported many years ago by Burden (1928). On occasion they enter salt water voluntarily, swim for distances of at least 1000 m to offshore islets (sometimes in areas subject to strong tidal currents), and dive to depths of at least 2 m. However, there is no evidence that they regularly traverse the straits separating the larger islands from one another, in spite of the implications of some earlier authors that they do so. Unless they *are* found along the south coast of Sumbawa, Sapi Strait serves as an effective barrier to westward dispersion, just as the more mesic habitats on Flores apparently limit the eastward dispersion of most individuals. As stated elsewhere (Auffenberg, 1980), I believe the present distribution of the ora on several separate islands is the result of a large-scale subsidence associated with Plio-Pleistocene volcanism on Flores.

Pleistocene.—During parts of the Pleistocene, when considerable water was maintained in continental glaciers, the sea level was much lower than it is today. Most geologists use the 100 fathom contour as the probable lowest ocean levels of the Pleistocene. During these periods Komodo was apparently separated from Sumbawa by the strait that isolates them now, but the gap was narrower.[1] Tidal currents through both Sapi and Lintja Straits were much stronger than they are now, perhaps similar to those in Selat Molo at present. If so, they were much more effective in isolating Komodo at that time.

Assuming relative stability of the Flores-Komodo area during the Pleistocene, the maximum land surfaces available to oras would have been during glacial times. For a sea level lowered 15 m, the minimum area available to oras is approximately 3000 km^2, or a 50 percent increase. At the 30 m contour the minimum area available is about 4000 km^2, or an approximate 100 percent increase. Thus, the range of the ora has probably been reduced about 100 percent during the last 20,000 years, irrespective of any changes in ecology or possible extirpation by early man.

Hooijer (1944) suggested that fossil varanids on Timor and Java are closely related to, if not conspecific with, *V. komodoensis*. Such a conclusion suggests considerable post-Pleistocene range contraction. However, recent analysis by Auffenberg and Franz (MS) suggests the fossil material is *not* to be referred to *V. komodoensis*, but to *V. salvator* (or a closely related extinct species). Therefore, there is no evidence that the present range of *V. komodoensis* is relictual.

Origin and History

Geological origin.—Zoologists who have commented on the evolution of the Komodo monitor have all presumed it evolved very recently. However, the fossil record clearly shows that the genus has a long evolutionary history, and morpho-

1. Burden (1928) suggested Komodo, Padar, and Rintja may all have been connected at one time but that transverse faulting separated them. If connected, they were more probably separated by submergence caused by adjacent massive volcanic extrusions on Flores in the Tertiary.

logic studies suggest that *V. komodoensis* is the most primitive species of the genus. Therefore one would expect this species to have considerable antiquity.

Probably the most important factor leading to the conclusion that *V. komodoensis* is a late evolutionary product is the common statement that the Lesser Sunda Islands are geologically young. Emphasis is placed either on the obviously recently elevated coral reefs found on the flanks of some islands or on the recent volcanism that has produced much of the present land surface of the archipelago.

There is no question that certain small islands and portions of larger ones have been formed or modified during very late geological time. However, it is also true that the geology of the entire area is exceedingly complex (Umbgrove, 1949), and that the region has been a zone of diastrophism and volcanism since the Mesozoic. Certain sedimentary rocks also suggest considerable antiquity, and ancient eroded surfaces prove that land of some type was available for a long time, though intermittently. In fact, a *succession* of land surfaces have been available in the Komodo area of the archipelago since its earliest developmental phases. It is also clear that most *Varanus* species are extremely adept at moving across marine barriers. Thus, while the remote origin of the *komodoensis* group may have been on one early island in the archipelago, succeeding generations could have moved over a succession of subsiding and emerging islands. There is, in fact, a large body of evidence suggesting that the ancestry of the *V. komodoensis* extends into the Miocene (Auffenberg, 1980 includes a complete discussion of the geological history and zoogeography of the area).

Geographical origin.—Sternfield (1924) was the first to suggest that the evolutionary center of the genus *Varanus* was in Australia. Later, Mertens (1934) drew the same conclusion while discussing zoogeography of the Sunda Islands, believing that *V. komodoensis* moved westward from Australia over Timor Laut and Wetar, Timor Roti, Sabu, Sumba, Flores, and Komodo. Although he stated that this probably happened in the Pliocene, he gave no evidence for it. This is the same route suggested earlier by Burden (1928), who honestly admitted that it was little more than a guess. However, after a careful analysis of all the living species of *Varanus* and a review of all fossil members of the group, Mertens (1942) concluded that the distributional center of the genus is in the Indo-Australian archipelago and *not* in Australia. Additionally, the most recent study on chromosomal evolution in *Varanus* (King and King, 1975) strongly supports the view that the Varanidae moved eastward to Australia. It is very likely that this occurred during the Miocene, when the Australian block is thought to have come in contact with the Indonesian island chain (see Auffenberg, 1980, for discussion).

Thus, it is very likely that the present distribution of *V. komodoensis* is at, or near, its geographical point of origin, and that this origin was in the Mio-

Pliocene. I believe that the present distribution of the ora on several separate islands is not the result of the lizard's dispersal from one island to another but of subsidence in the entire area (Auffenberg, 1980). This is probably correlated with Mio-Pliocene volcanism, which is associated with the collision of the Australian block and the Indo-Australian Archipelago.

4

Ecology

Macroecology

The general aspect of the islands on which the Komodo monitor exists is typical of arid, uplifted regions. Thus, the mountains all have steep angular slopes and alluvial fans, and streambeds and valley floors tend to be rocky, with shallow residual or alluvial soils. Vegetational communities are simple, the low rainfall and its seasonal occurrence producing open, semiarid types dominated by the savanna. The long drought and its accompanying fires deplete the already sparse vegetation and alter the physical conditions of the soil, making it susceptible to later erosion.

Most of the ecological data presented here were gathered from Komodo (Chap. 1), though additional material was obtained from Padar and Flores.

Edaphic Factors

Geologic.—Although climate is obviously responsible for altitudinal zonation of Komodo plants, latitudinal zonation is determined largely by geologic factors. The driest habitats are in lowland areas where the ground is impermeable and the slopes steep, as on the dense coralline limestones near Ntodo Klea and the igneous hilltops of many parts of the island. These areas are often sparsely clothed with grass. Under such conditions, detritus collects only at the break in slope at the foot of most hills and along intermittent stream courses. The permeable ground of these areas receives and holds runoff from the hills, supporting an open forest.

Most of Komodo is formed on an igneous base composed largely of fine-grained. slightly basic rocks, of which the chief mineral is plagioclase. Flanking this volcanic mass are lenses of tuffs, sandstones, and conglomerates with intercalated limestones, sandy shales, and clays. The eastern part of Komodo is composed mainly of steeply tilted beds of coralline limestone: Padar, Rintja, and Flores are very similar. Detailed geological accounts of the area are given by Tobler (1921), Ehrat (1928a, b), Brouwer (1942a, b), and Umbgrove (1949).

Soils of the Lesser Sunda Islands have not been extensively studied. They have little of no visible duff during most of the year. Leaf litter collects at the base of some perennial shrubs but generally disappears quickly.

Volcanic.—One of the chief volcanic belts of the world passes through the islands on which the ora is found. In many parts of the archipelago, slight earth tremors are felt every few weeks; one major and two minor shocks occurred on Komodo during our 11-month stay. As might be expected, some shocks are very destructive (Zollinger, 1850; Bickmore, 1868).

Aside from the immediate destruction caused by the explosions, quakes, tornadoes, and possible tsunami waves, the most important aspect of volcanism in the Lesser Sundas is probably the ejection of large quantities of ash. In the great eruption at Krakatoa, volcanic ash rose 40 km into the air and was dispersed around the world. Some floated for two years before it finally settled to the surface. When Tamboro erupted on Lombok, the ash fall reached an average depth of 0.6 m as far away as Flores (400 km east) and Java (420 km west). Most of the approximately 44,000 people that perished as a result of the Tamboro eruption did so because of the heavy ash fall (Bickmore, 1868). Zollinger (1850) pointed out that this same explosion modified the distribution of savanna and forest habitats on Sumbawa by starting many forest fires. Some of the resulting savannas remained for many years. Zollinger believed that the open, decimated forest communities caused less rain to fall on Sumbawa, produced drier and hotter surface temperatures, and ultimately caused rivers to remain dry for longer periods than before the eruption. These changes were said to be less marked in eastern Sumbawa, where the ash fall was slight. On Ternate almost all the vegetation died in 1862 as the result of an ash fall originating from an eruption 80 km away.

In addition to the effects of suffocation and fire caused by ash falls, the complete clogging of wells and water holes (Bickmore, 1868) must also have extremely important repercussions for wildlife and man. Moreover, soils are completely and quickly modified. And, lest the reader consider volcanism and ash falls as rare phenomena in this area, in the Banda Islands alone eruptions have been reported on an average of once every 21 years since 1586 (Bickmore, 1868).

The immediate and long-term effects of modern ash falls were summarized by Wilcox (1959), who reviewed several factors applicable to the study of past and present volcanic ash environments. Plants are smothered and mechanically overloaded in addition to being chemically attacked. Animals may be debilitated or may die, often months later (Wilcox, 1959; Thorarinsson, 1956).

Both immediate and long-term, as well as direct and indirect, effects of ash falls can also be documented on Komodo. On 4 January 1969 Kali Bara, located 300 km east of Komodo, near Ende, Flores, erupted. The southeast trade winds blew most of the ash northwestward, affecting the entire range of the ora. In western Flores the ash was said to have been over 30 cm deep; almost all the crops were crushed by the weight of the ash, and many other plants apparently suffocated shortly thereafter. In certain places on Komodo the ash attained depths of approximately 1 cm, and much of the grass, which had just begun to sprout, was killed. Thus, the food of most herbivores was greatly reduced. In at least the lowlands most game wallows were completely choked with ash, so that very little water was available for drinking. Deer were said to be stumbling around, apparently weak from lack of food and/or water, and predation by oras on large animals was probably heavy. However, it was quite likely a particularly difficult period for small oras, since insect populations, normally high during the wet season, were undoubtedly very low.

Climatic Factors

There is very little climatological data for this part of Indonesia. No weather data were available for Komodo, Rintja, or Padar, though some from Flores and Sumbawa have been used for extrapolation by Blower et al. (1977). During our study, an almost complete annual cycle was obtained from sea level at the base camp at Loho Liang, Komodo. Techniques are outlined in Chapter 1.

Precipitation.—The geographic range of the ora includes the driest areas in the Republic of Indonesia. In fact, the precipitation pattern in Loho Liang is the driest of any surrounding area for which such information is available, with a total of 650 mm of rain per year, compared to 787 mm for eastern Sumbawa and 815 mm for western Flores (Fig. 4–1). The typical rainfall pattern for this part of the world is heavy precipitation in the monsoon months of December to March. Most monsoon rains, often in the form of cloudbursts, last a few hours or less. However, cloudy weather associated with these rains often prevails for several days, due to the regional nature of the storms. There is less rainfall on Komodo than on any of the surrounding larger islands, partly because the January storms move southeastward and most of the water has already fallen onto islands northwest of Komodo, such as Java and Bali. In addition, Komodo receives no rain from the June monsoon, because little moisture has been picked up by the northwest-moving storms developed in high-pressure systems over Australia.

Though no rainfall data are available from higher elevations on Komodo, we know that rain fell there on at least twice as many days as it did in Loho Liang. In addition, these high peaks were commonly hidden by small cloud banks from late November to well into May. This would suggest that the higher forests of Komodo probably receive a minimum of 1000 mm of rainfall yearly.

In the Nggoer, Flores, area the summer monsoon is from November to April,

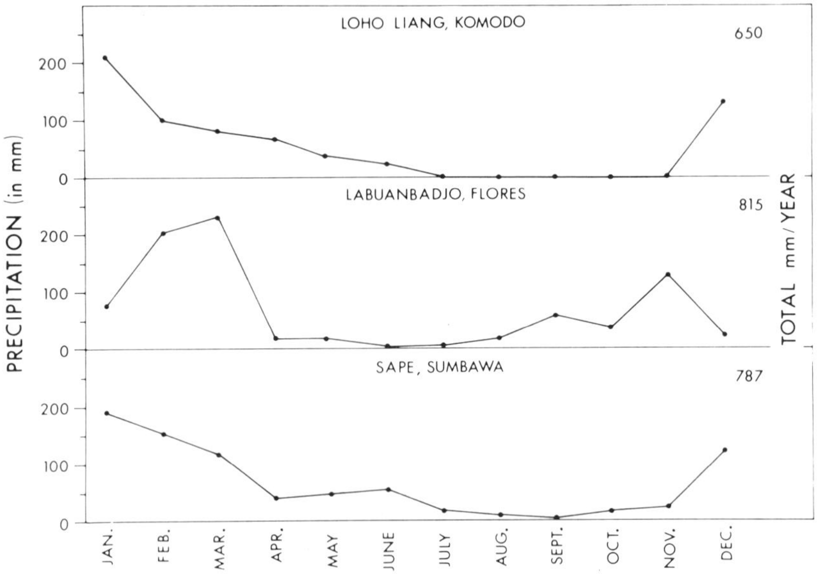

Fig. 4–1. Annual precipitation patterns in Komodo and surrounding areas.

though most rain falls from December through February—a pattern similar to that on Komodo. However, in the more mesic mountain forests several hundred meters above Nggoer, the monsoon lasts from October to April. Records from Wayarana Mission in eastern Mangarrai (el. ca. 400 m) show an average annual rainfall of 1,850 mm over a 4-year period.

Not all lowland areas of Komodo had as much rainfall as Loho Liang. Field experience and observation from the base camp both indicate that less rain fell in Loho Kalo, even less east and south of the Ntodo Klea complex. Padar received rain only rarely during the wet season, and I would estimate that total rainfall there is less than 375 mm.

Annual rainfall patterns on Rintja and western Flores are very similar to those on Komodo. However, the northern part of western Flores receives only a January monsoon, whereas the southern coast also receives a very small June monsoon.

Groundwater.—Free water is available in pools at all elevations on Komodo during the rainy season. The largest and longest lasting are those at high elevations in the Gunung Ara–Gunung Satilibo and Gunung Vau complexes. Here natural depressions, enlarged by boar and water buffalo to depths of 1 m and diameters as great as 25 m, often contain water all year long. Pools formed in streambeds at medium to high elevations are common, the largest often lasting until June, while the streambeds themselves rarely hold water for more than a few days after each rain. Wai Sadrap, near Kampung Komodo, is an exception, with flowing water carried in it until May or after. A few small springs produce water all year, though it is often unpotable during the last half of the dry season.

There are only three small water holes on Padar, all containing water throughout the year. Surface water is much more common on Flores, but within the range of the ora permanent streams are also rare. The reports of Hoogerwerf (1954a) and Darevsky and Kadarsan (1964) suggest that there is little surface water on Rintja.

Long-term precipitation patterns.—The results of the analyses of tree ring widths (Chap. 1) are shown in Figure 4-2. Only the extreme years were considered as wet or dry. The samples showed considerable correspondence and provided a total record of 103 years. Apparently the years from 1866 to 1885 were somewhat dry, while those from 1885 to 1919 were equally distributed between wet and dry. However, the period from 1919 to 1924 was quite wet, followed until 1943 (with the exception of 1935) by a long dry period. These are the years in which practically all the early expeditions came to study the ora. It was generally wet from 1943 to 1947, and trees of this period had twice the ring width of any other in the entire sample. The years from 1948 to 1953 were dry. During most of the following years the weather has fluctuated between dry and wet, with the latter more prevalent. Since 1965 it has been slightly dry.

Wind.—The prevailing winds are southerly, becoming northerly during the end of the dry season. By the early part of the rainy season, storm winds shift to the northwest, bringing the first hard rains. By the middle of February the monsoon season is at its peak. The average wind velocities at Loho Liang vary from

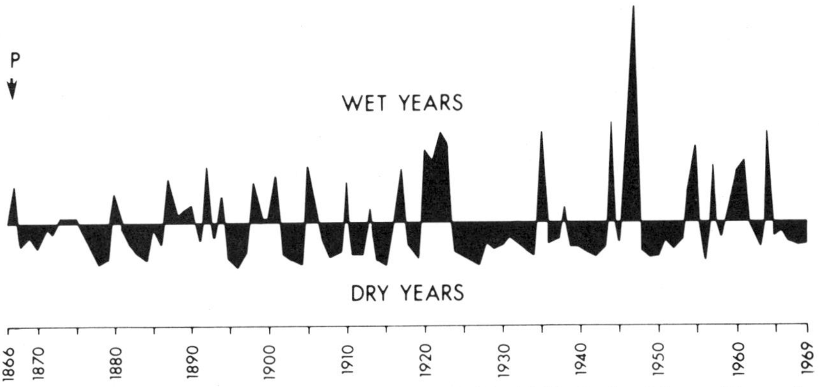

Fig. 4–2. Relative annual precipitation, 1866–1969, Loho Liang, Komodo; based on tree ring analysis, *Tamarindus indicus,* breast-high diameter 21 cm.

0.6 to 5.6 km/hr, being lowest during the monsoon. Maximum gusts at the base camp were estimated at 80–100 km/hr, usually southerly and accompanying rainstorms. The principal effects of air movement on animals and plants of Komodo were changes in evaporation and transpiration rates. However, for monitors, wind direction and velocity are both important in locating carrion (Chap. 9).

Barometric pressure.—Barometric pressure at sea level on Komodo is very uniform throughout most of the year (729.0–751.8 mm Hg) except during the early part of the monsoon season, when the average drops to 690.0 mm Hg. There is no correlation between pressure change and ora behavior.

Relative humidity.—Monthly mean relative humidity at the Komodo base camp varied from a low of 75.2 percent in October to a high of 86 percent in February. Oras tend to become less active at higher humidities and are only occasionally active during rain (Chap. 9).

Solar radiation.—In this part of the Sunda Islands solar radiation varies from 13,128 Cal/m^3 per day in June to 16,732 Cal/m^3 per day in October (Löf et al., 1966). Oras are exposed to highest solar radiation late in the dry season when most trees have shed their leaves, and it is at this time of year that behaviors are noticeably modified (Chap. 9).

Temperature.—The average weekly macroenvironmental temperatures at the base camp are shown in Figure 4–3. Maximum air temperature during the year was 43° C, minimum 17° C. Black bulb temperatures varied from 23° C at night and on cloudy days to 54° C after the short rainy season and before the beginning of the dry season. Highest temperatures tended to be at noon or slightly after. Average annual temperature at sea level on Komodo is 26.7° C. Though no continuous records are available for higher Komodo elevations, at Wayarana Mission (el. 400 m), Mangarrai, Flores, the average annual temperature is 22° C.

Vegetational Factors

There are two large regions of seasonally deciduous forest in Southeast Asia—the Indoburmese and the Sundanese. Vegetationally they are very similar (Van Steenis, 1935: Champion, 1936), though separated by the evergreen rain forest massif of mainland Southeast Asia.

Many systems of tropical vegetation analysis have been proposed since the important work of Schimper (1935), but none has been generally accepted (Richards, 1952). Excluding grasslands, most authorities recognize several major lowland tropical "climatic formations" on Komodo and adjacent islands. Three major lowland types are recognized: monsoon forest, savanna, and steppe. Most areas also possess small fringing mangrove forests and quasi-cloud forest at elevations of 500 m and higher. The distribution of major types is shown in Figures 4–4 and 4–5.

Though plant cover for each major plant association was investigated on several islands, the most comprehensive work was conducted in the Loho Liang study area.

Cover density.—Cover density and its obscuring qualities in each habitat are obviously important to both larger predators and their prey. These factors determine where an attack might best be launched or where it might be avoided by the prey.

When raised to its fullest extent, the head of an adult ora is approximately 40 cm above the surface, and when lowered the entire animal is less than 20 cm high. Table 4–1 shows the result of determining the visual obscurity (see Chap. 1 for techniques) of four plant associations near these levels. Note that foraging

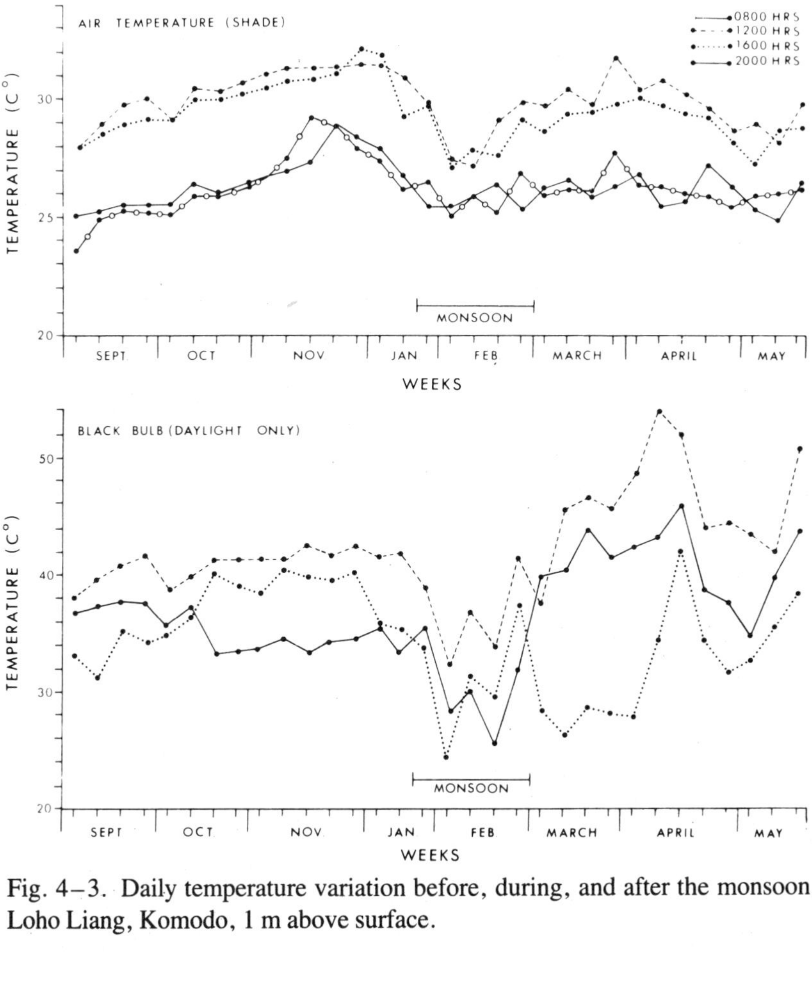

Fig. 4–3. Daily temperature variation before, during, and after the monsoon, Loho Liang, Komodo, 1 m above surface.

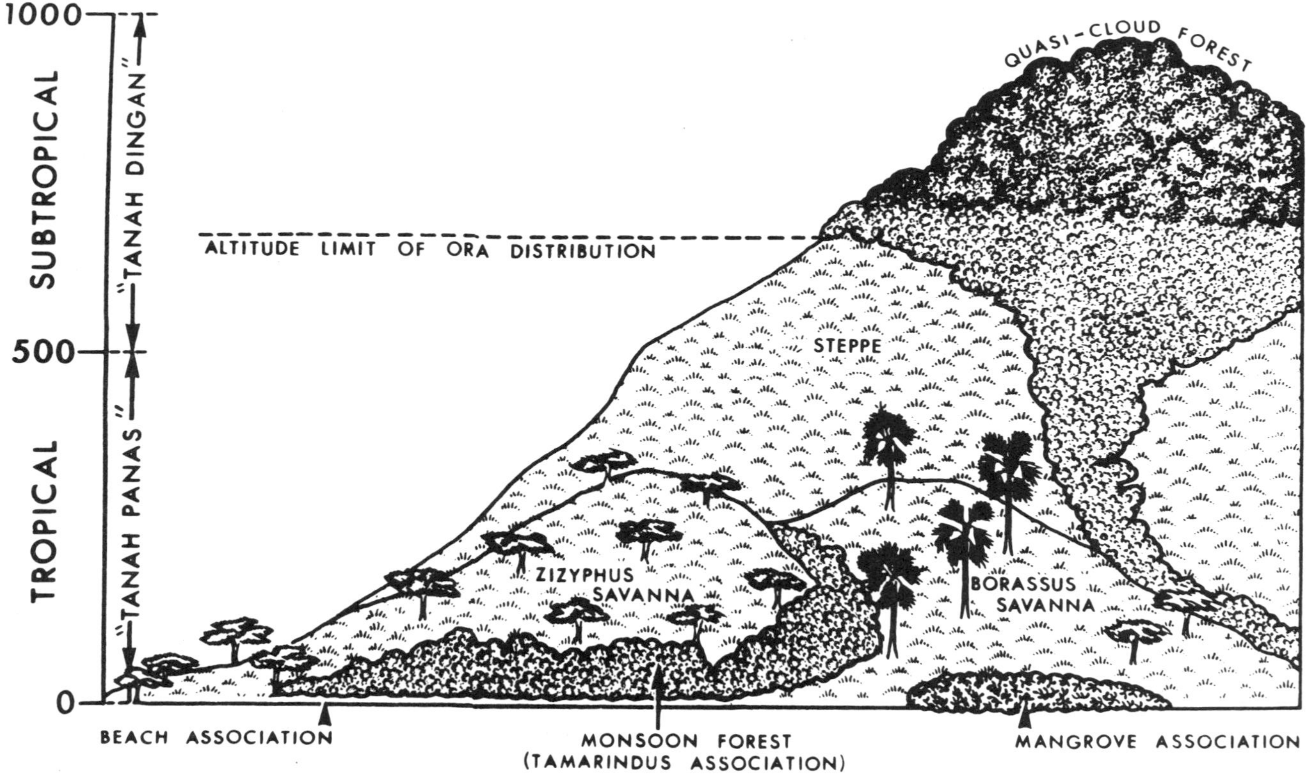

Fig. 4–4. Usual topographic distribution of major vegetation types on Komodo (elevation in m).

oras must extend their necks full to use distance vision in locating prey both in savanna situations and in the hillside monsoon association; they would not be able to do so in the other monsoon forest type, where distance vision can only be employed with the head held lower to the ground. Juveniles would be unable to use distance vision in prey location in any but the valley monsoon forest. Because much of the foraging takes place in ecotonal situations, the visual obscurity of these zones becomes particularly important to both the ora and its larger prey.

Principal Habitats

Principal habitats and their topographic relations are shown in Figure 4–4, distribution of major forest types in Figure 4–5, and diagrammatic vertical cross sections in Figure 4–6.

Quasi-cloud forest.—On the often cloud-bathed pinnacles and ridges above the 500–700 m elevation, these forests of the Lesser Sunda Islands are dense, shady, cool, and moist during most of the year (Fig. 4–4). *Tanah dingan* (cool ground), as it is called by the Sundanese, is a community often harboring a relictual flora and fauna of many endemic species (Auffenberg, 1980). Floristically, it is characterized by moss-covered rocks, rattan, bamboo groves, and trees generally not found at lower elevations, such as *Terminalia zollingeri*, *Podocarpus neriifolia*, *Uvaria rufa*, *Ficus orupacea*, *Calophyllum spectabile*, *Mischocarpus sundaicus*, *Colona kostermansiana*, and *Glycosmis pentaphylla*. On Komodo distribution of such forests is limited, though they are widespread on Flores.

Oras are rarely found here, being generally restricted to lower habitats. In the mountains of western Flores the vertical boundary of the animals is 250–450 m above sea level in a transition zone between the quasi-cloud forest above and the monsoon forests and savannas below. The transition forest is dominated by a mixture of xeric and mesic trees and shrubs (G. Wasedoo, Fig. 4–6). *Uvaria rufa*, *Pterospermum diversifolium*, *Bambusa blumeana*, *Murraya paniculata*, *Colona kostermansiana*, and *Diospryros malabarica* reach their greatest abundance on mesic slopes and often extend into the lowlands along permanent watercourses (Fig. 4–4). The ora is not typically found in this association, though *V. salvator* is common along streams.

Tropical savanna forest.—Savanna covers most of Komodo, Padar, and Rintja (Figs. 4-4, 4-5; Pl. 3). The open canopy allows a great deal of sunlight to reach the lowest levels, particularly when the leaves have fallen in the dry season. The understory is composed mostly of medium and tall grasses (0.5–4 m high). Fires are common during the long, dry season.

Two facies of a single savanna community (*Borassus-Zizyphus-Themeda* community) that sometimes blend into one another are recognizable on Komodo

Fig. 4–5. Distribution of major forest types on Komodo. Solid represents quasi-rain forest, leafy area represents monsoon forest, and grassy areas represent savanna types.

and nearby islands. One facies is dominated by the lontar palm, *Borassus flabellifer* (Fig. 4-6, Pl. 3), and is usually found in areas with perched water tables in a rocky, calcitic or dacitic, residual or colluvial lithosol or a reddish, lateritic clay. The *Borassus* facies is distributed from near sea level to approximately 400 m. It is not found on Padar, but occurs on both the north and the south coasts of Flores and is common on Rintja.

Zizyphus jujube and scattered *Tamarindus indicus* (Fig. 4-6, Pl. 2) are the dominant trees of the other facies of this savanna community, which is distributed from sea level to about 500 m. The soil, sometimes waterlogged for a few

Table 4–1. Average visual obstruction (from 100 m) at different heights above the surface in four ora habitats (Loho Liang, Komodo; N = 4 for each habitat)

	Plant associations			
Height above surface (cm)	*Borassus-Themeda* (hillside savanna)	*Zizyphus-Themeda* (valley floor savanna)	*Tamarindus-Hypoestes* (valley floor monsoon forest)	*Tamarindus-Cladogynos* (hillside monsoon forest)
151–180	No[a]	No	Yes	No
121–150	No	No	Yes	No
91–120	No	No	Yes	No
61–90	No	No	Yes	Yes
31–60	No	No	Yes	Yes
0–30[b]	Yes	Yes	No	Yes

a. No = no visual obstruction; yes = visual obstruction.
b. Adult oras can just barely lift their heads above this level.

weeks in the rainy season and often dessicated the rest of the year, is primarily a slightly halomorphic, azonal alluvial type, usually a black to gray sandy clay, rarely rocky or gravelly.

In general, the life forms of the *Zizyphus* facies are more varied than those of the *Borassus* facies. The grasses are similar and in some parts of Komodo, as along the north coast, the dominant trees of these two communities are completely intermixed. Appendix 4 lists the common plants of both savanna facies on Komodo, with remarks on the importance of each to the biology of the ora.

The tropical savannas of Komodo and their boundary woodlands constitute the major habitat of *V. komodoensis* (Pl. 3, top). However, this habitat may be limiting when trees become very scattered (Pl. 3, bottom). Most other animal life of the savanna varies seasonally. In the wet season (January–March) insects are very common, particularly grasshoppers. Birds are never abundant, except a few species of ground doves. Water buffalo, deer, and wild boar feed in the savannas, and deer sleep in the high, dense grass clumps, particularly in the foothills.

Grass productivity in most areas is fairly high. Fire usually occurs only once each year in most areas, and overgrazing is not a problem on Komodo.

Monsoon forest.—The other major forest association of Komodo is the tropi-

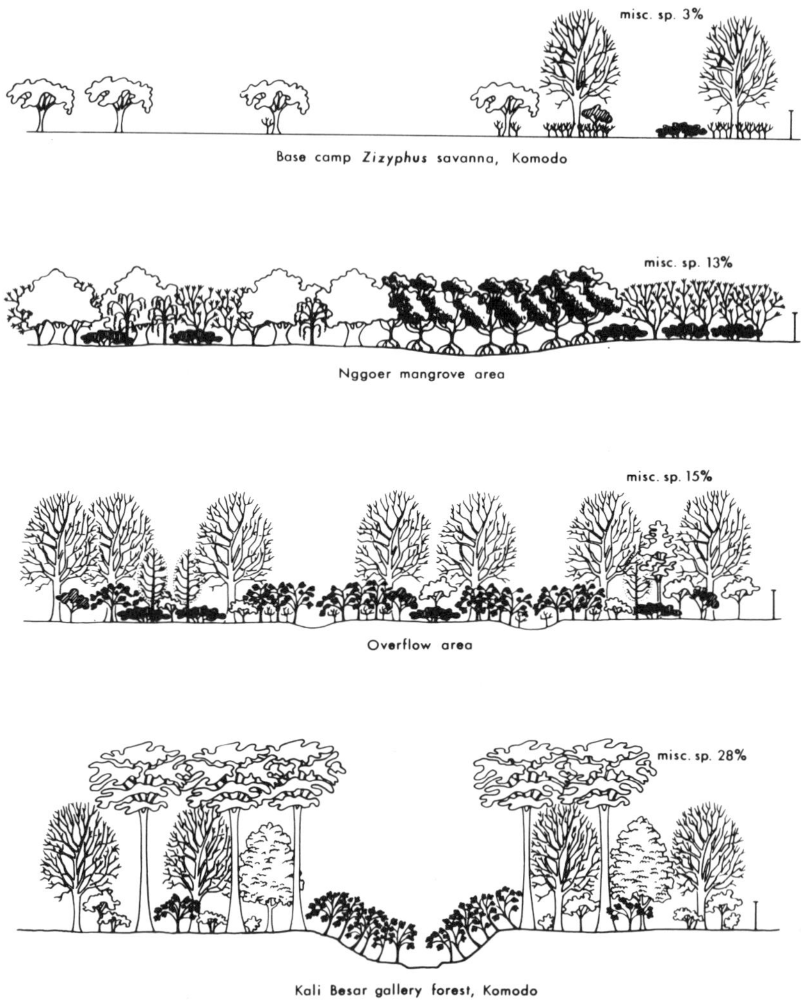

Fig. 4–6. Idealized profile diagrams of plant associations inhabited by oras on Komodo and Flores. Bar on right represents 2 m. Each symbol represents approximately 2 percent of total plants tabulated in transects. Key shows plant symbols (p. 60).

cal deciduous monsoon forest (Fig. 4-6; Pl. 3). It is composed mainly of phreatophytes, though it includes some xerophytes. Like the savanna forest it is largely fire resistant, and many of the trees are almost leafless in the dry season. Some dominant trees, like *Tamarindus indicus,* occur in both associations, but the monsoon forest canopy is higher and more continuous (Fig. 4-6). The crowns commonly touch and undergrowth may be dense, though the surface level is

Fig. 4–6 Continued

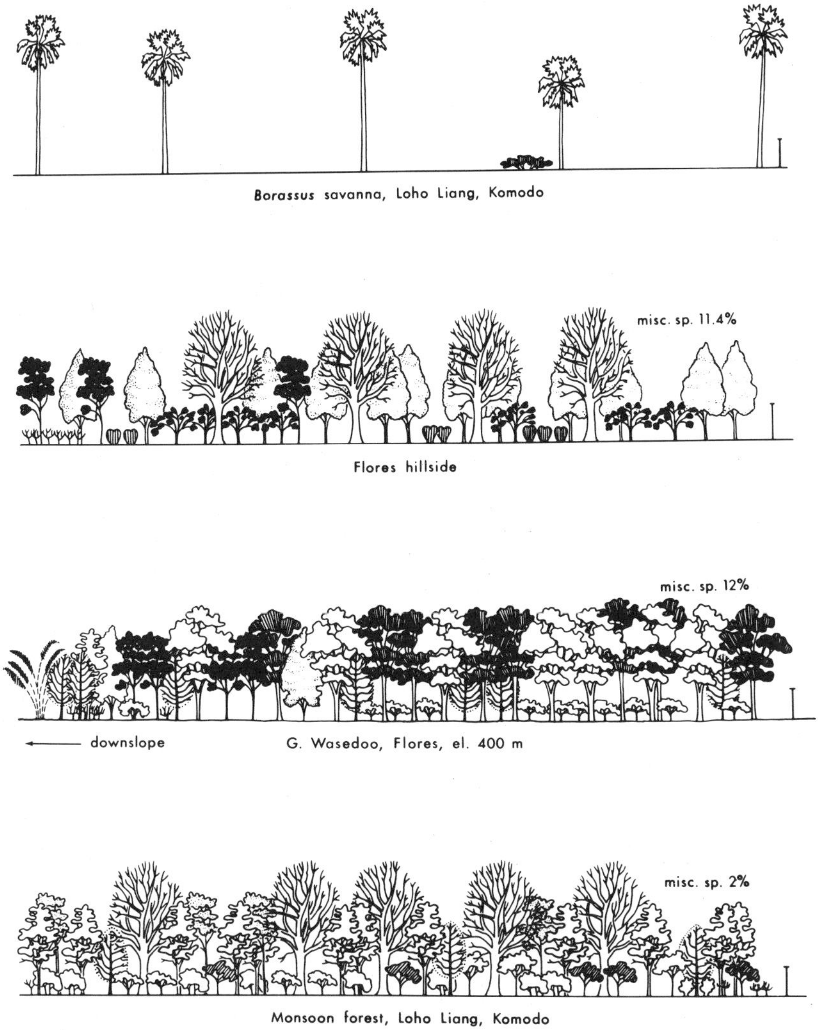

(Continued)

usually fairly open. *Tamarindus indicus* and *Sterculia foetida* are larger than any other trees and may have a breast-high diameter of as much as 80 cm.

On Komodo, monsoon forest covers considerably less of the total area than does savanna (Fig. 4-5). It is found at all elevations below approximately 500 m; that is, the almost continually dry, hot region known as *tanah panas* (hot ground). On Komodo, the forest may be a gallery type found along stream courses (Fig. 4-4; Pl. 3). A bandlike strip of monsoon forest often extends along the base of hills. A list of the dominant floral elements of the Komodo monsoon forests is provided in Appendix 6 and for western Flores in Appendix 8.

Two weakly developed facies of the *Tamarindus* monsoon forest community are here recognized: *Tamarindus indicus–Jatropha curcas* and *Tamarindus indicus–Cladogynos orientalis*. The latter is best developed in drier areas. The former is salt tolerant and occurs in slightly wetter areas, often on sandy soil behind the low storm dunes. The *Tamarindus–Jatropha* facies often produces tangled thickets that are important in the biology of the larger oras near the coast, for deer and boar are hunted there. The rivulet beds along which the thickets are formed are common ambush sites and are often the only routes by which large animals can move through the dense vegetation.

Ecotonal zones are very important in the behavioral ecology of the ora. They are the preferred nesting sites of many species of birds and are areas in which basking and other major activities of many reptiles occur, particularly arboreal species. Foraging oras frequent the ecotones, and ambush sites are usually located along game trails in these zones.

Fig. 4–6 Continued

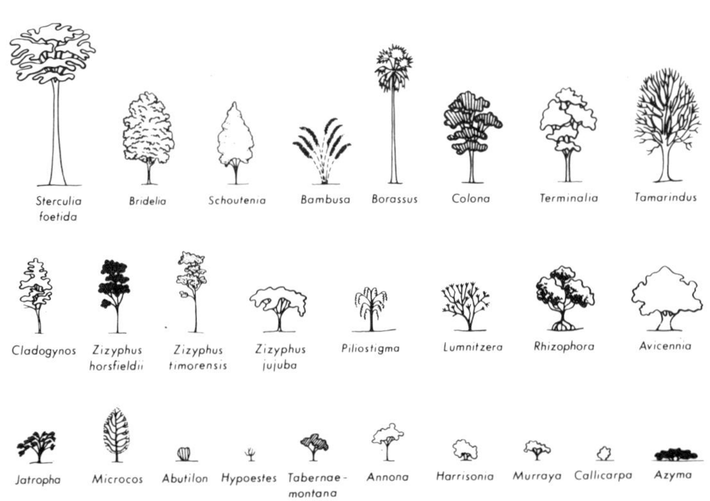

Pl. 3. *Top,* savanna association, *Borassus flabellifer* facies, Loho Kalo, Komodo. Monsoon forest trees restricted to small stream banks. *Bottom,* monsoon forest clothes some hillsides and follows small stream courses. The savanna association in the immediate foreground and on the hillsides represents the *Zizyphus jujube* facies.

On Flores the near-sea-level monsoon forest association includes several communities not found at low elevations on Komodo. Of these, the most important is probably that dominated by thorny bamboo (*Bambusa blumeana*). Oras are found here, but densities are apparently low. However, the usual occurrence of this community on the base of steep hillsides on Flores means that most oras and their ungulate prey must pass through it when moving from one elevation to another. On Komodo this community is restricted to higher, steeply sloping hillsides on Gunung Ara and Gunung Satilibo.

Plant succession in monsoon forest is an important factor in the behavioral ecology of the ora. Two major disturbances destroy or modify existing plant communities at low elevation in these islands–fire and shifting agriculture. Fire is important on all the islands. However, gardening is extensive only on Flores; there is none on Padar and very little on Komodo and Rintja. Land is cleared by slash and burn methods. Except for coconut plantations, garden lands are generally used on a 5-15 year rotational plan. Within 8 to 10 years these plots are usually dominated by *Hypoestes malaccensis* or *Abutilon indicum*. Other common shrubs include *Solanum junghulnii*, *Callicarpa sappan*, *Azyma sarmentosa*, and *Microcos paniculata*. The most common trees are *Annona squamosa* and *Tamarindus indicus*. The grasses are mainly *Setaria verticillata*, *Brachiaria ramosa*, and *Digitaria adscendens*, though the most spectacular and important is *Rottboehia escalbata*—a species that reaches heights of 3.5 m during a single wet season. These thick plots are favored by deer and boar, because of both the cover afforded and the variety of food plants. They are also extensively used by oras for hunting, particularly in the afternoon. Repeated annual burning forms a derived grassland (*padang*) that is an obviously deteriorated habitat. In this case, the densities of both ora and prey species become depressed.

Mangrove forest.—The oriental region possesses approximately 30 mangrove species (Chapman, 1970), of which 4 occur on Komodo, where they are flooded once daily by the tides. Differential saline tolerance leads to horizontal zonation of the species along the shoreline.

Rhizophora mucronata usually grows well within the tidal zone, closest to the open water. Impenetrable thickets are formed by the tangled buttress-like roots. On many islands this association forms an almost unbroken fringe along the entire coastline, but on Komodo it is restricted to the east coast near Sabita. This zone is only occasionally inhabited by oras; we have seen them here, presumably hunting for crabs, on Padar, Flores, and Komodo.

Behind the *Rhizophora* zone, in appropriate areas, is an association in which *Lumnitzera racemosa* is dominant. Though not tall, this species grows very densely, forming large thickets, occasionally broken by small *Cyperus javanicus* marshes. The land is less flooded than seaward, but birds and reptiles are not common. However, wild boar are sometimes very abundant. Thus, oras, particularly larger ones, are often found in the *Lumnitzera* community.

Further inland, in certain parts of Soro Masannga Bay, there is still another mangrove community composed mainly of *Avicennia alba*. This species is least tolerant of salt water flooding and is often mixed with species grading into monsoon forest (Fig. 4-6). The terrestrial animal life is quite varied, and the community is not limiting to oras. The mangroves *Ceriops tagal* and *Avicennia alba* are also found on Komodo, but both are very restricted and are thus not important from the standpoint of ora ecology.

Mangrove associations are much more extensive and complex on Flores. Five genera occur near the mouth of Nanga Löok, each showing distinct zonation. *Sonneratia* is usually found on sandy or muddy flats just above the lowest tides. *Bruguiera* is found in one small cove near the mouth of Nanga Löok. The small areas in which both genera are found make them relatively unimportant in the lives of the local oras. However, the remaining mangrove species (*Exoecaria agallocha* and *Lumnitzera littorea*) are widely distributed and often form large stands. Here wild boar and monkeys are common, and oras spend considerable time hunting them (see Chap. 9).

Pandanus-Spinifex community.—This is a hot, dry, sandy, saline, and windy zone immediately behind the beach. It is generally not important to the oras, though they occasionally sleep in particularly thick stands of *Spinifex littoreus, Crotalaria retusa,* or *Barleria prionitis*.

Perhaps the most important vegetative feature of this zone is the shade produced by certain plants (*Jatropha curcas* and *Thespesia populnea*) whereas it is otherwise scarce. Shade is particularly important to oras hunting the beach wrack, where they are exposed to high solar radiation. It is this radiation level and the absence of both cover and appropriate prey that make the beach community generally unsuitable for oras.

Steppe.—The great variation in occurrence and composition of grassland types has generated a vast literature, especially in respect to their formation and continuance in tropical areas. Because much of Komodo is composed of savanna and steppe, the question of development becomes important in the ecology of the Komodo monitor.

There are two major viewpoints regarding grassland formation in Southeast Asia. One is that in most tropical regions, areas dominated by grasses are demonstrably secondary, arising from the destruction of deciduous forest as a result of burning or shifting agriculture; the second is that the distribution of grasslands is entirely climatically and edaphically determined.

Part of the confusion and uncertainty regarding grassland formation in the Lesser Sunda Islands is undoubtedly based on the often repeated error that the grasslands of these islands are composed of *alang-alang* (*Imperata cylindrica*) (e.g., Horst, 1926; Lallemant, 1929; Pfeffer, 1959; Kern, 1968). Actually, the extensive grass-covered areas within the range of the ora are dominated by *Themeda* and *Chrysopogon*. *Imperata cylindrica* normally occurs on heavy, often seasonally waterlogged soils. On Komodo it is found only in the moist garden area of the main village and in a few small stands in the mountains (Hoogerwerf, 1954a), and on Padar there is only one small stand adjacent to a permanent spring. On Flores, where conditions are generally more mesic, *Imperata* is more widely distributed, and is usually associated with old gardens (*ladang*). Periodic burning of these plots produces a pure alang-alang grassland, called padang, which is of little agricultural value. Most of the padang areas on Flores have formed within the memory of present-day villagers, and the ladang to padang sequence is well known to most of them. Alang-Alang grassland associations can also form after catastrophic destruction of monsoon forest, as occurs during volcanism. Such padang areas may remain for a long time, even without fire (Zollinger, 1850).

The *Themeda-Chrysopogon* grassland community covers extensive portions of the ora's range. Unlike *Imperata* stands, it usually occurs on the well-drained soils of hills, as well as in valleys. None of the villagers has seen any change in the character or distribution of this grassland type, whether burned or not. Furthermore, it is in this community that the relictual grassland species, such as *Vipera russelli* (Auffenberg, 1980), occur, and it is this community, and not the one dominated by *Imperata,* with which the ora is most closely associated. I believe the *Themeda–Chrysopogon* community is a natural relictual grassland with a climatic and edaphic basis.

With the exception of the mesic forests, all the major vegetative types within the ora's range are fire adapted. Besides man, many natural phenomena cause vegetation ignition: lightning, volcanism, rock falls (Phillips, 1965), spontaneous combustion, and even thermophilic bacteria (Phillips, 1936). Of these, the most common is lightning (Komarek, 1966, 1967). Several studies have made it clear that lightning potential over many parts of the world is extremely high, providing a firm meteorological basis for fire-controlled communities (Komarek, 1966, 1967).

Though several types of thunderstorms produce lightning fires, nearly all peak loads occur during general type storms which take place over a wide area and literally create waves of fires (Komarek, 1967). Most are produced when pulsating general storms occur after a long drought, particularly if the storms pass over an area in which the forests or grasslands contain a quantity of flammable material. The pre-monsoon period in the Lesser Sundas meets these conditions perfectly.

In addition, the Komodo topography favors ignition, since mountainous areas are more prone to lightning fires (Kourtz, 1967) than are lowland areas. Volcanic eruptions also spawn much lightning (Bickmore, 1868).

Though the entire question of such fires in the drier parts of tropical Asia deserves study, that lightning is an important factor in the formation and maintenance of savanna and steppe in the Lesser Sunda Islands can hardly be disputed.

This does not, of course, deny that man starts fires, or that he has not been an important factor in shaping the ecology of the area. His role as a pyromaniac in many parts of the world is attested to by the destruction of much original habitat by burning.

In the Lesser Sundas, fires are deliberately set for agricultural purposes, to make walking through the savanna easier, to improve grazing, and mainly to drive game.

The interplay of animal associates, vegetation, and fire has been examined in relatively few parts of the world. Most work has been done in Africa, practically none in Indonesia. As far as can be determined, oras seem little affected by fire, in spite of the unsupported statements of others. The villagers state they have never found dead oras in recently burned areas, and we found none with what could be considered fire scars. I presume oras escape fires either by running ahead of them or by entering burrows. Studies by Phillips (1965) and others show that large mammalian predators are not affected by even large fires; they in fact hunt near them in both derived and wooded savannas. Hoofed animals may travel long to moderate distances to return to burned savanna or grassland in order to browse and graze in and along their margins, even within the ora's range, where deer, buffalo, and goats are often found grazing on the new grass and browse produced after extensive burning. However, as in Africa and elsewhere, wild hogs tend to move away from burned areas and do not return for long periods of time.

Microecology

Near-Surface Environment

Diel thermal patterns.—The climate of the area immediately adjacent to the monitors is considerably different from that described by general macroclimatic data. Except for hatchlings, the ora's world rarely extends more than a meter above the ground. This narrow zone, from the substrate to just above the animal's head, is characterized by little ventilation and great radiation. These two conditions account for much of the ora's thermal behavior. While a thin boundary zone about 1 cm thick immediately above the surface is extremely important in determining the diel behavior of small poikilotherms, the most important climate in respect to ora activity is in the intermediate zone, which is about 80 cm thick. In this zone there is a steep vertical temperature gradient, and reversals in temperature patterns occur from the daylight (insolation) to nighttime hours (radiation). These changes from warming to cooling processes are exceedingly important in the diel behavioral patterns of the ora.

The transition from the insolation to radiation patterns occurs from 0500 to 0700 (Figs. 4-7, 4-8), when the intermediate layer is consistently lower than stations above the oras (who are generally inactive during these early hours). The boundary layer warms rapidly at about 0700, while at 10 and 100 cm the warming process does not begin until about 0800. The rapid temperature change within the intermediate layer from 0800 to 0900 clearly controls the beginning of the ora morning activity period (Chap. 5). At about 1000 an on-shore breeze causes great temperature fluctuation on Komodo; this increased wind velocity, coupled with the increase in eddy diffusion caused by upward movement of warmed boundary layer air, scatters the readings over a wider range.

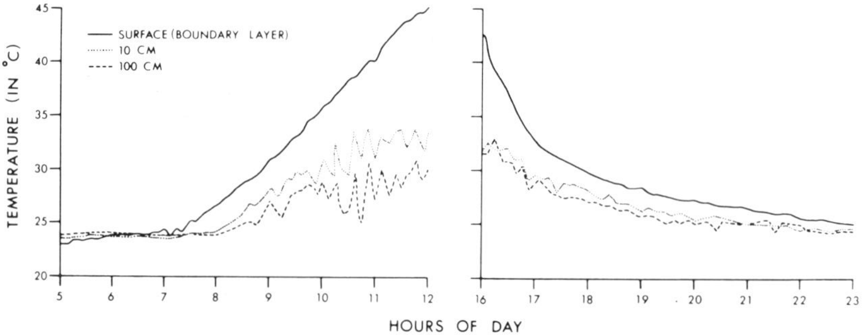

Fig. 4–7. Incoming (*left*) and outgoing (*right*) radiation patterns at three different levels on hill 19, Loho Liang, Komodo, Nov. 15, showing inversion from 0600 to 0700 hrs and thermal instability from 1000 to 1100 hrs. Compare with ora activity patterns, fig. 5–1.

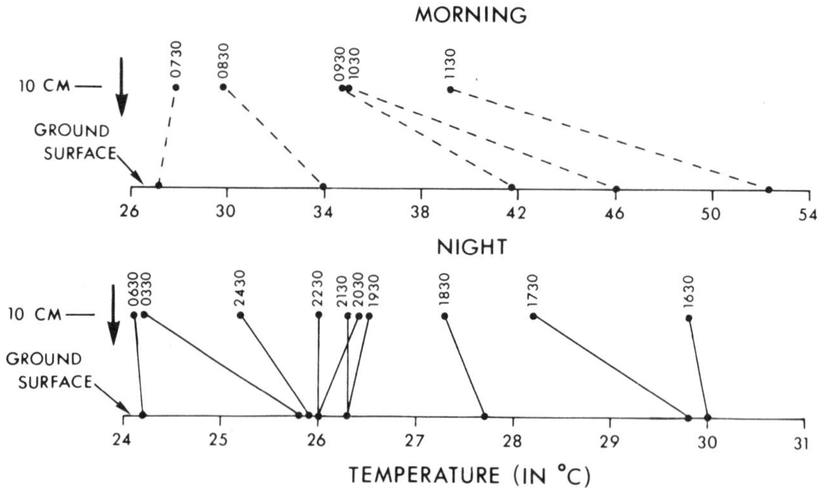

Fig. 4–8. Typical progressive thermal changes within 10 cm of the surface in the mornings and evenings.

Nocturnal cooling, even close to the surface where oras must function, begins before sunset (about 1800) and lasts till after sunrise (about 0600). Rate of nocturnal heat loss in ora habitats is highest in the early evening (Fig. 4-8), though the coolest parts of the entire day are just before dawn. The greatest temperature change occurs in the intermediate zone occupied by the ora. At night this zone contains the coolest air layers, with warm, lighter layers generally above the ora (Fig. 4-9). Similar results have been reported in xeric habitats near Poona, India (Ramanathan and Ramdas, 1935). Though not typical of mesic temperate areas (Geiger, 1950), the pattern may be common in xeric tropical habitats, as on Komodo.

Figure 4-10 shows the generalized temperature layers in the zone within which

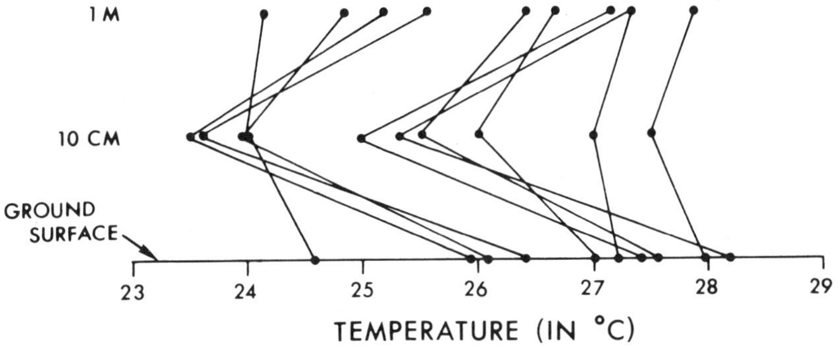

Fig. 4–9. During the night the coldest places in open areas are within 10 cm of the ground surface, that is, within the height of an adult ora over the surface.

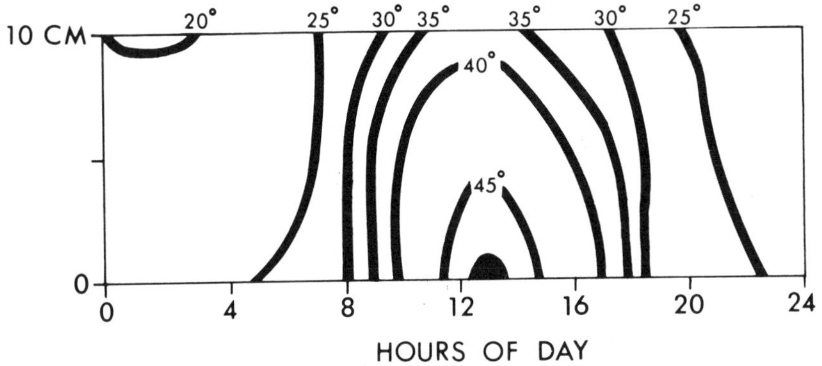

Fig. 4–10. Isotherm distribution within 10 cm of ground surface in an open environment. Optimal temperatures occur between 0900 and 1800 hrs, comparing favorably with diel ora activity patterns (compare with fig. 5–1).

the ora is operative. During the morning the bottom of a cold air layer is located about 10 cm from the ground surface, while a hot air layer occurs in the boundary layer after midnoon. (The importance of this thermal structure in regard to ora activity is discussed in Chapter 5). During this time the boundary layer temperature often rises above 50°C. Cloudy weather during the January monsoon may cause decided flattening of the usual diel temperature curves from 0 to 10 cm, particularly at night.

Topography.—In a large valley system, the downward movement of cooler air develops a well known characteristic evening "down-valley wind." It is of local occurrence and to a great extent determines the microclimate of the region affected by it. In every recorded instance the evening resting spots (Chap. 5) of adult oras were outside the areas influenced by down-valley winds.

Just as the difference between mountain and valley produces its own downslope winds by night, it often produces an upslope wind by day; its strength increases in proportion to insolation and the steepness and barrenness of the slopes. Upslope winds are important when oras are locating carrion. Under some upslope wind conditions, carrion can be scented from distances as far as 5 miles. Since wind direction is consistent for long periods, certain sites are more important than others in locating carrion by scent (Chap. 9).

In flat areas or straight valley systems, thermal stratification is usually rather simple. However, it becomes very complex in the hilly country present throughout the range of the ora. As in lowland areas, the evening outgoing radiation pattern in hilly areas results in the formation of a cold air layer just above the ground. Initially this stratification is of equal vertical extent at all points so that the cold air of the higher areas is the same distance above the surface as the cold air over lower ground. However, the density differential causes cold air from higher elevations (of even a few centimeters) to flow slowly to lower places and be replaced by the warm air forced upward. From the standpoint of ora behavior, it is clear that at night concave land forms such as valley bottoms are usually cooler islands, so that the coldest places at equivalent elevations on Komodo are valley savannas. The complex nocturnal temperature stratification in valleys and on hillsides is of particular importance to ora activity cycles and nightly shelters utilized.

Figure 4-11 is a diagrammatic cross section of the Loho Liang Valley. On the entire valley surface, the divide, valley walls, and floor, the intermediate layer cools off at night at the same time as the air in the boundary layer and a great reservoir of warm air fills the lower portion of the valley. The near surface environments of broad divides, such as Banu Sambi, remain relatively cold, because there is little or no downward cold air flow. However, narrow divides, such as Gunung Rankang and Nggoko Djarang, and parts of the hill slopes are warm.

Plants and ora behavior.—Vegetation also plays an important role in micro-

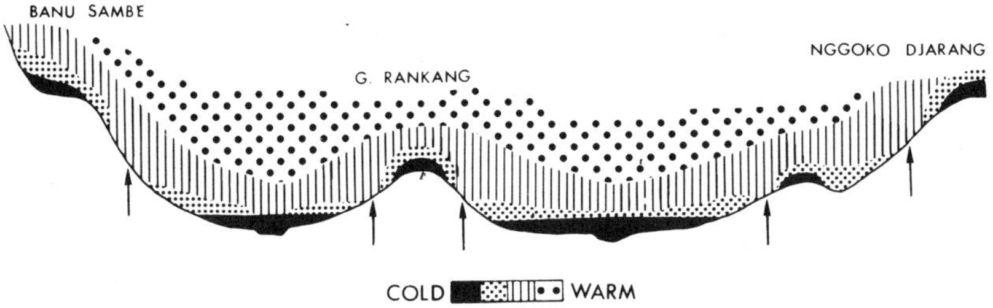

Fig. 4–11. Generalized diagram of distribution of evening cold and warm air masses, Loho Liang, Komodo.

ecology at the level of even adult oras. The 10-cm high diel thermal pattern at different times of the year in different microhabitats is illustrated in Figure 4-12. Note that the highest daytime temperatures (other than at the sand surface) throughout most of the year are in the grass and the coolest in the shaded but open brush. However, during the monsoon season these relationships are reversed. A 10-cm high thermal transect near the western coast of Flores is provided in Figure 4-13. Open beach associations are warmest; progressively cooler ora microhabitats are savanna (in grass), hillside bamboo thicket, seaside scrub, and under *Zizyphus* trees in the savanna, with the coolest being the shaded, open riverine forest.

Where convection is restricted by plant growth, stratification of hillside climates is better developed. On slopes free of vegetation, vertical stratification is often destroyed by winds and convection. In general, the closer plants are to one another and the larger the stand area, the cooler the lowest air layer during the day. As shown in Chapter 5, the precise activity particularly of large oras is constantly adjusted to these minor differences. Highest temperatures occur consistently at the surface where vegetation density is low. At night the air under plants is consistently warmer at ground level than above it.

In any microecological consideration of the ora, great significance must be attached to the ecotonal environment between monsoon forest and *Zizyphus* savanna, because it is in this transitional habitat that oras spend most of their time.

The climate at the ecotone is a result of the area's location between forest trunk-space and open savanna environments. The influence of trunk-space environment predominates on the outer edge, the savanna environment on the inner edge. The edge of the monsoon forest has the same effect on the immediate environment as an abrupt change in elevation. It is often in a wind shadow and, according to the direction it faces, it catches and accumulates or withholds heat from the savanna.

In most cases oras spend the night immediately inside the stand edge

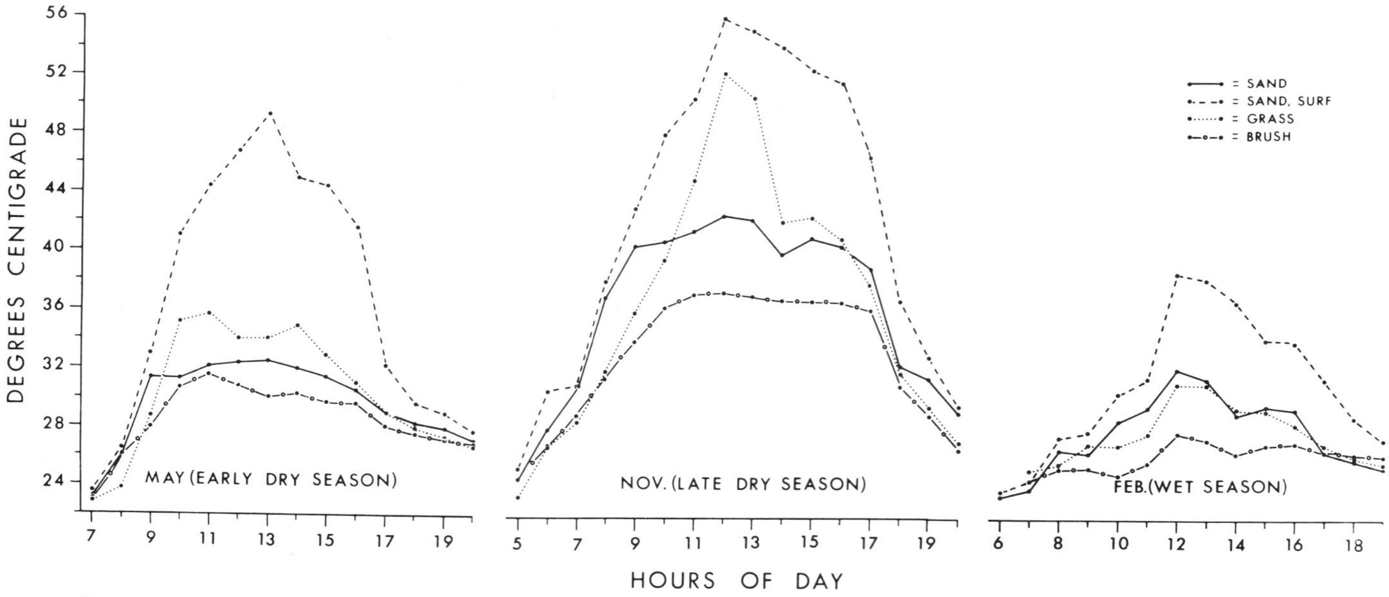

Fig. 4–12. Annual thermal patterns in different habitats at 10 cm above surface (except sand surface).

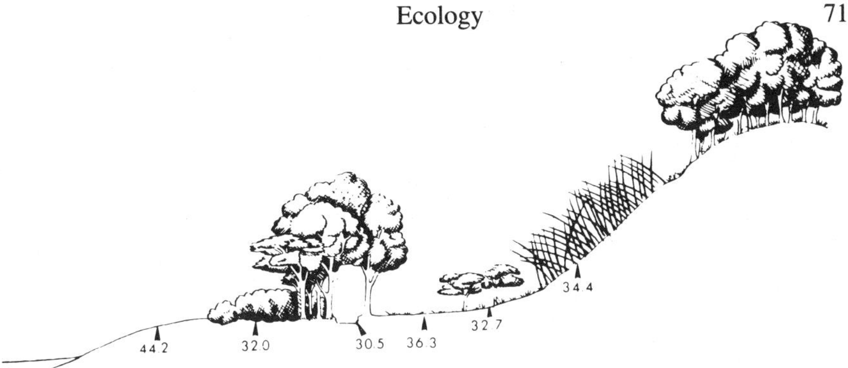

Fig. 4–13. Thermal transect (10 cm above surface), western end of Flores, constructed from averages of simultaneous readings at 1400 hrs, June 5–9, 1971.

(Chap. 5). Here, under the overhanging trees, half the sky is cut off, so nocturnal heat loss by radiation from the surface of either the soil or the ora is only half of what might be expected under the open sky of the savanna. During the day the forest hinders air currents, creating a dead air space near the surface. In this band, daytime surface temperatures are sometimes very high (Fig. 4-14). On the other hand, trunk-space temperature within the adjacent monsoon forest is generally low, and there are often strikingly different temperatures on either side of the actual stand border, though separated by only a few meters. The regular resting sites of large resident oras are always found in the ecotone, within a short distance of the high and low temperatures characteristic of this immediate area.

A series of thermal transects were made along lines passing through the regular resting sites of adult oras on hills 19 and 23 and along other lines running some distance on either side of them. These data show that the thermal gradients are steepest along those transects passing through the ora resting sites (Fig. 4-14). The importance of this steep gradient, as well as that of the "warm slope" phenomenon produced on hillsides at night in both nocturnal and diurnal thermoregulatory activity of large oras, is further discussed in Chapter 5.

Burrow Microecology

During this study 213 burrows were investigated with regard to temperature, humidity, soil moisture, and light penetration. Samples were drawn from the burrows along heavily shaded banks of Kali Besar, monsoon forest and hillside savanna of hill 19, and the open, exposed steppe at the top of Gunung Kalo.

The daily variation of light penetration into different parts of the burrows (mouth, midway, and terminus) was measured by a photographic method dependent on the amount of reflected light captured on film. The light in each of 5

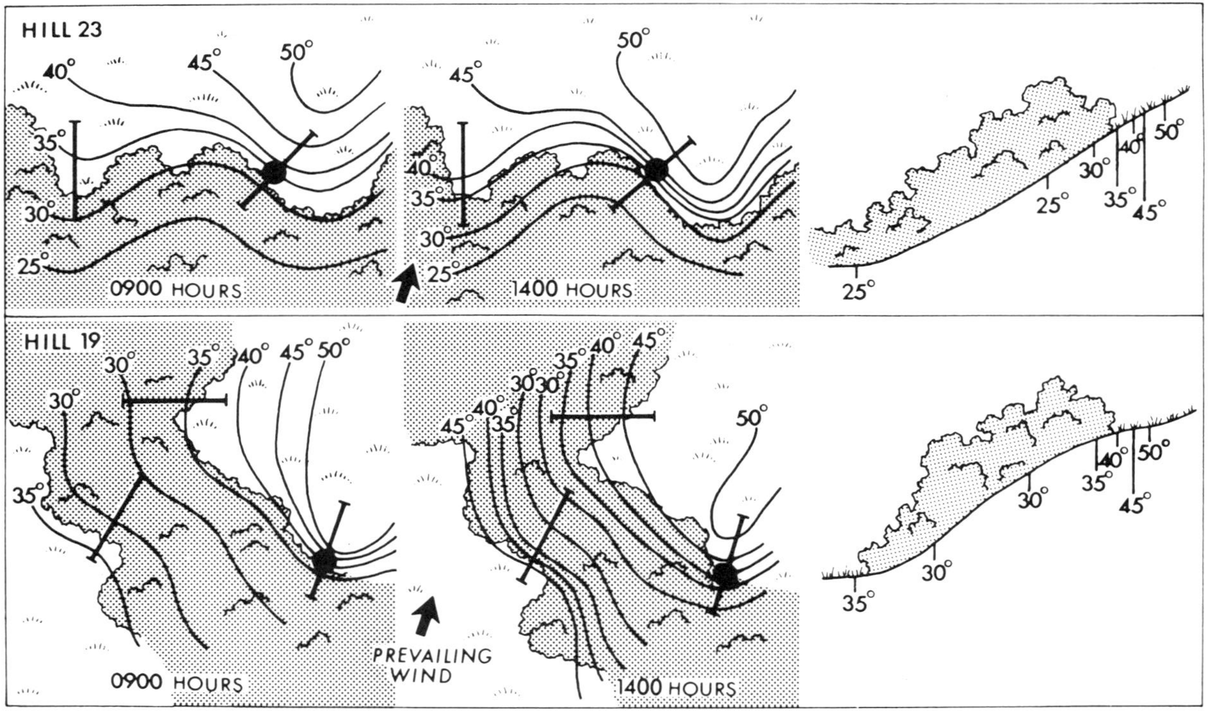

Fig. 4–14. Isotherms on hills 19 and 23, Loho Liang, Komodo. Bars represent thermal transects constructed. Dots are traditional resting sites. The left and middle sections are plan views, and those on the right are cross sections through the resting sites at 1400 hrs.

Table 4–2. Average seasonal differences between temperatures (in ° C) inside (mid-burrow) and outside ora burrows

Place and time	Wet season		Early dry season	
	Inside	Outside	Inside	Outside
Kali Besar				
0800–1000	0.7	–	–	0.3
1100–1300	0.2	–	–	2.5
1400–1600	–	0.3	–	2.3
after 1700	0.2	–	1.8	–
Hill 19				
0800–1000	0.4	–	0.3	–
1100–1300	–	3.9	–	3.0
1400–1600	–	2.9	–	1.8
after 1700	1.1	–	1.2	–
Gunung Kalo				
0800–1000	–	0.5	0.3	–
1100–1300	–	5.8	–	7.0
1400–1600	–	5.7	–	4.5
after 1700	0.2	–	0.8	–

Note: All differences are listed for the warmest site only.

burrows was measured in the morning, midday, and late afternoon over an extended period of time. Not surprisingly, light is brightest at midday, less bright in the morning, and darkest in the late afternoon. Burrow entrance direction is obviously important in determining light intensity levels in the holes. Almost all the light entering the burrows is reflected rather than direct illumination. There is no correlation between burrow use by oras and burrow light intensity.

Approximate soil moisture in the 12 sample burrows was obtained by weighing quantities of soil, drying them in the sun until completely friable, and then reweighing. Seasonal variation was from 18.0 percent (S.D. 8.1) in the wet period (N = 107), to 11.1 percent (S.D. 2.6) in the dry period (N = 63). During the monsoon period the soil in the hillside holes is wetter than that in the protected creek bed.

Relative humidity in burrows was obtained by means of an Atkins pistol type hygrometer provided with an extension sampling tube (N = 170 samples). Not surprisingly, the burrows are characterized by high atmospheric humidity (75.3–88.9 percent). It is generally highest from late afternoon to evening, and there is no significant difference in humidities of burrows located in creek beds or hillsides, or during wet and dry seasons. In general, humidities immediately outside the burrow at hole mouth height are about 10 percent lower than those well inside the burrow.

Burrow soil and air temperatures (N = 170 samples) show relatively little fluctuation, with variations due to obvious structural differences; the major microecological changes occur near the mouth where microclimates are transitional

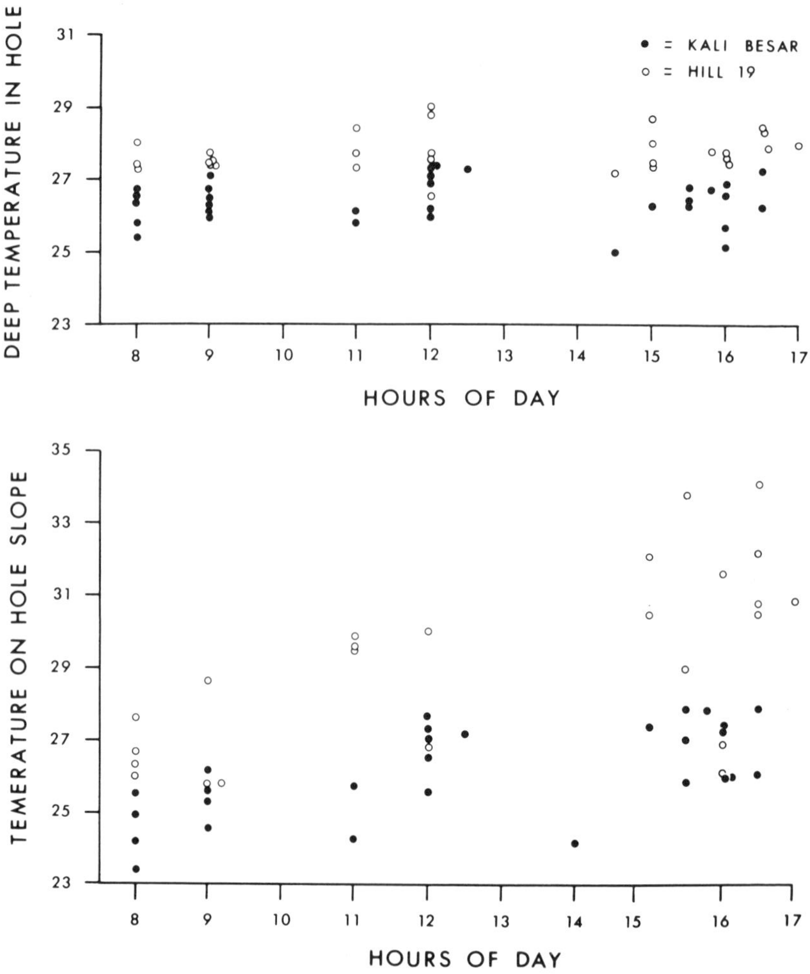

Fig. 4–15. Comparison of temperatures inside and outside creek bed (Kali Besar) and hillside burrows, Loho Liang, Komodo, showing the ameliorating effect of temperatures in both situations, as well as the slightly warmer temperature conditions in the hillside burrows.

between open surface and ground climates. As expected, the burrows of large oras have more air movement and are less uniform climatically than those of smaller individuals. However, the larger mass of adult oras completely negates this factor when they occupy their burrows.

In general, terminal and middle burrow temperatures during the monsoon season are about 2°C cooler than during the dry season in all habitats studied. The greatest seasonal differences are found between the warmer hillside steppe holes and the cooler conditions in the Kali Besar creekside burrows (Table 4-2). Within even a single day, hillside burrows are consistently cooler than creekside burrows are (Fig. 4-15), reflecting the climatic differences in the two habitats. Except for a short time in the morning, when outside temperatures are proportionately low, creekside burrows are almost isothermic with the intermediate air layer outside the burrow for most of the day during the monsoon. In the dry season these same holes are significantly cooler than intermediate layer air temperatures during the middle of the day and warmer at night. Hillside holes on the monsoon forest border follow the same pattern, but the midday differential is much greater, particularly when the rainy and dry seasons are compared. Thus, the hillside burrow temperatures during the rainy season form an entirely different pattern from that of the gallery forest holes. On the open steppe hillsides the afternoon differential between outside and inside temperatures is even greater, though morning and evening in and out temperatures tend to be isothermic. Burrow use is described in Chapter 5.

5

Activity and Movement

Activity Cycles

Seasonal cycles.—The two principal annual seasons in the Lesser Sunda Islands are characterized by major changes in precipitation that result in a moderate monsoon rainfall from January through March and a long dry season of 7 to 9 months (Chap. 3).

Attenborough (1957) has stated that oras are inactive in the monsoon season, particularly during rainstorms. This is not consistent with our observations. Young and adult oras were seen walking about and feeding on carrion during rainstorms on January 19, 25, 31, and February 20. Though precipitation was particularly heavy and lasted several hours on January 25, a 2.8-m ora walked at least 1 km through the heavy downpour. Ora social interaction during rainstorms was also observed. On January 31, oras 23W and 19W intercepted one another on hill 19 during heavy precipitation, and circled, licked, and touched each other with their snouts (ritualistic courtship, see Chap. 8) for approximately 10 minutes.

However, though oras are sometimes active in the rain, the majority will leave exposed situations for nearby cover. Of 32 instances in which we watched oras feeding at the beginning of a rainstorm, in only 4 did they continue to feed after the first 5 minutes of precipitation. However, these 4 animals fed throughout the entire storm.

While Attenborough thought oras were inactive in the wet season, Darevsky and Kadarsan (1964) assumed they were inactive during the dry season. Our data show that although many Komodo reptiles have a quiescent period, oras do not. In fact, the ora's annual activity pattern is remarkably uniform in view of the drastic seasonal changes. It is certainly true that other species of monitors are inactive when conditions warrant it: *Varanus griseus* is inactive for several months in the North African winter (Werner, 1904), though in India it remains active all year (Smith, 1932). In South Africa, *V. niloticus* is inactive during the cool season from April to August (Cowles, 1930); *V. gouldi* hibernates from May to August in the temperate parts of Australia (Green and King, 1978). Sometimes inactivity is associated with size; young *V. griseus* often aestivate longer than mature individuals (Thilenius, 1897). There is no evidence for size-correlated seasonal activity in *V. komodoensis*, nor did I find any evidence supporting seasonal habitat preferences such as presumed by Darevsky (1963) and Kern (1968).

Oras were observed a total of 1013 times (Table 5-1), with the highest frequency occurring during the wet season. When only sightings at carrion baits are calculated (Table 5-2) there is no significant difference in number of oras/day, regardless of the season. Table 5-3 shows that it is the foraging individuals not at

Table 5–1. Seasonal pattern in ora activity (all observations combined)

Month	Number of ora sightings
September–October	196
November–December	178
January–February	283
March–April	198
May–June	104
June–July[a]	54
Total	1013

a. Loök-Nggoer area, Flores. All other data from Loho Liang, Komodo.

Table 5–2. Seasonal activity pattern of oras on basis of number of individuals seen at carrion bait per day per month

Month	Range of ora sightings per day	No. of baiting days	Total no. of ora sightings	$\overline{X}$ oras sighted per day	Standard deviation
September–October	0–15	37	125	3.4	±0.7
November–December	0–14	31	80	2.8	±0.4
January–February	0–15	41[a]	113	3.6	±0.8
March–April	1–9	34	112	3.2	±0.3
May–June	0–8	29	54	2.0	±0.4

a. Multiple baits offered in different areas on some days.

Table 5–3. Total monthly sightings of oras not at carrion

Month	Number of ora sightings
September–October	61
November–December	98
January–February	170
March–April	86
May–June	50

carrion that determine the seasonal pattern. Activity levels are lowest in the early part of the dry season and increase into the end of the dry and throughout the wet season. This higher foraging activity is probably related to the abundance of weakened prey and the higher mortality of prey species during the end of the dry season, as well as to the increase in all general prey classes at the beginning of the monsoon.

Daily cycles.—The factors that influence seasonal activity are also those that determine daily patterns. The diel cycle is, however, obviously broken by at least one major period of inactivity—sleep.

Several authors have stated that the ora, though primarily diurnal, is also seen abroad at night. Lallemant (1929) believed oras sometimes hunted large prey at night, perhaps even lying near or in water holes on bright, moonlit nights (he provided no supporting data). Pfeffer (1959, 1965) stated they are occasionally nocturnal based on noises heard in the surrounding forest at approximately 0100 hr. (No one acutally saw an ora. Piazzini [1960], apparently camping with Pfeffer at the time, did not mention the incident, though he stated that "on occasion it will even go hunting at night." Komodo villagers who accompanied the group to their campsite at Poreng described the incident as one in which oras unaccountably continued to feed on the bait until after dusk—very late by their own cognizance. The noise in the nearby forest many hours later was presumed to be an ora on the basis of the admittedly unusual situation during the early evening.) Darevsky (1963) stated they are "not seldom" active on bright, moonlit nights (but provided no evidence).

As far as is known, all monitor species are usually diurnal, though Hagen (1890) stated that *V. salvator* often enters hen houses at night. In reporting on the diel activity pattern of *V. griseus* Werner (1904) observed that it often begins to forage before sunrise, though considerable light is already present, and Cloudsley-Thompson (1965) showed that the diurnal cycle in *V. niloticus* is determined largely by water retention factors.

Though our study team members were often abroad in the evening, gathering information on the nocturnal use of burrows, recording miscellaneous ecological data, and collecting other material and specimens incidental to our work, we never found one shred of evidence that oras are ordinarily nocturnal. In addition, all the Komodo and Flores villagers with whom we spoke insisted that the ora is completely diurnal. Father Erwin Schmutz, a gifted naturalist of Nunang, Flores, whose parish includes the extreme western Mangarrai District where the ora is found, has no evidence that oras are abroad after nightfall. Almost all the information I obtained from experienced zookeepers, as well as from published accounts, make it clear that captive oras are completely diurnal. The only exception known to me was a specimen in the San Diego Zoo, that was repeatedly lured from between the walls of an enclosure (where it had secreted itself) by placing carrion in the cage nearby. The ora apparently left its retreat to feed on the carrion only at night, and there was no evidence that it ever roamed about the enclosure beyond the food.

On Komodo we found that newly captured oras often exhibit a pattern different from the one they assume when they have become accustomed to their quarters. After having been introduced to a large activity pen, some oras would continue to attempt escape as late as 2330 hr. In several instances oras kept in our traps escaped after nightfall, so that they had clearly been moving about in the dark. However, all these nocturnal movements were by newly captured animals, and I agree with Mertens (1942) that activity in captivity often bears little resemblance to that in the field.

Though carrion bait regularly remained at all our observation sites during the night as well as the day, there was not one instance in which the oras came to the bait after nightfall or before sunrise. Pfeffer's statement (1959) that monitors move at night was based on their observed entry into holes early in the morning and on his assumption that the lizards were returning from a nighttime hunt, rather than an early morning sojourn. As will be shown below, oras are already wide awake and well into their daily activity patterns by 0800 hr.

However, in the Lesser Sundas moonlight may be very bright—particularly near the end of the dry season. During full moon we could walk through the savannas in the middle of the night and be able to see clearly standing deer and hogs and sometimes even moving snakes. Therefore, though we have no evidence that oras do so, it is probably physically *possible* for them to move effectively on bright, moonlit nights.

On several occasions we had the opportunity to watch oras during the night, when they were trailed to their resting places and regularly checked for several hours. It is clear that the animals usually fall into deep sleep at approximately 1930 hr. While they were sleeping we sometimes obtained tick samples from their skins, attached plastic bands to their toes, moved their feet and tails, took temperatures from their skin surfaces, and otherwise slightly manipulated them.

However, some individuals remained awake until 2330 hr. At these times they would lie in the grass or in their holes, with head outside the mouth of the burrow and eyes wide open, staring into the black surrounding forest.

Diel activity patterns were analyzed on the basis of data collected in several different ways. One method utilized captive animals. Several medium size oras were maintained in a 15 × 8 m enclosure (Fig. 1-8) provided with appropriate shelter. Only one individual was placed in the pen at a time. Observations on the diel pattern after a period of acclimation indicate that under these conditions oras emerge from shelter between 0430 and 0615 hr ($\overline{X}$ emergence time = 0441 hr). However, they remain near the evening shelter, basking until between 0615 and 1330 hr ($\overline{X}$ = 0852), depending on incoming radiation patterns. During the remaining part of the day the captives generally moved constantly through the sunlight mosaic in the pen. They sought nocturnal shelter between 1718 and 1852 hr ($\overline{X}$ = 1804).

The second method utilized a 1000-m-long disturbance strip (see Chap. 1) designed to provide data on the movement of oras through the savanna. These data confirm the conclusions based on activity pen observations; i.e., that the active diurnal phase generally begins between 0800 and 0900 hr and ends between 1800 and 1900 hr. There is, however, some evidence for a slight shift to both an earlier beginning and later ending time from November through the early part of the monsoon season (discussed below).

The third method utilized observations on the arrival and departure times of all lizards at the carrion baiting sites, as well as the times when free-ranging oras were observed in the field, irrespective of other factors.

There is good correspondence between the peak activity times recorded by all four methods (Table 5-4). In the case of the expressed total observation hours for each method used, it is important to point out that the longer observation period of diel activity in the pen are due to convenience. In the case of free-ranging oras, the earliest recorded movement periods were obtained during continued tracking of individuals for several days. It is obvious that for the entire year the average morning peak lies at 1000 hr, and the afternoon peak at 1530 hr. The statements of several earlier workers that oras are not active until 1000 hr (e.g., Burden, 1928; Broughton, 1936; Hoogerwerf, 1954a) are almost entirely based on observations of the arrival of the lizards at carrion. If the carrion is distantly located from the evening shelter of an ora, or if wind conditions are poor, it takes considerably longer for the animals to locate the carrion. The diel pattern in respect to microecological factors is discussed below.

The bimodality in diel activity patterns varies seasonally, (Fig. 5-1) being strongest during the dry months (September to early January, and May to July). It is less marked during the rainy monsoon season from the middle of January through April (abnormally long the first year of our study). During the extremely wet period of February, the curve is nearly unimodal. The observed seasonal changes are dependent on microecological factors (discussed below).

While many oras were kept in captivity during our initial year of study, the best activity sequence is of ora 50W (14 kg, 1.97 m), who was observed continuously throughout each day from April 12 to April 25.

Like free-ranging individuals, the captive awoke long before sunrise. Anticipatory behavior placed the ora in an approximate position for basking prior to sunrise. Basking time in early morning sun varied from 15 min to 3 hr and 15 min ($\overline{X}$ = 1 hr 50 min). Thus, in most cases, the monitor basked until approximately

Table 5–4. Diel activity pattern in Komodo oras, based on four different data sets

Data set	Peak activity times AM	Peak activity times PM	S.D. in minutes AM	S.D. in minutes PM	Activity period limits	Sample size
Activity pen	1013	1520	±23	±46	0530–1940	7
Savanna trail	1001	1517	±31	±43	0600–1900	92
Bait sites	0945	1541	±28	±30	0715–1810	386
Free ranging	1010	1537	±51	±46	0515–1840	118

0900, and until much later on cloudy days. Basking periods were longest when shade temperatures were lower than 35°C. There was a better correlation of activity pattern with shade than with air temperature.

When the behavior of captive specimens is categorized into basking, moving, or resting in the sun, shade, or under shelter, it is obvious that morning and afternoon patterns are considerably different. Basking constitutes 72.4 percent of all activity from 0600 to 1200 hr, whereas it constitutes only 32.6 percent of activity from 1200 to 1800 hr. However, casual movements in the sun are more common after 1200 hr (9.5 percent) than prior to that time (5.7 percent). Activity in the shade (regardless of type) is much more common post-1200 hr (57.9 percent) than pre-1200 hr (21.9 percent). The emphasis of the morning behavioral pattern on basking and the afternoon pattern on shade-seeking suggests that thermoregulation is the determinant in these behavioral cycles. Figure 5-2 shows a typical record for ora 50W during a single day in the activity pen.

Haulout time in free-ranging *V. varius* varied from 0600 to 0632 hr (Stebbins and Barwick, 1968). In captive *V. komodoensis* average haulout time is 0700 hr. The first appearance of *V. varius* at the openings of tree hollows occurred over a narrow overall time range of 25 min (mode = 13 min) and, as in our captives, seemed independent of body and air temperatures. Total variation in haulout time for captive *V. komodoensis* was very similar, being 31 min (mode = 12 min).

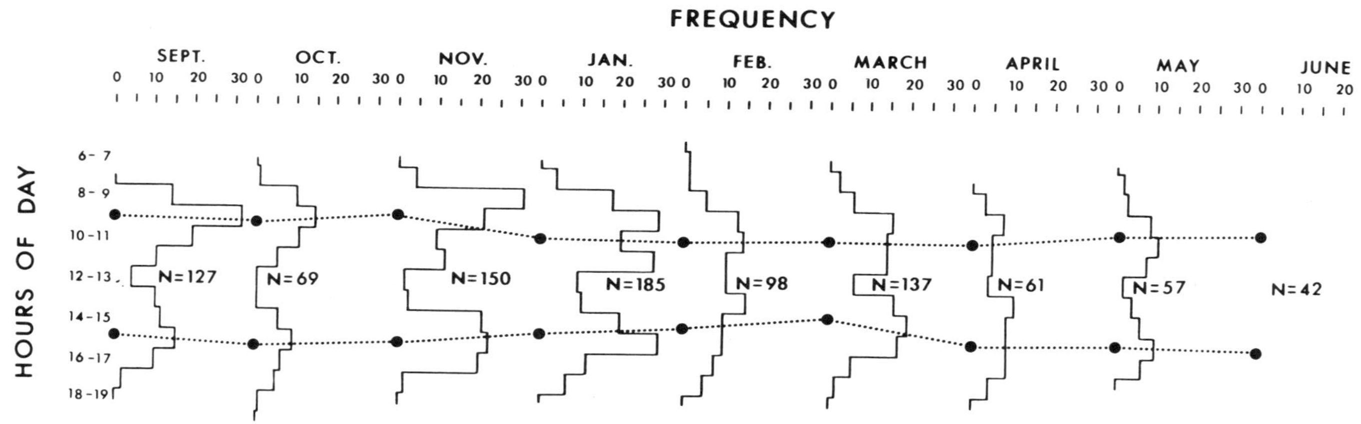

Fig. 5–1. Seasonal diel activity pattern, Komodo and Flores observations combined (N = 926). Dots represent morning and afternoon averages, showing slight shift toward unimodality during the monsoon season.

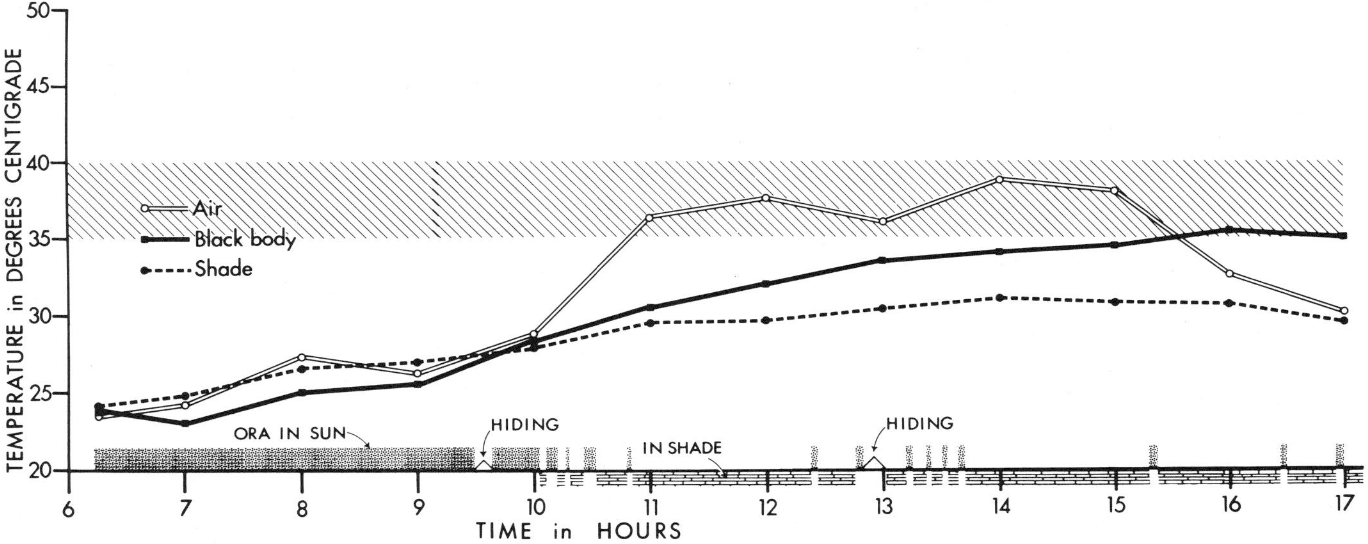

Fig. 5–2. Typical ora diel activity pattern, based on captive 50W, maintained in large outdoor pen on Komodo (see text). Basking is largely an early to mid-morning activity. Most late morning to evening activity takes place in the shade. The specimen retired into its subterranean shelter at 1710 hrs. The cross-hatched strip across the top represents overall range in operative deep body temperatures of free-ranging oras, and shows the general relation between typical active phase of the diel pattern and the late morning rise in air temperature.

Temperature Relations

Body temperatures and activity.—The ability of reptiles to regulate their body temperatures behaviorally has been well documented (e.g., Bogert, 1949, 1959; Soule, 1963; Schmidt-Nielsen, 1964; Heath, 1965; DeWitt, 1967). However, many workers have shown that though reptiles can choose the ceiling of their preferred range by behaviorally utilizing solar energy, they have less choice at the lower threshold values because they lack the cooling mechanisms characteristic of most mammals and birds. Therefore, they can cool themselves only to the level of the coolest segment of their environment.

Body size is an important parameter of heat exchange in all animals, since it influences heating and cooling rates by affecting surface area and heat capacity, as well as the rate of heat production. Though varanid lizards generate more metabolic heat than do most lizards (Bartholomew and Tucker, 1964), it is unimportant in maintaining a temperature differential with the environment. A more important thermal factor in *V. komodoensis* is that the large size potentially confers on it a great thermal independence, even during quiescence. Goodman (1971) has shown that the critical maximum body temperature (CT Max) is directly related to the highest air temperatures experienced within a reptile's habitat, and that these environmental thermal extremes are important in the thermal preference evolution of reptiles. Heliothermic lizards show a very high correlation between preferred body temperature (PBT)[1] and the highest temperatures ever recorded in the immediate habitat. It is, however, important to point out that the actual level of preferred body temperature is not entirely a function of high air temperature extremes. In the ora these extremes are modified by burrowing or panting, behaviors that meliorate any temperature exchange. Temperature lowering by water evaporation, if effective at all, is probably too expensive in terms of water loss to animals living in xeric habitats (Schmidt-Nielsen, 1964). It is obviously the interaction of behavioral patterns, circulatory shunting mechanisms, and physiological tolerances under the selective pressures of high temperatures that determine the level of the preferred body temperature. Brattstrom (1968) has shown that in *V. varius* vasomotor control of heat retention is an important function in the endothermy reported in this lizard earlier by Bartholomew and Tucker (1964).

Rand (1967) and others have shown that the body temperatures of many species of lizards (generally small) are almost always close to, and usually above, the air temperatures in their immediate vicinity. This is not true in the Komodo dragon. The few data previously available for these lizards (Darevsky and Kadarsan, 1964) suggested that their body temperature is equal to or less than air temperature. However, these data do not permit a critical evaluation of the influ-

1. I follow all other authors in use of "preferred body temperature" despite the work of Licht (1965), who suggested the term be restricted to experimental studies only.

ence of lizard size on thermoregulation, because the temperatures were taken in the morning after the animals spent the night in a trap, and no information was given on the size of the lizards. The only other ora temperature data is provided by Lederer (1942), who stated that a large specimen successfully maintained in the Frankfurt Zoo for 16 years was kept between 20 and 30°C and was most active at 25–30°C. At 36°C this specimen's respiration rate increased.

On Komodo we obtained thermobehavioral data on free-ranging *V. komodoensis* by means of telemetry (Chap. 1) and physiological probes. Cloacal temperatures were taken within one-half hour of the oras' entrance into our traps. Simultaneous measurements were taken of the shaded air temperature at 10 cm above the ground surface and soil temperature of the trap in the sun to obtain a rough measure of the radiant heat loss in the oras' environment at that time of the day.

Between 1000 and 1700 hr the cloacal temperatures of 28 oras varied from 32 to 40°C, 10 cm shaded air temperatures varied from 28 to 32°C, and surface soil temperatures varied from 46 to over 53°C (the upper limit of our equipment) (Fig. 5-3). Thus, the cloacal temperatures of oras during the day were approximately intermediate between the two local environmental temperatures available. Equilibrial cloacal temperatures are independent of body weight (Fig. 5-4), although they are somewhat higher than field temperature measurements of small Australian varanids (Stebbins and Barwick, 1968).

Three individuals were also utilized for establishing CT Max (necessarily under field conditions). All started gaping between 10 cm shaded air temperatures of 39.6–40.0°C. Correlated with this last observation is the fact that there are few measurements of cloacal temperatures greater than 40°C. CT Max is established as approximately 42.7°C.

We also estimated thermal conductance in two *V. komodoensis* by construction of cooling curves based on continuous telemetric monitoring of both ambient and deep body temperatures (McNab and Auffenberg, 1976). The apparent cooling constants of the 6.7 kg and 35 kg individuals are only 7 percent of those expected on the basis of weight-dependent equations established in studies of smaller lizard species. In fact, after nighttime cooling, deep body temperatures of a 35 kg Komodo monitor are 4.9°C higher than ambient temperatures. This means that large Komodo monitors are apparently capable of maintaining an appreciable differential between body and air temperatures by physiological means, primarily by reduced thermal conductance. Brattstrom (1973) has shown that the skin of *V. varius* (a related species) does not appear to be very effective in reducing heat loss, and the same is undoubtedly true of *V. komodoensis*.

As further protection against low body temperatures at night, some large Komodo monitors utilize thickets and burrows. A reduced conductance by burrow use permits large monitors to begin basking in the morning at a higher body temperature (i.e., 28–30°C) than would otherwise be the case. Because of this,

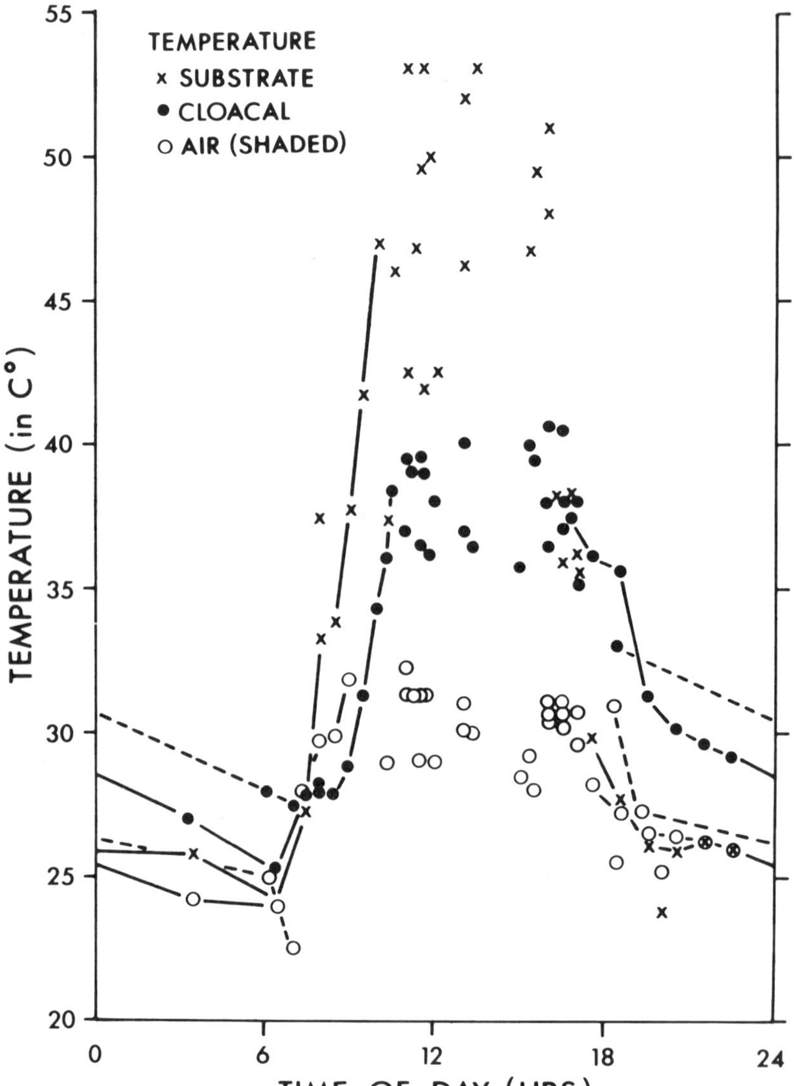

Fig. 5–3. The body temperatures of *Varanus komodoensis* during the day are intermediate between shaded air and substrate ambient temperatures (after McNab and Auffenberg, 1976).

adults spend only about one-half hour in basking activity before beginning their morning foraging pattern. Thus, the basking time required to raise cloacal temperature to active levels in large monitors is reduced as much as 1 or 2 hours by starting the day with a body temperature already 5°C higher than air temperatures. During initial morning basking, increase in deep body temperatures varied from 0.11 to 0.18°C/min ($\overline{X}$ 0.14°C/min).

Clearly, large size has allowed the Komodo monitor to maintain a body temperature more independent of ambient temperatures than is typical of smaller species. The relative uniformity of ora body temperature must surely extend the activity period into conditions in both temperature and time that are unacceptable to the smaller ectotherms of this area, thus allowing adult oras to approach activity patterns typical of mammalian carnivores.

Evaporative cooling and hyoid panting.—Warburg (1965) and Templeton (1960) have suggested that evaporative cooling is not one of the usual mechanisms to which species of desert lizards resort when subjected to heat stress. However, Bentley and Schmidt-Nielsen (1966) found that water loss in the same lizards increased proportionately at high body temperatures. Crawford and Kampe (1970) suggested that evaporative cooling in lizards probably becomes effective only at high ambient temperatures (above 44°C).

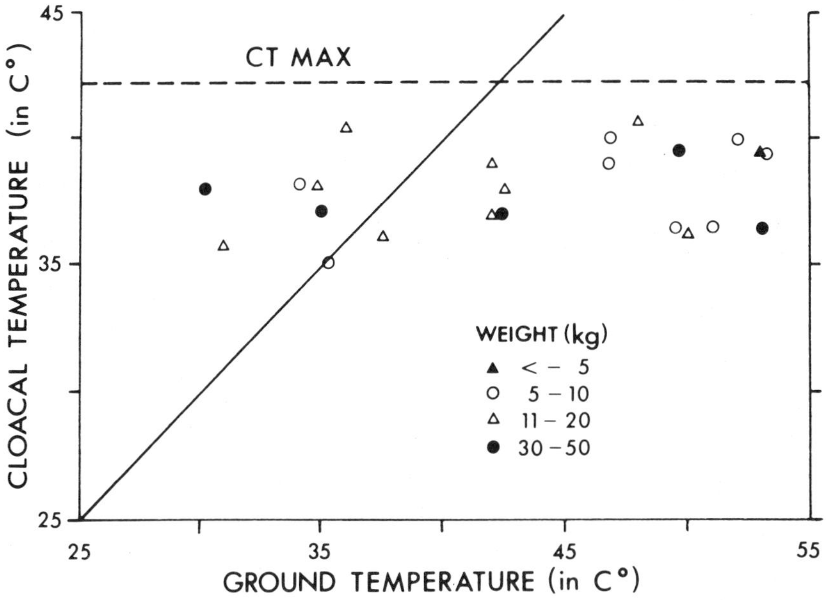

Fig. 5–4. Equilibrial cloacal temperatures of the ora are independent of body weight (after McNab and Auffenberg, 1976).

Since thermoregulation begins long before the onset of panting, evaporative cooling by means of panting is usually not an important factor in maintaining a thermal difference in ambient and body temperatures. However, that hyoid pumping, or panting (for whatever reason), is of importance in oras under thermal stress is obvious since it usually begins at about 40°C (Fig. 5-5). The differences among dorsal skin, cloacal, and buccal temperatures were determined in 28 specimens. These data show that buccal temperature was 2.4°C lower than cloacal and 1.0°C lower than that of the dorsal skin surface ($\overline{X}$ cloacal 38.0, $\overline{X}$ dorsal skin 36.6). Mean ground surface temperature in the sun (to measure maximum heat load) was 42.7°C, and at 10 cm above the surface it was 34.8°C.

In at least one desert lizard, *Sauromalus obesus,* Norris and Dawson (1964) found unusually large abdominal lymph sacs and nasal salt glands. They suggested that the glands probably permitted the animals to cope with saline foods. Templeton (1964) has shown that the glands compensate for potassium and sodium loading. In spite of its role in excretion, the combination of the large lymph sacs and nasal structure may enhance evaporative cooling in lizards during critical periods in xeric habitats. It is not known whether nasal glands are present in the ora.

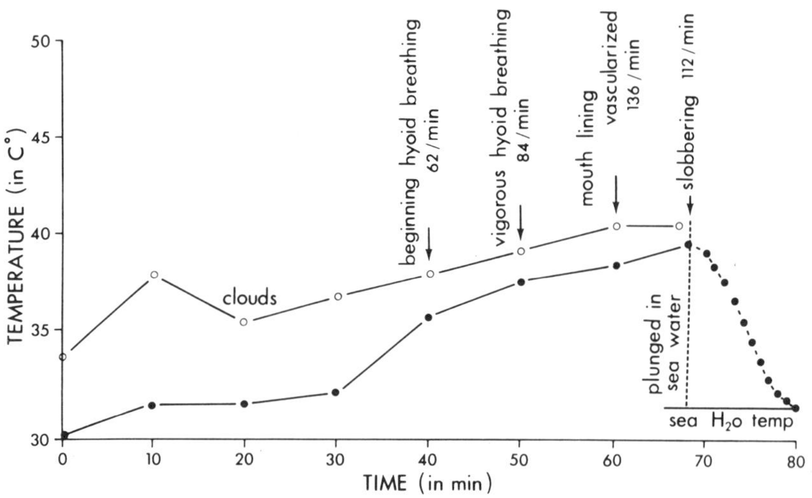

Fig. 5–5. Relationship of ora cloacal temperature (solid dots), ambient 10 cm temperature, hyoid breathing (= pumping), and respiration rate taken from record of tethered ora (13 kg). Hyoid pumping in oras generally begins about 40°C cloacal temperature, often with vascularization of buccal area. Salivation is common as cloacal temperatures reach 40°C.

Table 5–5. Respiration rate per minute in free-ranging oras

Ora number	Time (shade temp. in °C)			
	1120 (30.0)	1140 (30.5)	1150 (30.8)	1200 (31.0)
8W	90	90	108	114
19W	96	80	116	84
23W	114	100	72	120
34W	90	90	96	108

Temperature and respiration.—The usual respiration rate of oras at normal operative body temperatures varies from 66 to 95 breaths per minute. Heat stress causes higher rates (Fig. 5-5). Under free conditions, the rate in oras exposed to high heat loads (such as feeding on carrion in the sun) may climb to 120/min; under restrained conditions it may rise to at least 161/min when near CT Max. There is no correlation between ora size or sex and respiration rate.

Though respiration strength and rate are generally positively related to body temperatures, rate changes are often sudden and erratic, particularly during a rapid rise in ambient temperatures (Table 5-5). In this case ora 19W had been feeding on carrion under the open sky from 1010 to 1020 hr. The accumulated heat load apparently not only caused it to move into the shade, but to begin panting and to increase its respiration rate. That the heat load at this time was significant was indicated by the widely open epiglottis (sometimes producing popping sounds) being thrust anteriorly from its usual position, and increased vascularization of the tongue sheath and mouth lining. Body positions during this time varied from Ventral Adpression to the Alert Position (see Chap. 6). The hyoid apparatus is often vigorously brought into play during panting phases and apparently pumps great volumes of air in and out, perhaps resulting in greater heat loss during panting than has been demonstrated in smaller lizards.

Temperature and digestive activity.—Cowles and Bogert (1944) suggested that the digestive functions of some reptiles are considerably influenced by the efficiency of their thermoregulatory behavior. Additionally, Regal (1966) has shown that in at least some reptiles digestive efficiency may be a determinant in thermoregulatory activities. Thus, in reptiles, the utilization of external heat sources probably varies as a function of changes in digestive state. The survival value of such responses is clear. Elevated body temperatures undoubtedly accelerate enzymatic and mechanical digestive functions. In view of the large meals sometimes ingested by oras, it is particularly important that digestion continues at high temperatures, because there is a tendency for carrion food to putrefy when body temperatures remain low. Regurgitation and even death are common sequelae in large captive reptiles (Oliver, 1955). The ora can hardly be expected to be an exception, and, as will be shown below, the animal goes to great lengths to maintain a uniform body temperature, particularly during the digestive phase.

Spatial Relations

Activity range.—Tracking studies on Komodo have shown that adults may move as much as 10 km/day, though the mean is 1.8 km/day. This large area makes defining the activity range, particularly of adult animals, troublesome. Additionally, workers are not in complete accord as to what part of the total observed movement should be included in the term (see Milstead, 1961; Rand, 1967; and particularly Wilson, 1975). The interspecific or even seasonal variation in behavioral patterns leads to at least some kind of emendation for the term in almost every species studied. By activity range in the ora, I mean the total area in which the individual will most likely be found, including all of its burrows and sleeping or thermoregulatory sites. The range is comprised of two usually zonal regions—the *scavenging* and *foraging areas,* of which the former is much larger. The latter includes a *core area,* where most of the activity takes place. All are approximations, and together they form a graded series in size, but not necessarily in function (Auffenberg, 1978). Their shape and size are dependent upon topography, prey location, and density, as well as on an individual's social status, size, and sex. Very young oras do not have a scavenging area, for they rarely feed on carrion.

Residents and transients.—While some oras may utilize the same activity range for many years (= residents), others apparently move about over a much larger area, passing through the activity ranges of several oras within a period of several weeks or months (= transients). Only a few of the more than 300 individuals observed in our Komodo study area were regularly seen. Most remained in the area from only a few weeks to several months (Table 5-6). Of 46 individuals sighted during the year at the Base Camp, only 7 were regularly seen in the area during the entire year. Of these, none was less than 1.5 m long, and only 2 were less that 2.0 m (Table 5-7). These 7 were clearly residents. General locations of core areas of oras resident in Loho Liang and Loho Kalo are shown in Figure 5-6.

Of the remaining oras, only 31 were seen several months of the year, and of these 26 were seen a total of only 5 days or less. Of these 15 were smaller than 1.5 m and only 3 could be assigned residency in any of the surrounding study blocks. Thus, 23 were casual visitors, entering the study area at one end and slowly passing through it. Once they left the valley, we never saw them again; they were clearly transients as far as the study was concerned. Observations in western Flores in 1971 support the same pattern.

Therefore, ora populations seem to be comprised of many variously transient

Table 5–6. Sightings of oras near the base camp area, Komodo, from July 1969 to June 1970

Ora number	Sex	Total days seen	Present in base camp area		
			July–Oct.	Nov.–Feb.	March–June
Total length between 2.5 and 3.0 m					
19W	M	68	X	X	X
Total length between 2.0 and 2.5 m					
8W	M	22	X	X	X
23W	F	34	X	X	X
32W	M	2	0	X	0
34W	M	10	X	X	X
28W (4Y)	M	19	X	X	X
22W	F	6	X	X	0
33W	M	6	0	X	0
37W	M	3	0	X	X
50W	M	1	0	0	X
14W	M	2	X	0	0
Total length between 1.5 and 2.0 m					
11W	F	17	X	X	0
44W	F	8	0	0	X
47W	M	3	0	0	X
35W	M	4	0	X	0
3W (45W)	M	4	0	0	X
3Y	M	1	0	X	
11Y	M	1	0	X	0
24W (29Y)	M	4	X	0	X
25W	F	2	X	0	0
2W	M	5	X	0	0
5W (10Y)	M	9	X	0	0
13W	M	8	X	X	X
46W	M	1	0	0	X
36W	M	3	0	X	0
43W	M	4	0	0	X
29W	M	5	X	X	X
Total length between 1.0 and 1.5 m					
18Y (18W)	F	2	X	0	0
26W	M	10	X	X	0

(Continued)

Table 5–6—*Continued*

38W	M	2	0	X	0
41W	M	1	0	0	X
10W	M	8	X	X	0
6W	F	6	X	0	0
16W	F	2	X	0	0
27W	F	4	X	0	0
31W	M	7	X	0	0
30W	M	7	X	X	0
40W	F	9	0	0	X
42W	M	9	0	0	X
20W	M	2	X	0	0
7W	M	10	X	X	0
1W	M	2	X	0	0
49W	M	1	0	0	X
17W	M	3	X	0	0
Total length less than 1.0 m					
21W	M	4	X	0	0
1Y	M	2	X	0	0
2Y	F	1	X	0	0

individuals and a few residents. Almost all residents are large adults, while transients vary from hatchlings to an animal that was nearly the longest recorded in this study (Table 5-7). Very few of the transients were adult females. The transients (including many mature males) represented a dynamic faction in the local population, constantly entering new areas and interacting with resident and other transient individuals.

General features of resident activity range.—The activity range may include habitats at any elevation from sea level to about 500 m. While not imperative, it nevertheless commonly includes slight elevations—most frequently the spurs of higher hills or mountains. However, most of the area is generally flat to rolling. Precipitous regions, particularly large-scale areas, are almost always excluded. Near the sea it may include offshore islets, sometimes even reefs and bars. While most of the terrestrial portion usually encloses monsoon forest and/or savanna habitats, also commonly included are mangrove swamps, open beach, steppe, and *Pandanus* thickets.

Scavenging area.—The fact that oras regularly feed on carrion and that they are able to recognize it over great distances provides this species with a space utilization strategy different from that of any other lizard studied to date. The size of the potential scavenging area is determined mainly by vagaries of wind currents, topography, and ora locomotor speed and endurance. Though scavenging areas may extend in any direction from an individual's core area, they are usually most extensive in an upwind direction. Furthermore, they usually include part or all of the activity ranges of several neighboring residents. The total size is not as clearly related to ora length as are the sizes of the foraging and core areas.

Perhaps most important is that the scavenging area affords the greatest potential for intraspecific coactions, for they are largely shared areal resources. The carrion baiting sites are described in Chapter 1, and their locations are shown in Figure 1-9. Analysis of the exchange of marked individuals (Fig. 5-7) among the five baiting sites in the study area shows several patterns that are probably representative of the ora movements throughout this species' distribution. The lizards rarely move east-west from one valley to another; only 1.8 percent of the marked oras on either side of the hilly range (el. ca 400 m) are known to have crossed it. The greatest movement was up and down the valleys. Movement between the Base Camp and Inaloah is high (15.8 percent), because both sites are located close to the large, dry creek bed often used by oras moving through the eastern part of the valley of Loho Liang. The fairly high exchange between the Vai Nggulung and Base Camp sites (10 percent) is probably due more to prevailing wind direction than to any other factor. The next highest exchange is between the mouths of two major creek beds, one at the Base Camp and the other at Nanenepi (8.2 percent). This movement rate is much higher between sites located along the same creek bed (Nanenepi and Vai Nggulung). Apparently, the movement rate between the creek beds in this area is high because many oras forage in the mangrove swamp and overflow area between the mouths of these intermittent

Table 5–7. Size and sex of transient and resident *V. komodoensis* in the southern part of Loho Liang, Komodo, 1969–70

Size classes (total length, mm)	Sex	Residents' Frequency	Transients' Frequency
1.1–1.5	M	0	8
	F	0	4
1.6–2.0	M	2	10
	F	0	1
2.1–2.5	M	3	3
	F	1	0
2.6–3.0	M	1	0
	F	0	0
Total		7	26

streams. There is very little exchange between the same two creek beds upstream at the Vai Nggulung and Kali Inaloah sites (5.1 percent), probably because of the intervening southward projected spur of the large mountain called Nikipolak. Furthermore, scent from either of these areas would only rarely be blown in the direction of the other.

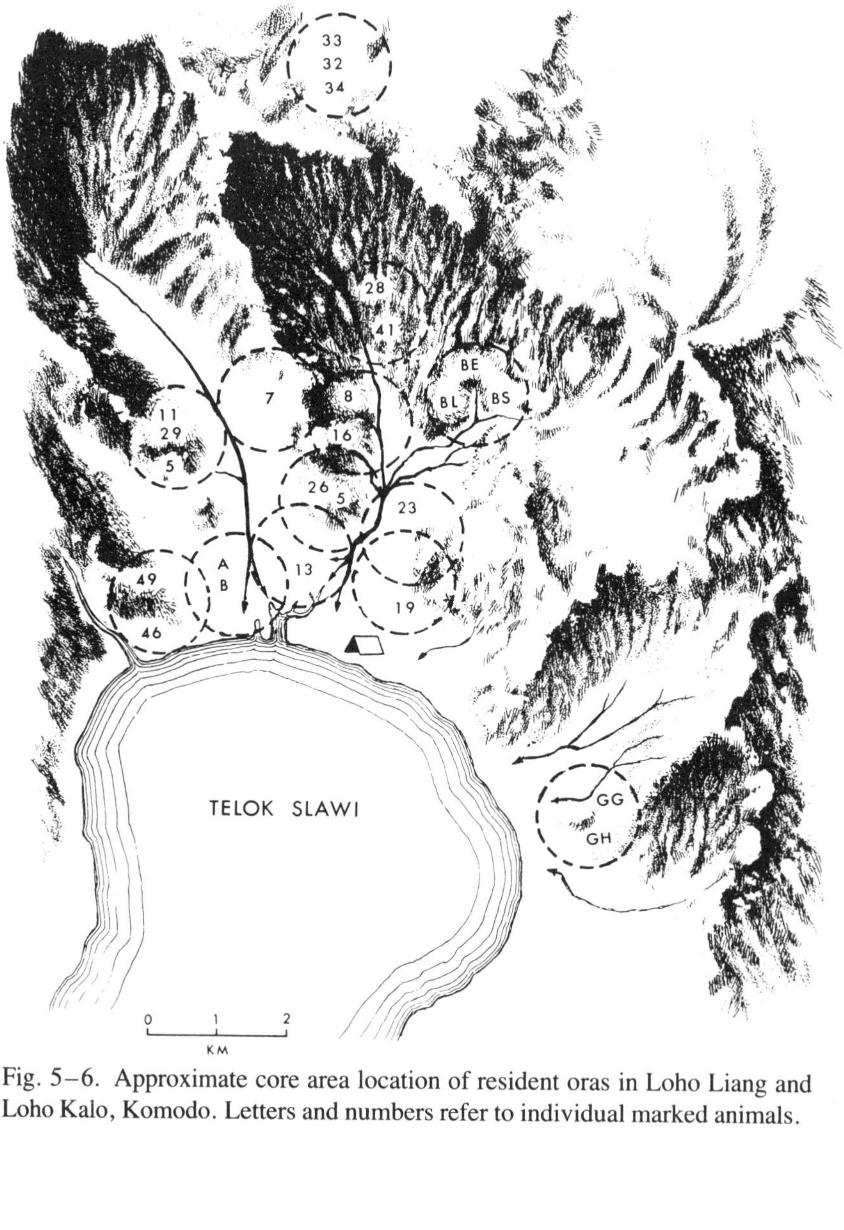

Fig. 5–6. Approximate core area location of resident oras in Loho Liang and Loho Kalo, Komodo. Letters and numbers refer to individual marked animals.

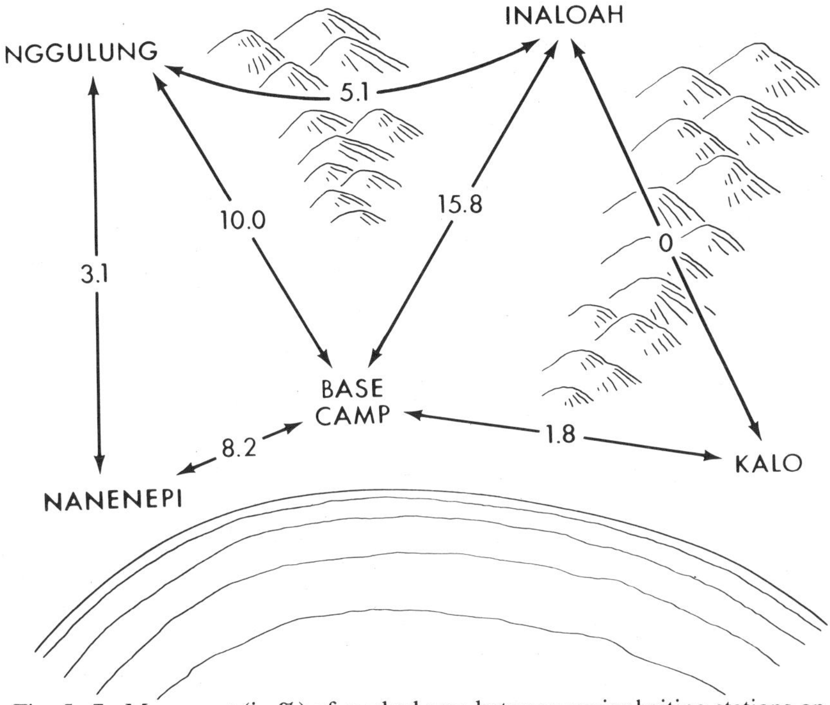

Fig. 5–7. Movement (in %) of marked oras between major baiting stations on Komodo. See text for explanation.

Physical and biotic features are least important in the shape and size of the scavenging area, since the location of carrion is usually fortuitous. However, the behavior of most dying wild animals is such that death frequently occurs in and on the edges of thickets, usually on level to undulating surfaces. Precipitous mountains are rarely involved, though cliff bases may be. Beaches are an exception to the generally more or less random distribution of carrion. Here dead marine animals are often concentrated in and near the beach wrack zones, so that scavenging areas often parallel the water edge. Prevailing wind currents and valley shape and orientation determine the size and shape of most scavenging areas—producing acute hemiconic areas in valleys paralleling prevailing wind direction, and more or less obtuse subtriangular areas in valleys perpendicular to prevailing wind patterns.

Foraging area.—Of the several types of activity areas recognized in ora movement patterns the forage area is the easiest to define. Typical examples are shown in Figures 10-8 and 10-9. There is a better correlation between ora size and foraging area than scavenging area (Table 5-8). For hatchlings the foraging area may include only a small group of large dead trees, and for resident adults it may encompass an area of several square kilometers. In general, the foraging area can be much more precisely located than the scavenging area.

The mean foraging area of resident adults (2.1–3.0 m, N = 23) is 4.2 km^2, expanding from 2 to 6 km in all directions from the central core area. This area is much larger than has been reported for any other lizards (see Green and King, 1978, for review of varanid studies). Turner et al. (1969) provided an equation for the relationship between home range and body weight: $A = 171.4W^{0.95}$. When used for adult oras, the estimated range would be 2.4 km^2, or 1.7 times smaller than that actually observed. The observed home ranges of *V. gouldi* (Green and King, 1978) also exceeded the predicted values by a similar amount. Further work with varanids may suggest a modification in the formula for the highly predaceous, larger varanids.

While including practically any habitat within the ecological species limits, the foraging area tends to be concentrated in level to rolling areas of monsoon forest and savanna. Coastal foraging areas may contain beaches, dunes, islets, and possibly even reefs or shoals to a depth of 2 m. The most consistent determinant is apparently the distribution of live food, which may vary seasonally. Also important are the sites from which prey movement may be monitored and the interrelationships of territorial boundaries established by individual oras. However, there are other resource classes consistently found within each foraging area. Thus, it will generally include nests or nesting areas, such as the sandy beaches used as nest sites by sea turtles; the large permanently located nests of megapode birds; the seasonally located nests of jungle fowl, etc. It will also include prey shelter sites, such as particular thickets where deer tend to sleep, rock slides and fallen trees, standing dead trees, burrows, beach wrack, and rarely garbage or latrine areas near human habitation. For large oras it also includes ambush sites along game trails, near water holes, and rarely in domestic livestock pasturing areas. Some trails are made by oras, by humans and their domestic animals, or by wild animals, and some are dry creek beds and other openings—any area consistently used for easy and rapid movement thoughout the region.

Core area.—Most studies of carnivorous lizards have shown that individuals spend approximately half their time in an area that is fairly small compared with the total area used during the course of a day or a year. Kaufmann (1962) and Rand (1967) called such regions core areas. In the ora study 50 percent of all the site activity records for particular individuals falls within a very small part of the total known activity range of resident individuals. This core area is not only that part of the activity range of the resident that is most heavily used, but it is also that part of the range which least overlaps the activity ranges of other residents.

Table 5-9 shows the approximate size of ora core areas. Figure 5-8 illustrates the relationship between foraging and core areas for an adult resident ora. The foraging area (where known) is from about 5 to 28 times larger than the core, and

Table 5–8. Ora size classes, forage areas, and proportionate forage areas

Ora size classes (total length, m)	Average forage area (m^2)	Forage area per body area[a]
1.1–1.5	7,824 (N = 6)	29,731
1.6–2.0	247,380 (N = 5)	940,044
2.1–2.5	1,411,368 (N = 14)	5,363,198
2.6–3.0	4,813,000 (N = 9)	18,289,400

a. Total length × greatest width.

Table 5–9. Ora size classes, core areas, and proportionate core areas

Ora size classes (total length, m)	Average core area (m^2)	Core area per body area[a]
1.1–1.5	338 (N = 6)	1,303
1.6–2.0	9,341 (N = 5)	16,526
2.1–2.5	48,000 (N = 4)	41,744
2.6–3.0	168,000 (N = 3)	85,714

a. Total length × greatest width.

the scavenging area is as much as 125 times larger (N = 15 adults). Resources generally include thermoregulatory and/or sleeping sites (often combined or adjacent), including specific thickets, hillsides, rockslides, and burrows; commonly used ambush sites and trails to and from them are also included; monitoring sites from which carrion can be scented are important in larger individuals. This combination generally places the core area near spurs at the foot of larger hills and mountains. Distribution of prey species, or the sites from which prey movement can be monitored, and the interrelationship of territorial boundaries established by the individuals comprising the ora population itself may also be important. Typically the core area may also include fairly specific dunging sites. Water holes are also frequently included. Whether females return to specific

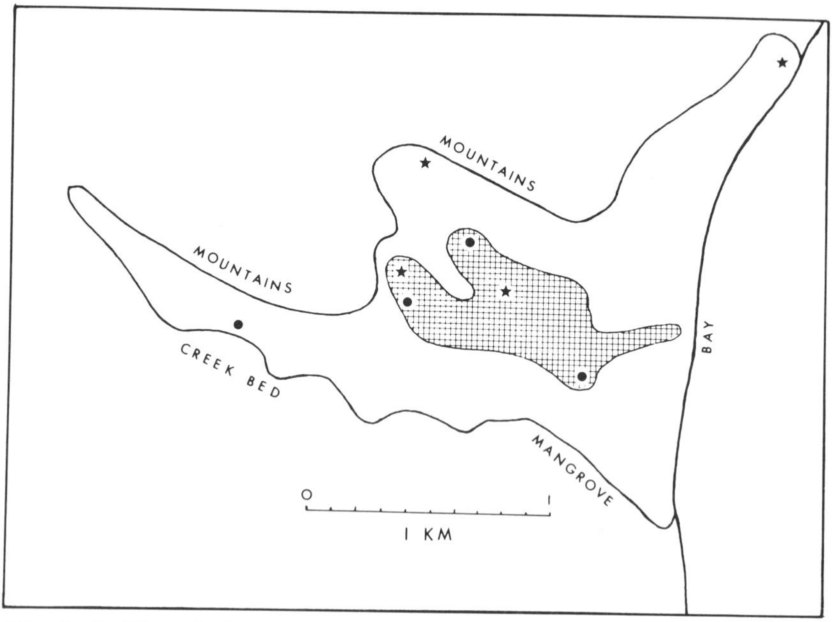

Fig. 5–8. The relationship between core (cross-hatched) and approximate limits of the foraging area (borderline) for ora 19W (2.8 m long) in Loho Liang, Komodo. Stars show the location of this individual's thermoregulatory sites; dots represent burrows known to have been used during the study period.

nesting sites each year, or if the nests are even generally located within the core area, remains to be shown.

Relationships between components of activity range.—The size, shape, and location of activity ranges for all individual oras, regardless of their size, is influenced by the distribution of the habitat requirements of the particular size group being considered, plus location of other lizards of the same or larger sizes. Thus, oras are not distributed randomly with respect to one another. For example, residents of the same sex are rarely found in the same activity range (except at carrion), and it is apparently rare for more than two resident adults of the opposite sex to occur in any core area.

Though the core areas of adult males are mutually exclusive, males are often seen within one another's overlapping scavenging ranges. Under most circumstances, the fact that carrion occurs within the foraging or core areas of a particular adult male apparently does not discourage intrusion by other individuals of any sex or size. The core area is ostensibly most related to shelter (including thermoregulatory sites); the foraging area is related to prey density; the scavenging area is related to the capricious location of a large dead animal. In habitats appropriate with respect to both shelter and prey, each core area has a high hill or knoll, which is extensively utilized by the resident individual for monitoring and shelter purposes (Fig. 5-8) and which can thus be considered the center of the activity area.

Depending on local conditions in regard to both biotic and abiotic factors, the ora tends to move over the core and foraging areas fairly regularly, though not necessarily on a daily cycle (Fig. 10-9). However, the outer edges of the foraging area are much less frequently utilized than the core area (Fig. 5-9). During the scavenging foray, the normal, rather cyclic activity patterns within the foraging area are abandoned and oras may be drawn to a large carcass from distances as great as 8 km, completely overrunning one another's activity ranges, and even central core areas. Ora aggregations at carrion are thus partly comprised of individuals unrelated through normally overlapping or adjacent activity ranges. The meeting of strangers is the rule, with a resulting high level of social interaction (Chap. 12).

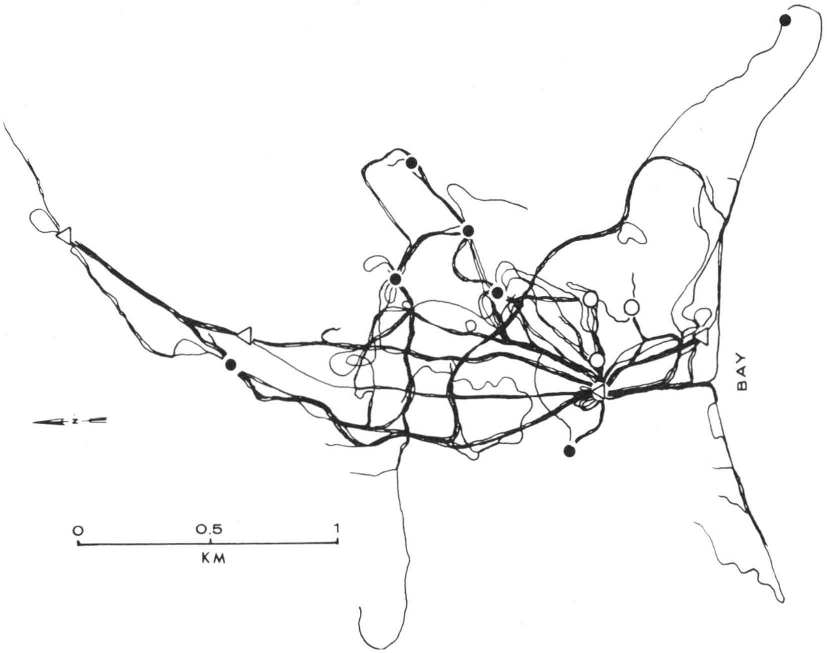

Fig. 5–9. The movement pattern of a resident, adult ora (19W) in Loho Liang during 1969-70 showing the tendency to use certain paths repeatedly, the infrequency with which the peripheral areas are visited, and the tendency to move between shelter (solid circles), water pools (open circles), and food sources (carrion baiting stations shown as open triangles).

When carrion is all eaten, at least the larger specimens comprising such groups are somehow able to return to their former core areas. However, young oras apparently continue a more or less transient life, moving in somewhat random fashion from the immediate area of the carrion outward.

Homing and orientation.—Orientation within the activity range of resident oras is undoubtedly based on clinotaxic, olfactory, and visual cues. In all areas in which oras occur, broad-range orientation must be dominated by the obvious long-range visual and geotaxic cues offered by the mountainous relief surrounding almost all the valleys. Less specific, intermediate range factors include clinotaxic factors. Thus, upward leads to a generally cooler environment, and sentinel, basking, or traditional resting sites; downward generally refers to deeper shade, water, and higher food and ora densities. Mountainous skylines offer almost continuous visual cues in the generally open habitats. Large, dry creek beds and elongate savannas form natural, easily followed pathways extending over many miles, clearly shaping the chief intermediate range visual orientation factors in long distance movements for man and beast alike. Regional scents may also play a role at this level, but data are lacking. The importance of scent in carrion location is discussed elsewhere.

Scent is clearly important on a close range, immediate level. Its utility in the precise location of carrion and fresh kill is obvious. Trailing proficiency tests using both dead and live, moving prey are discussed in Chapters 9 and 10. An unusual relation to carrion by juveniles is the peculiar behavioral pattern called *Rubbing* (Chap. 6) in which trail marking with carrion scent may be the most efficient form of carrion relocation. However, more important in view of the persistent use of mutually exclusive and widely separated core areas is homing ability, especially from carrion located in an unfamiliar area outside the normal activity range. Tracking observations of individuals over long distances, both going to and returning from carrion baits, strongly suggest that pheromones may play an important role in homing, though there is no proof of this.

Sixteen oras, all known residents, were tracked for varying distances by means of their spoor as they moved to the carrion. Almost every segment of their path was clearly imprinted in either the soft volcanic ash that covered most of the island or the fine sand of the dry creek beds (Chap. 1), providing an accurate record of their convoluted paths as they moved to the carrion. The same individuals were later trailed back towards their core areas from the carrion bait sites. Both the incoming and outgoing trails were clearly visible, making direct comparison possible. The distance between the two trails was measured at 1 m intervals and an average of these differences constitutes the ingoing and outgoing trail deviation/m (Table 5-10). Smaller oras tend to have greater trail deviation/m than larger ones. However, it is remarkable that all the deviations are so small. The average deviations/m are 8.8 cm for specimens longer than 2 m, and 11.6 cm for those shorter, with an overall average of 10.6 cm. The return trails often follow precisely the incoming ones, through shortcuts via visually obvious, wide foot trails, game trails, or dry creek beds exist (Fig. 5-10). This close correspondence in movement patterns, plus the fact that the tongue is constantly used to touch the ground surface, suggests that a pheromone is utilized in homing. Such trailing correspondence has been noticed over distances as great as 810 m, and there is no reason to suspect it does not occur during the longest and most convoluted trails all the way back to near the point of origin (as far as 11 km when moving to and from carrion; see Chap. 9).

Resident and nonresident oras show different movement patterns, whether in or out of their own foraging ranges. This is most easily demonstrated by compar-

Table 5–10. Deviations per meter in trails of oras to and from carrion

Ora size classes (m)	Number of observations	$\overline{X}$ deviations per m (cm)	Standard deviation
1.1–1.5	3	10.7	±3.1
1.6–2.0	8	12.5	±1.8
2.1–2.5	7	8.8	±2.1

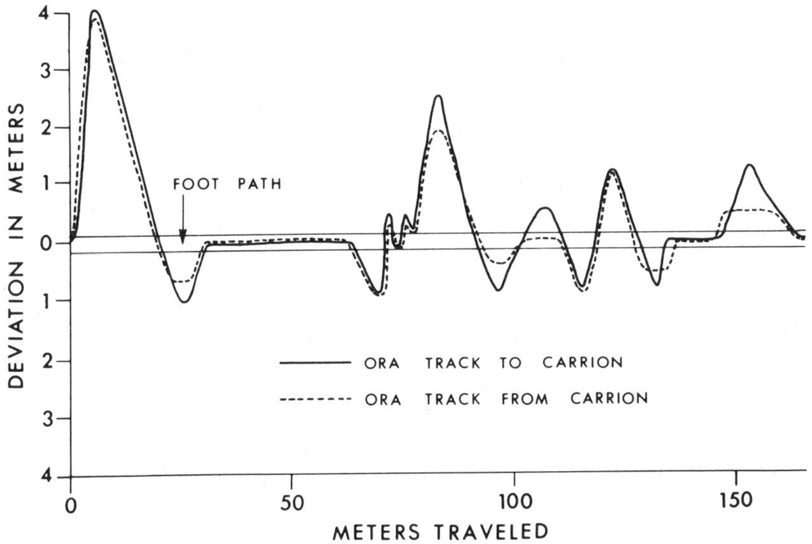

Fig. 5–10. Ora orientation is apparently more dependent on scent than on visual cues, as shown by the close correspondence of the tracks to and from carrion by a single ora, in spite of the strong visuality of a wide footpath (straightened for simplification) passing in the correct direction.

ing the average trail deviations ($\overline{X}$ = 11.8 cm in residents, 9.4 cm in nonresidents, differences between averages significant at 2 percent level). That residents show a higher deviation is probably related to their familiarity with local topographic and vegetative details; the homing residents often take advantage of shortcuts, and may even pass through or over an immediate obstacle in order to enter an open area beyond sight.

Resident oras regularly use certain trails in preference to others that go in the same direction. In general, when traveling longer distances between potential foraging sites, the widest, most open, natural or produced pathways are used. These are most commonly dry creek beds, human paths, important game trails and beaches, and other areas naturally devoid of vegetation. Because they are often used by all the resident and many of the transient oras, some trails become virtual ora highways, probably due to the ease with which they are traveled and the points they connect.

Residents often left the major trails at particular places on their way to known feeding, resting, and other specific sites. Some trails were consistently used by only certain oras—always large residents. Frequently, an entire sequence of specific trail segments was repeatedly used in foraging movements of resident adults. Two or three circuits were often remarkably consistent, though separated by many months duration. Details of an earlier track, measured from adjacent logs, rocks, and other topographic features, were rarely more divergent than 0.5–1.0 m from a later track made by the same individual, even though the topography and vegetation would have allowed considerable variation.

Most ora trails were indistinct and not easily recognized, but the repeated use of trail segments by specific individuals sometimes resulted in very distinct trails. Most of these segments were 50 m or less in length, about 1 m wide, and noticeable only when the lower level vegetation was rather dense and pushed down or aside. The paths lead to frequently used ambush sites, traditional resting sites, or burrow complexes. In general, these trails are very similar to those observed near core and nesting areas of adult resident crocodilians. A particularly obvious ora trail, easily seen from adjacent hills, is located on the north slope of Gunung Soro in Loho Kalo. It is approximately 300 m long, free from all vegetation, and extends from near the bottom of the slope to a hole cluster two-thirds of the distance to the top. The trail does not follow the easiest contours, but extends straight up the slope face at an angle of about 45°. To my knowledge it was never used by any animal species other than *V. komodoensis*.

Although some trails were used by oras of all sizes, some were used exclusively by one individual, suggesting that other oras did not know the paths existed and/or avoided the area close to the residents' core area (where most of the trails in this category are located). Individual pathways within the core area of 19W were mainly associated with its more frequently used ambush and thermoregulatory sites.

Shelter Characteristics and Use

The major shelters used by oras are burrows, natural cavities, and overhanging vegetation. Of these, hollow trees and burrows are cited as the most common in many *Varanus* species (Smith, 1932; Mertens, 1942). *V. komodoensis* frequently enters both. Specimens from 75 to 150 cm total length are occasionally found in hollow trees, often climbing to heights of 15 m to enter openings. Smaller specimens from hatchling size to about 75 cm spend most of their time in tree hollows and under loose slabs of bark, particularly of *Tamarindus indicus* (Chap. 10). The most common shelter of oras over 150 cm in total length is a burrow, often referred to in earlier reports of *V. komodoensis*. Oras use holes made by rodents and civet cats, the hollows made by wild boar in overhanging banks, or (on Flores) the long straight holes dug by porcupines; they also seek shelter in openings provided by rockslides. However, the most commonly used burrows are those dug by the lizards.

Burrows.—The length of the burrows excavated by *Varanus* species depends on several factors, of which species behavioral traits and soil conditions are most important (Lucas and Frost, 1896; Zakhidov, 1938). Those of oras are relatively short, generally 0.75–1.25 m in length. They are concentrated in two major habitats: along dry creek beds and on relatively open hillsides. There is no significant difference in burrow length in the two habitats (Fig. 5-11).

Creekside holes are almost always excavated in the nearly vertical clay banks, often under tree roots (Fig. 5-12; Pl. 4). Their height above the river bed varies from 0 to 3.5 m ($\overline{X} = 1.5$, $N = 85$), depending largely on bank height and conformation. The farthest that burrows are located from the channels is 8.5 m.

Burrows along all creek beds in the Loho Liang study area are nonrandomly distributed (Fig. 5-13), being concentrated in a long belt between 10 and 60 m in the middle section of the stream courses. The 10 m contour in Loho Liang often crosses the creek beds where the banks are about 1 m high. From this point the holes are found upriver to about 60 m, where rocks and boulders in the bank become so common that digging is difficult. Burrows are never found in periodically flooded or mangrove areas, but they are sometimes dug along the banks of ancient abandoned creek channels and oxbows. Ora burrows were also found along dry creek beds on Flores, though there were none along Nanga Loök, a creek containing water all year long.

Ora holes are rarely found on the flat valley floor. Only two such burrows were found on Komodo. However, there was a small hole cluster (described below) on the relatively low storm dunes near Nggoer, Flores (sea water sometimes enters these burrows during exceptionally high tides). On Komodo two ora burrows were found approximately 1.5 m and 2 m above the beach in a vertical clay bank eroded by the sea.

Hillside burrows (Pl. 4) are common, usually concentrated near the ecotone

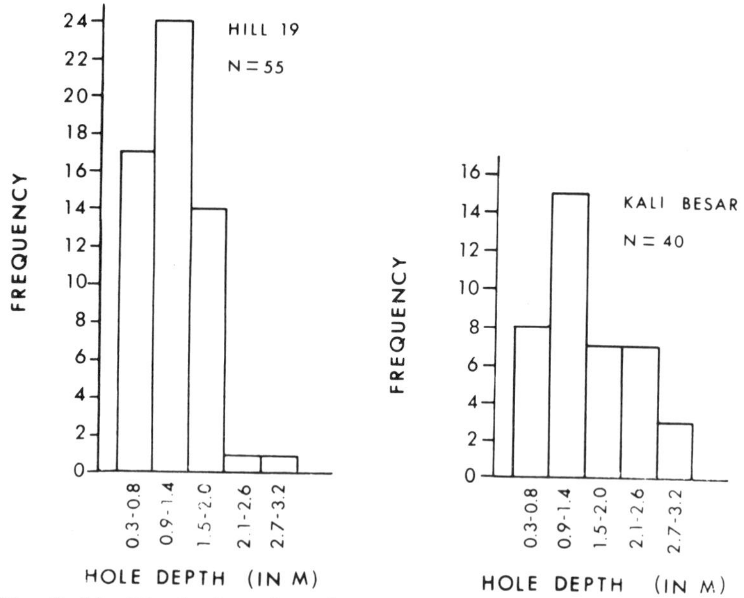

Fig. 5–11. Distribution of ora burrow lengths on hillside (hill 19) and along a creek bed (Kali Besar).

between valley monsoon forest and hillside steppe or savanna (Fig. 5-14), often along the top of the debris slope. They are most commonly excavated under boulders (48 percent) or tree roots (46 percent), and are less frequent against low vertical banks or earth slides (6 percent). Convenient vertical banks are sometimes produced by rooting wild boar, or even by continued use of certain game trails. There is great variation in the burrow position above the valley floor (5–80 m, $\overline{X} = 13$ m on hills 19 and 23, Loho Liang, N = 27). Most hillside burrows are located on a slope about 30° ($\overline{X} = 32.1°$, range 16°–53°, N = 27). Though most creekside and hillside holes have a straight, parallel-sided bore, internal shape varies greatly (Fig. 5-15).

Dunn (1927b) stated that most holes on Komodo face south, and Pfeffer (1959) reported that three of four holes he investigated on Rintja face southeast, the few remaining ones west or northwest. Both workers suggested that the direction in which the burrow faces is probably correlated with some environmental factor. However, our observations make it clear that burrow-facing direction is slope dependent, so that burrow directions on flat surfaces exhibit a higher coefficient of variation than those on hillsides (Table 5-11; Fig. 5-16).

Hole clusters are known from several localities on both Komodo and Flores, particularly on hillsides. The number of burrows in each cluster (ignoring multiple openings in boulder rubble) varies from 5 to 18. In every instance the soil in which hole clusters were found was of a type more easily excavated by the oras than were surrounding soils. In elevation hole clusters varied from 3 m to nearly 300 m. Several clusters were associated with old landslide areas (Banu Sambi, Gunung Kubur, Fig. 5-16; Gunung Kalo, Figs. 5-17, 5-18, 5-19; Gunung Klinta, etc., all on Komodo). A small hole cluster occurs just above the high tide line on the beach at Nggoer, Flores (Fig. 5-16). Hilmi Oesman (personal communication) suggested that these clusters may be ora nest sites, but excavation of many holes in search of remains of old eggshells provided no evidence of such use. However, a recent study of *Varanus bengalensis* in India (Auffenberg, unpubl. data) suggests that in at least this species the females return to specific places to lay their eggs. Over several years this behavior produces groups of holes similar to those described above. That some hole clusters of *V. komodoensis* have a very long, though probably intermittent, life is suggested by their

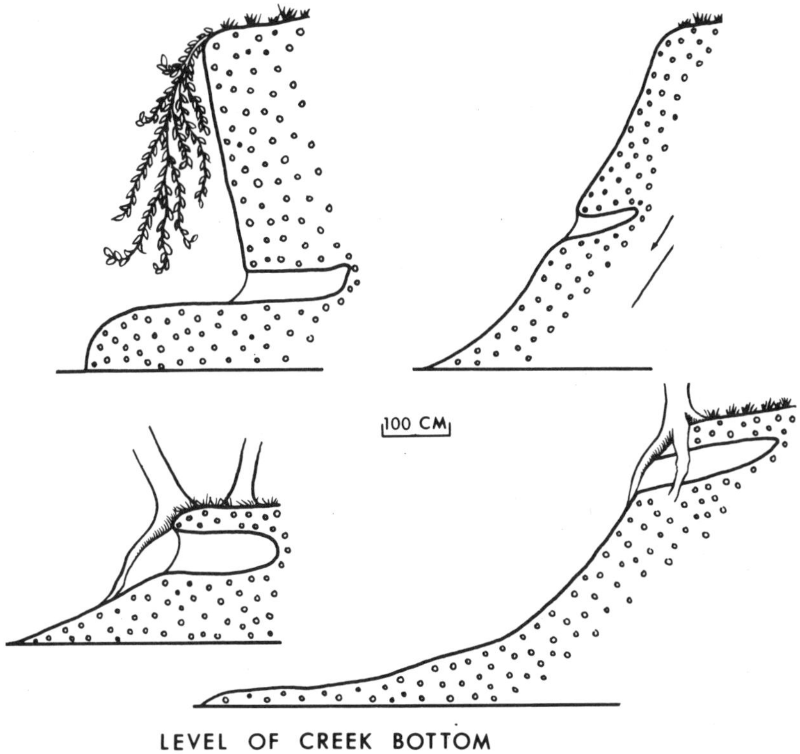

Fig. 5–12. Bank and burrow cross sections along Kali Besar, Loho Liang, Komodo.

Pl. 4. Variation in ora burrows, Komodo. *Top,* part of hole cluster, Gunung Kubur, in uniform sandy loam. *Bottom,* hole 31, hill 19, in boulder-strewn savanna.

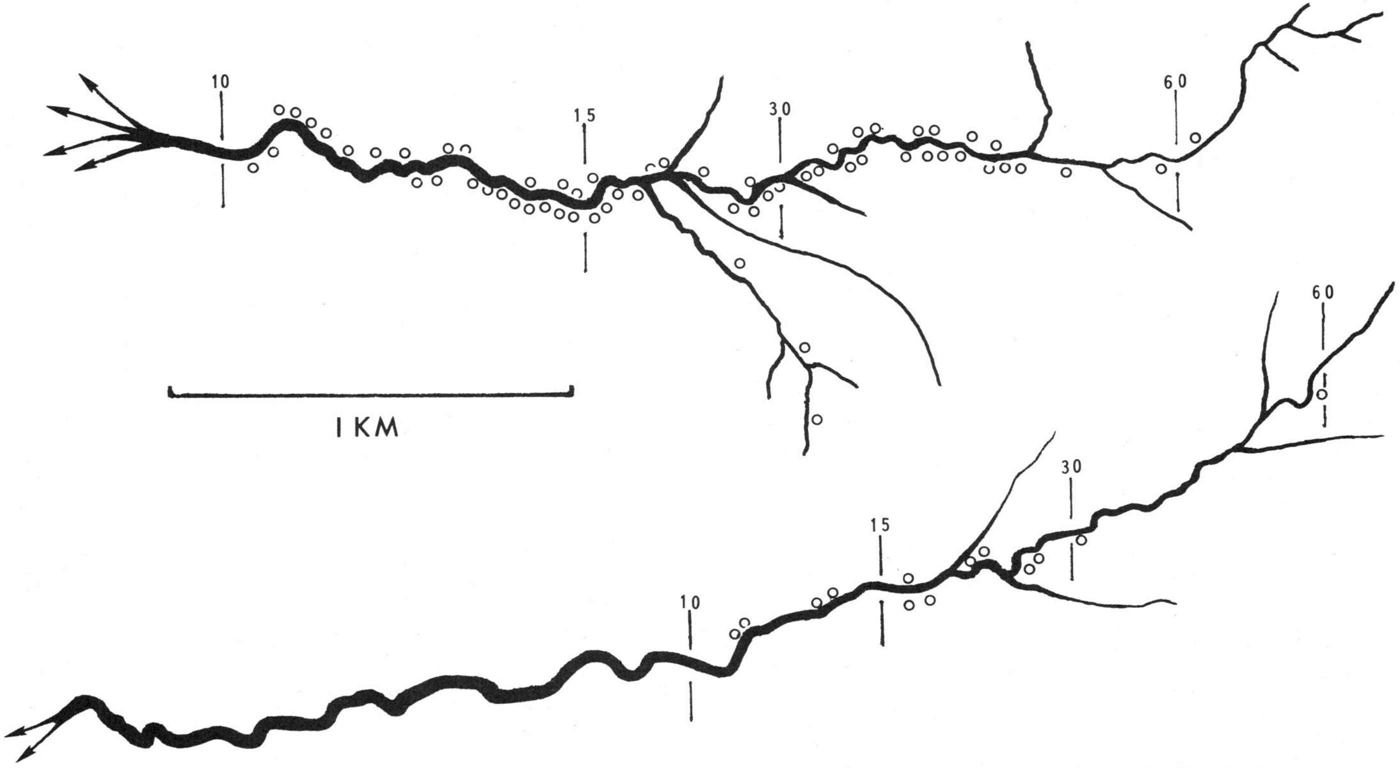

Fig. 5–13. Burrow distribution (circles) along dry creek beds. Numbers represent contour intervals (in m) above sea level. *Upper*: Kali Besar. *Lower*: Vai Nggulung, Loho Liang, Komodo.

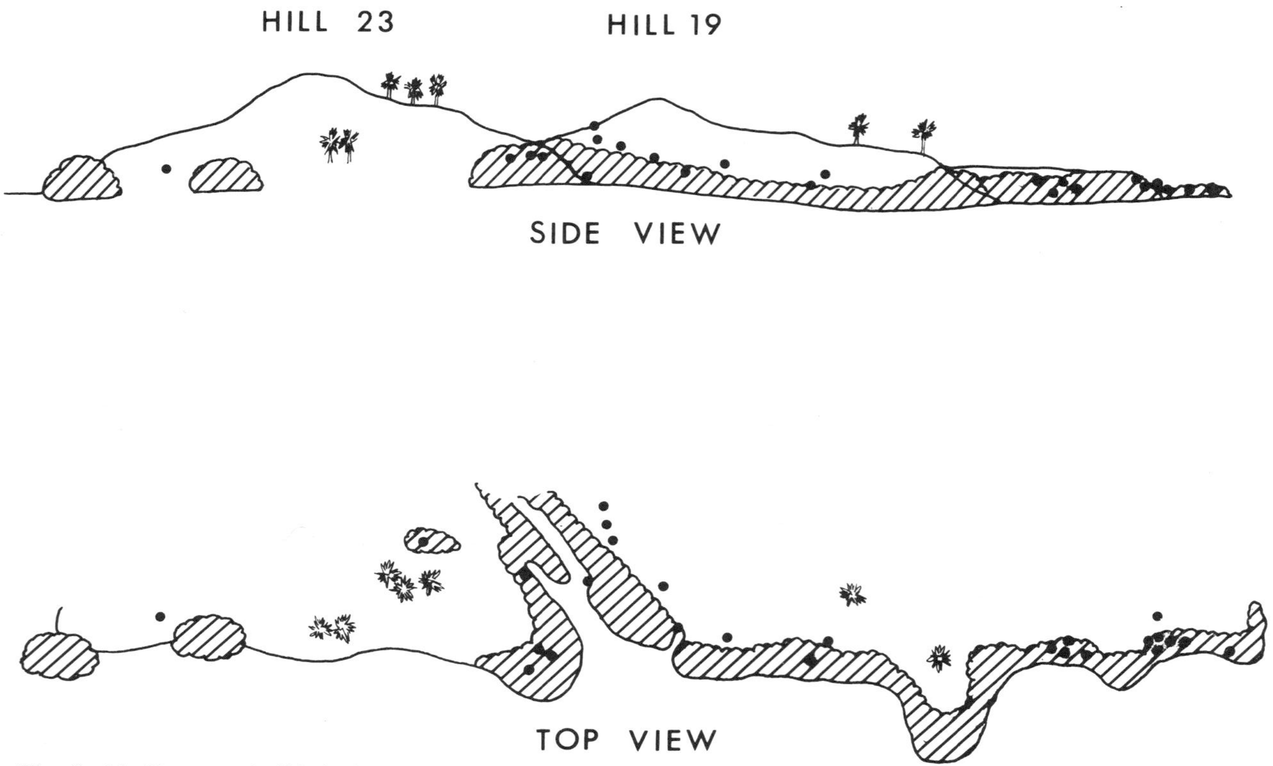

Fig. 5–14. Burrows (solid circles) are usually concentrated at the ecotone between monsoon forest (cross-hatched) and savanna (clear).

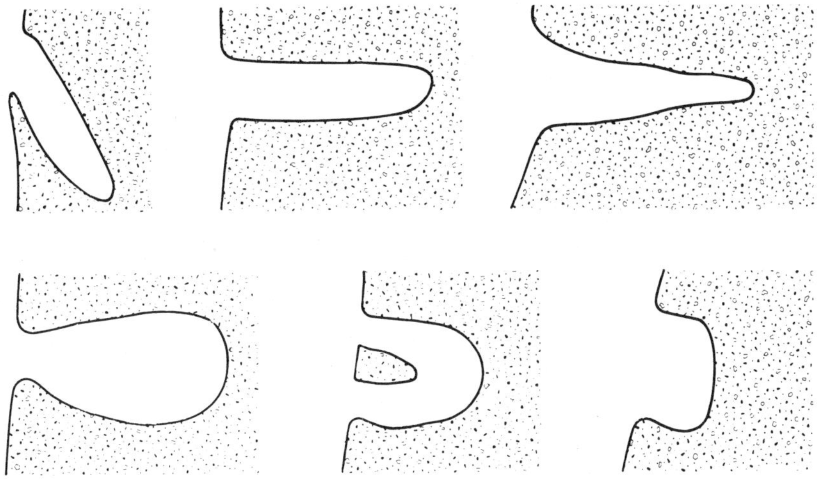

Fig. 5–15. Variation in internal shape of ora burrows, top view.

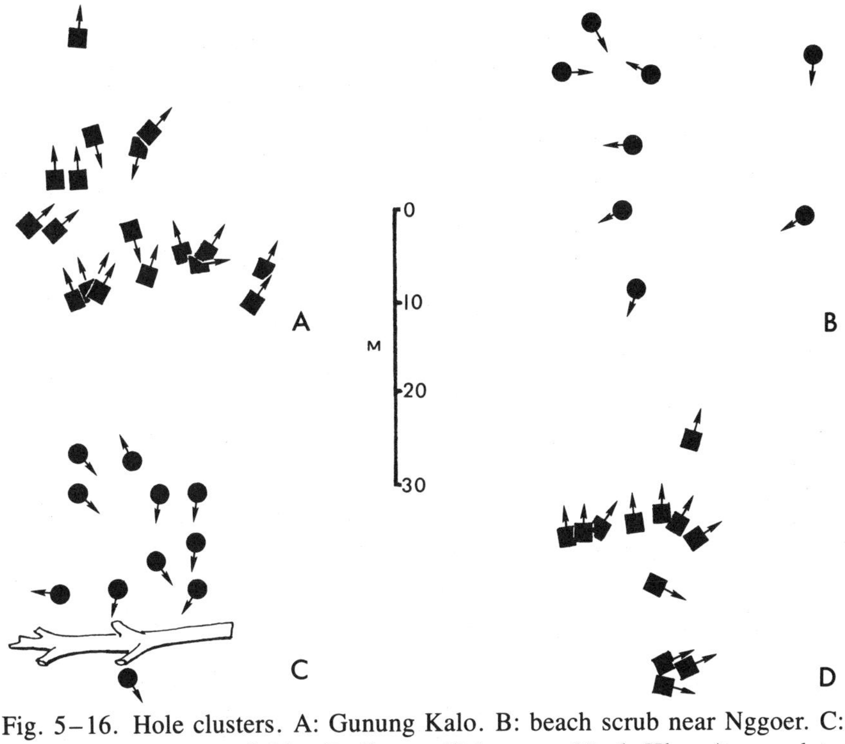

Fig. 5–16. Hole clusters. A: Gunung Kalo. B: beach scrub near Nggoer. C: Gunung Perangka, near Sabita. D: Gunung Kubur, near Ntodo Klea. Arrows show direction in which burrow mouth faces.

Table 5–11. Coefficients of variation (CV) in burrows' facing directions compared to $\overline{X}$ surface slope

Place	$\overline{X}$ slope (in °)	CV	N
Kali Besar	87	11.3	36
Gunung Kubur	62	30.7	12
Gunung Klinta	51	32.2	9
Gunung Inselung	38	41.3	6
Gunung Kalo	37	44.9	20
Peranka	28	43.7	11
Banu Sambi	18	52.2	13
Nggoer (Beach)	2	60.5	8

appearance, as well as by the statement of a reliable Nggoer villager that the Flores beach cluster has been used by oras for at least 30 years.

Kern (1968) has suggested that the concentration of burrows along creek beds and on hillsides reflects differential seasonal use, determined by climatic factors. Tables 5–12 and 5–13 show the results of having checked for the seasonal occupancy of ora burrows on hillsides 638 times and along creek beds 471 times. These data clearly show that during all parts of the year hillside burrows tend to be used more frequently than creekside holes (79.3 percent and 20.7 percent, respectively, when corrected for slightly differential visitation). There is no evidence that the use of creek bed or hillside holes is seasonally correlated. However, holes of both types are used most commonly in the dry season (June–November, 80.9 percent of all burrow usage), infrequently in the wet season (December–February, 19.1 percent), and never during the equable period of March to May. Table 5-14 shows identical seasonal results when usage data are corrected for number of times burrows checked/month. Of the total number of burrows used (Table 5-14), there is no significant difference between day and night usage in the wet season, though diurnal use is significantly high in the dry season.

Burrow use during the wet season may be related to generally lowered ambient daytime temperatures. However, usage during the dry season may be related to water conservation, ameliorating heat loss during the generally cool nights, or increasing heat loss during the middle of the day by using the walls of the burrow as a heat sink. During the dry season, evening temperatures are 0.8–1.8°C warmer than the outside nighttime temperatures and 1.8–7.0°C cooler than the outside daytime temperatures (Table 4-2), compared to 0.2–0.7, and 0.3–5.8°C, respectively, for the wet season. Thus, mid-burrow temperatures show greatest differentials with outside temperatures at night as well as the day during the dry season. This potentially greater ameliorating effect is probably largely responsible for higher usage during the dry season.

However, in general burrows are infrequently used at any time of year (only 41 of 1109 times checked) and play a minor role in the lives of most individuals. The large number of holes within the activity ranges of resident oras precludes the use of all of them at any time. Furthermore, though some holes are used more than others by some residents, few are regularly utilized. Of 16 hole uses recorded in September, only one hole was used 3 times during the month by the same ora; of the 41 total usages during the year, on only 7 occasions did the same oras use the same holes at least twice during a single month. Moreover, different oras (resident and transient) occasionally used the same holes. Thus, hole utilization can hardly be called either regular or habitual.

This intermittent hole use pattern is also suggested in *V. griseus* (Grijs, 1899),

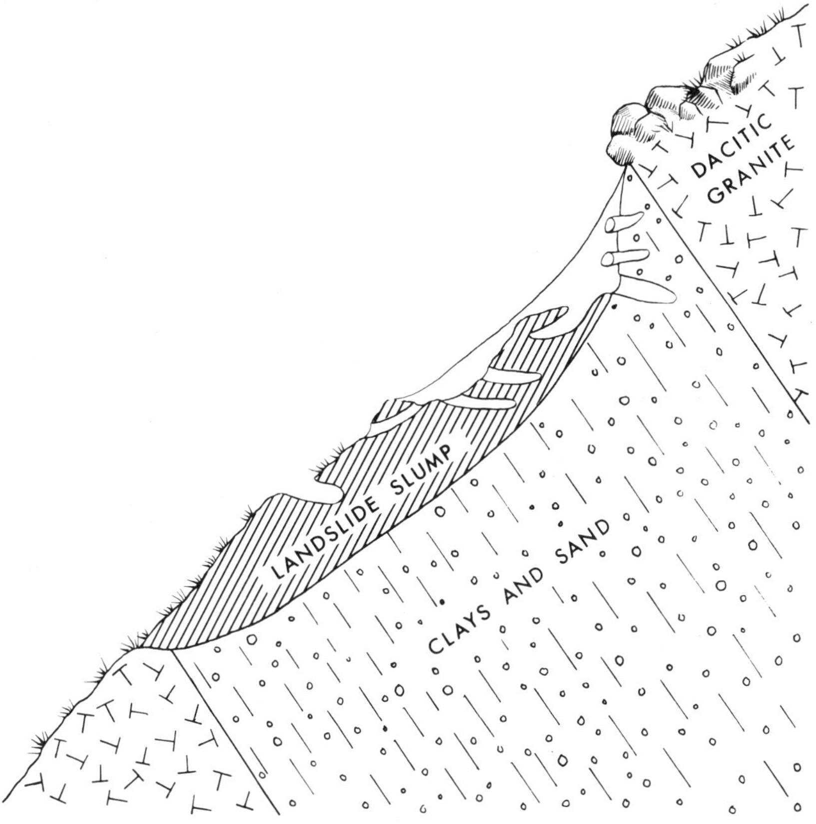

Fig. 5–17. Idealized cross section of Gunung Kalo hole cluster, Loho Kalo, Komodo, showing concentration of burrows in landslide slump.

though *V. niloticus* (Cowles, 1930) may regularly use specific burrows. Pfeffer (1959) showed that during 10 nights an ora on Rintja used its burrow only 5 times. Hoogerwerf (1954a) concluded that the holes are infrequently used, and Burden (1928) suggested that the ora may not always use the same hole but may use several. Holes are normally slightly lengthened terminally with each use. The greatest lengthening documented by a single use is 25 cm.

When found in holes, oras usually are alone. However, we twice found two oras in a single burrow at the same time. On one of these occasions an ora was

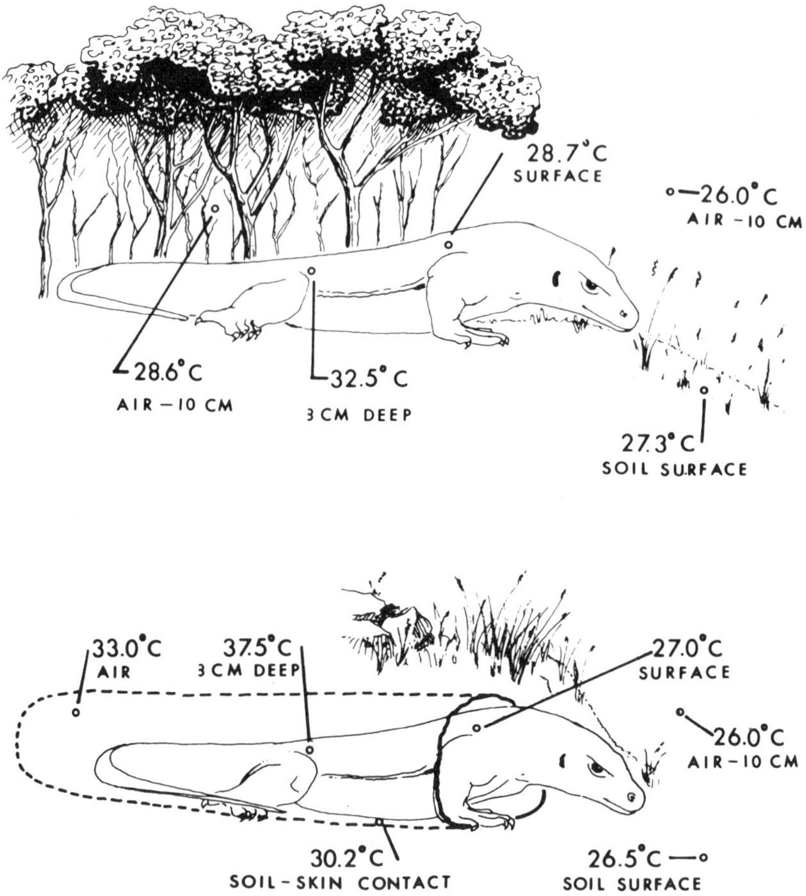

Fig. 5–18. Typical thermal relations outside and inside occupied burrow (below, Nov. 26, 2300 hrs, hill 23) and thermoregulatory site (above, Nov. 1, 0300 hrs, hill 19).

Fig. 5–19. Cross section of typical location of traditional thermoregulatory site (X) at junction of hillside forest and savanna.

seen to enter a burrow, remain there for a short period, and then dash out, followed by a second individual intent on biting the first.

Geckos (*Gymnodactylus d'armandvillei*), rodents (*Rattus diardi* and *R. rattus*), land snails, many insects, and even wild boar are frequently found in the burrows; none of these is an obligate.

In general, the burrows are relatively short (considering adult ora length), with an average length of 1.5 m. The ora is normally doubled inside the hole (unlike *V. bengalensis* [Mahendra, 1931], which is reported to sleep with its head at the terminus of the hole) so that the tip of the tail and head are near the burrow mouth. Normally the bend is near the base of the tail, so that the length of the body and head, plus the radius of the doubled base of the tail, equal the burrow length. The head of the lizard is much closer to the opening than is the tail tip. However, during the warm evenings or in the morning the ora may lie with the anterior portion of its body outside the hole. In small diameter burrows, oras do not bend in hairpin fashion but are pressed against the walls with the head pointed posteriorly and legs stretched posteriorly (as reported in captives by Pfeffer, 1959). Honegger and Heusser (1969) interpret the sleeping position of a captive *V. salvator* as largely thigmotropic, probably also the case in the ora.

Table 5–12. Seasonal use of burrows by oras

Month	Number of different holes checked	Number of times holes checked (N = 1109)	Hill holes (N = 638)	Creek holes (N = 471)	Number of times oras used holes (N = 41)	Number of oras involved	Hill holes used	Creek holes used
September	65	235	186	49	16	13	14	2
October	75	132	80	52	6	5	4	2
November	75	94	51	43	4	2	2	2
December	0	0	0	0	4[a]	?	4	0
January	54	147	29	118	2	2	1	1
February	102	152	88	64	1	1	1	0
March	59	86	39	47	0	0	0	0
April	86	125	80	45	0	0	0	0
May	32	59	37	22	0	0	0	0
June[b]	21	51	20	31	3	2	3	0
July[b]	28	28	28	0	5	5	5	0
August	0	0	?	?	?	?	?	?

a. Established by sticks placed in burrow mouth during absence from island.
b. Flores data. All other data from Komodo.

Table 5–13. Percent utilization of burrows by oras ([number of uses] ÷ [number of burrows checked per month] × 100)

Month	Percent	Number of times burrows checked
September	6.8	235
October	4.5	132
November	5.3	94
December	?	0
January	1.4	147
February	1.0	152
March	0.0	86
April	0.0	125
May	0.0	59
June[a]	5.9	51
July[a]	17.8	28
August	?	0

a. Flores data. All other data from Komodo.

Table 5–14. Diurnal and nocturnal use of burrows in wet and dry seasons

Season	Total	Diurnal use (%)	Nocturnal use (%)
Wet	7	58	42
Dry	34	83	17

In the hairpin position, the burrow lumen is often almost entirely filled, increasing the heat sink soil effect under one climatic regime, yet providing an insulative air space under another. Subsurface soil temperatures near sea level on Komodo are near 27°C, and body temperatures are between about 35° and 40°C. The burrow serves as a heat sink when the body and outside ambient diurnal temperatures are high, or as a temperature ameliorant when nocturnal temperatures are low. In the case of the latter, a 4–8 degree differential between outside and inside temperatures is maintained throughout the night in an occupied burrow, considerably slowing body heat loss. Differential temperatures between the air of occupied burrows and deep areas of the ora's body ranged from 0 to 6.5°C throughout the night, and the equilibrium between soil and deep body tempera-

Table 5–15. Nocturnal temperatures (°C) of burrows' surface soil and deep body temperature of resident ora (48 kg)

Time	Surface soil temperature	Deep body temperature
1700	30.3	38.2
1900	30.1	37.2
2100	28.9	36.3
0100	28.4	35.8
0300	28.0	35.2
0500	27.6	34.8
0700	27.3	34.5

tures was rarely attained before sunrise (Table 5-15). Rate of heat loss is apparently further regulated by how much of the body is outside the burrow. Thus, large oras sometimes lie with as much as their shoulders outside the burrow mouth (Fig. 5-18) during part or all of the night. During the middle of the day the inside burrow temperatures ($\overline{X} = 27.5$°C) are often as much as 8°C lower than outside temperatures 10 cm above the surface. Therefore, the burrow may be used as a heat sink during the daylight hours, particularly during the warmest part of the year.

Water conservation is probably one of the most important functions of the burrow. The relative air humidity inside and outside the burrows shows a differential of approximately 10 percent in both the dry and wet seasons (Table 4-2). While not important in the wet season, the differential is probably very important during the dry period. Under some circumstances oras are said to spend the night in water-filled game wallows (Lallemant, 1929), though we failed to observe any. Lederer (1942) reported that his captive sometimes spent the night in its pool.

Traditional resting sites.—The thermal characteristics of these extremely important sites within the range of most (all?) large adult oras have been discussed in Chapter 4; their use during the digestive pause is covered in Chapter 9; and their importance as territorial scent posts was discussed earlier in this chapter.

Each of these sites is usually approximately 0.7 m wide and 1.3 m long. Scooped free of most debris, the bare earth becomes a very shallow body pit. The core area of each resident adult in Loho Liang includes three to five such sites. In every instance they are found above the valley floor (10–150 m) and at the junction of forest and savanna vegetative associations. The usual ecotonal location is shown in Figure 5-19, the typical placement in respect to surrounding

Fig. 5–20. Location of thermoregulatory sites (X). *Top*, hill 23. *Bottom*, Gunung Rankin.

Table 5–16. Nocturnal thermal relationships (°C) at thermoregulatory sites

Time	Ambient temperature (10 cm height) Under open sky	Under canopy at site	Deep body tempera-tures
1800	41.2	35.3	40.0
1900	30.0	33.2	39.3
2000	29.2	30.6	39.0
2100	26.0	30.0	37.5
2300	25.1	28.8	36.3
0100	24.2	26.7	35.0
0300	23.3	26.2	32.5
0500	22.8	25.6	30.3
0600	22.7	25.5	30.0

Note: telemetry of ora 34W, 33.5 kg.

features in Figure 5-20, and their position within the core area of one resident adult in Figure 5-8.

Each of these features plays a role in determining the thermal properties of the resting sites, of which the most important is the formation of the steep thermal gradients (Fig. 4-14). These gradients are apparently extremely significant in thermoregulatory activities (Chap. 4), serving to provide a wide spectrum of ambient temperatures in a minimum area. The dead air space in the open savanna upslope from the forest edge plays an important role in morning basking, while the body pit under overhead vegetation reduces heat loss to the open sky at night. That this is less effective in reducing nocturnal heat loss is shown in Figure 5-19. However, large oras, because of their great mass, lose heat slowly in even this relatively open nocturnal environment (Table 5-16), so that deep body temperatures are at a nearly operative level at morning haulout time. However, heat loss in smaller oras is much more rapid, thus explaining why individuals less than 1.5 m long were never observed spending the night under the open sky or at traditional thermoregulatory sites. While few smaller oras were found during the night (N= 12), those we did find were in burrows, tree hollows, or in dense foliage in trees or bushes.

Vegetative cover and hunting.—Vegetation type is not only important in thermoregulation, but it has a major function in determining general movement through the environment, as well as specific hunting techniques, largely because of its obscuring qualities. Although it is clear that some plant species limit the vision of adult oras, others do not, and it is therefore not surprising that ambush sites for large prey are most often found in ecotones dominated by *Themeda* rather than by the high growing *Solanum* or *Hypoestes*.

The remarkable change in vegetation density from dry to wet season, particularly in regard to grasses and vines, modifies general movement patterns as well, for regions that usually are easily passable for adult oras during one season may be nearly impassable during others; other areas exposed to high insolation, and perhaps avoided during the dry season, may be thermally comfortable during the wet season, etc. These seasonal cover differences are apparently more important to the larger than to the smaller oras.

Optimum foraging strategy.—Selection favors those oras who forage for food most efficiently. Since the ora is a refuging predator (e.g., returning to a specific burrow or resting site each day; Hamilton and Watt, 1970), it is under selective pressure to minimize not only the time needed to find its prey but also the commuting distance back to the refuge. Based largely on theoretical considerations, Morrison (1978) has shown that search time for bats is minimized by the use of search paths with low turning rates, as these improve the chances of searching over areas nearer the refuge. Further, his models suggest that the relative importance of searching versus commuting depends on the number of return visits a predator can expect to make to a known prey site.

The models predict that for oras which habitually return to a specific hunting site, total foraging time is minimized by a search path that leads not to the nearest prey, but one that is well known and that can be expected to produce food, even though more distantly located. Therefore, many of the high prey densities within the commuting radius may remain unknown to, and so unused by, resident adult oras. This may help to explain why the foraging ranges of some adult individuals can be so broadly overlapping.

6

Individuals

Individual Characteristics

Morphology.—In addition to ontogenetic and sexually related differences, some oras possess distinctive individual characteristics usually most noticeable in large adults.

A commonly distinct variable, characteristic among the largest specimens, is the snout color. In these probably very old individuals the supralabials, infralabials, and sides of the snout are flecked or clouded with a metallic, light-greenish sheen. Others are sometimes diffusely marked with pale greenish-yellow or yellowish-pink, particularly on those parts of the skull subjected to abrasion (supraocular ridge, canthus, and preocular area).

Injuries.—With the exception of the literature on turtles and crocodiles, few data are available regarding injuries in reptiles. While crocodiles and turtles often lack part or all of a limb, the tail, or even portions of the snout or lower jaw, such is not the case in oras. Of the almost 200 individuals observed on Flores, Padar, and Komodo, and 87 additional museum specimens (representing few juveniles) studied, only one exhibited a major injury. This was "Stumpy" (21W), a small male with a body length of 53 cm, who was missing a significant part of its tail. Hoogerwerf (1954a) also reported seeing an ora on Komodo with most of its tail missing. Although such loss in males is undoubtedly most commonly caused by the tail being bitten off during fighting, humans are also occasionally responsible. For example, Komodo villagers report that the photographer with a recent expedition deliberately severed the tail of an ora for no apparent reason.

Though few oras are found with severe wounds or losses, many large specimens of both sexes possess old, healed scars; this is more predominant in males than in females—a common correlation among many species of large reptiles. It is suspected that in crocodilians such scars are caused by territorial fighting among mature males (Cott, 1960). Mertens (1942) provided a similar explanation for the more heavily scarred male varanids. Most scars on females are probably the result of courtship and breeding. An analysis of the scar patterns on the pair of adult oras in the Jogjakarta Zoological Park is particularly enlightening. Both specimens were obtained when immature and their present scars were accrued in captivity. The male (much larger than the female) has no scars, but the female has many, presumably the result of wounds received during the annual courtship in which they engage. Of the 11 scars on the female, approximately one-half are scattered all over the body, and the other half are concentrated on the tail tip (Fig. 6–1). That this is a typical pattern in adult females is corroborated by the field data: the tail tip is more regularly bitten in wild females than any other body part, and more females lack the tip than do males ($N = 23$, χ^2 less than 5 percent).

Wild, mature males exhibit a higher concentration of wounds in the shoulder and dorsal thorax regions than is exhibited in the Jogjakarta pair and in free-living, mature females (Fig. 6–1). Many of the scars are the long, slash-like wounds that oras inflict on one another when fighting. However, the wound patterns on oras 18W and 27W, after both had been killed by 19W, showed no concentration of slashes on any particular part of the body (Fig. 6–1).

In addition to slash wounds, many wild specimens receive punctures, often with a rip or slash extending from them. This type of wound has not been seen in juveniles, nor is it found in the Jogjakarta captives. The punctures are all about 1 cm in diameter, and I believe they are made by the tusks of wild boar. The concentration of wounds in the shoulder area (Fig. 6-1) is probably because the typical sideward-directed, defensive thrust of the boar brings the tusks into contact with this part of the lizard's body, particularly when the back or a hind leg of the boar is seized.

Temperament.—Review of the literature makes it clear that there is considerable variation in temperament of both captive and wild monitors (Mertens, 1942). Some individuals become tame very quickly (Hediger, 1934), though when moved to new quarters (particularly outdoors), some specimens may become aggressive (Ditmars, 1910; Mertens, 1942; and DePrato, MS). Von Rathgen (1894), Andres (1904), Werner (1904), and Lederer (1942) have all remarked on the ability of varanid lizards to recognize and come to only certain keepers, leading some authorities, such as Knauer (1911), to the conclusion that monitors are unique in their ability to adapt to new situations. Procter (1929) claims great intelligence for *Varanus komodoensis*.

The literature shows no agreement on the temperament of capitive oras. Indeed, opinions are often opposite. Some have described the ora as "excitable," "unpredictable," and "vile tempered." Others have called it "silent," "morose," and even "phlegmatic." The last is particularly true of animals maintained in captivity for long periods. DePrato (MS) believed that in captivity smaller oras were more aggressive than the large adults. Komodo villagers corroborated this for free-ranging animals as well. Pfeffer (1959) suggested that the oras of Flores and Rintja have a temperament different from those of Komodo (which I could not substantiate). Some individuals were decidedly aggressive. Within the Komodo study area, the most dangerous was 34W. The unique temperament of this animal is described in greater detail below.

The spectrum of behaviors exhibited by members of a feeding group to my presence varied from excitement and flight to seeming disinterest. In general, larger animals were more nervous than smaller ones at a carrion carcass, possibly due to a higher hunger level in young animals. Attenborough (1957), H. Oesman (personal communication), and S. Holmes (personal communication) all state that feeding adults can often be approached carefully within 1–2 m. This is also our experience, particularly if one crawls on stomach or hands and knees.

In spite of considerable individual variation, smaller oras in a free state are basically shy and nervous. Flight distance may be as great as 100 m when the ora is first able to see an approaching human over the savanna. Outwardly most large wild specimens are placid and not easily excited. Many flee in a rather slow

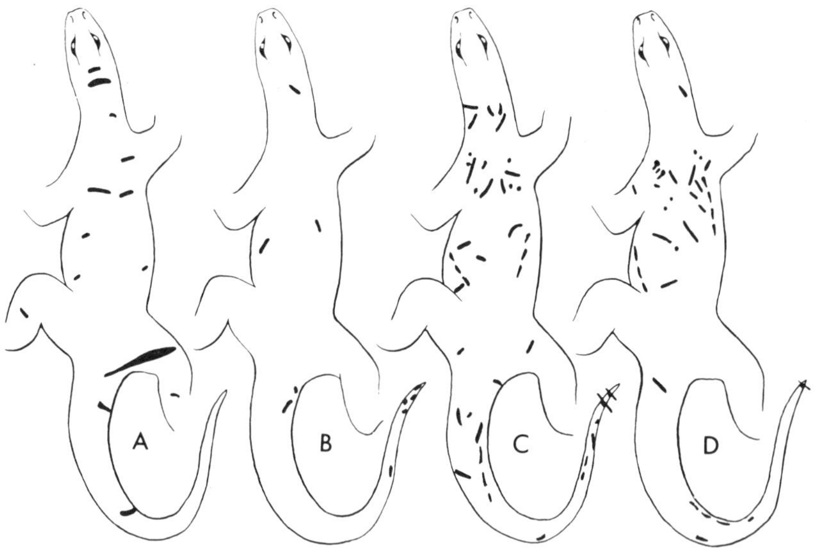

Fig. 6–1. Scar patterns in *V. komodoensis*. A: 1.5-m specimen killed by larger individual in wild. B: 1.5-m female specimen in Jogjakarta Zoo. C: composite scar pattern of wild-caught females (N = 27). D: composite scar pattern of wild-caught males (N = 51).

undulating trot rather than the impulsive, headlong dash of particularly the smaller individuals.

However, when cornered in the stockade traps, captive oras were often exceedingly aggressive. On several occasions specimens 1–1.5 m in length that had climbed partially up the vertical stockade walls and had reached the overhanging lip at the top leapt at me with mouths open, hurling themselves as much as 1 m or more off the wall. Larger individuals, usually shy and docile when free-ranging, would often become aggressive in the stockade. While we attempted to noose them, certain individuals would approach us in a very aggressive fashion, with the throat dilated, the mouth opened slightly, body compressed laterally, tail in striking position, and hissing. Some, such as 19W, 24W, and 8W, knew the position of the entrance trap and when confronted inside would open their mouths and move straight toward us, apparently to force us away from in front of the entrance. There is no question that they would have bitten us had they had the opportunity.

The most obvious and perhaps most intriguing individual behavioral traits are related to feeding, particularly the tendency of certain individuals to attack remarkably large prey or humans. Individual variation in frequency with which large prey, such as water buffalo, are attacked is discussed in Chapter 10. Human attacks are discussed in Chapter 12. In both cases, it is clear that oras that regularly attack large prey or humans are not behaving normally when compared to other individuals. A short case history of such peculiar conduct follows.

When we first arrived on Komodo the villagers stated that all the oras at Poreng and Gunung Vau were particularly aggressive. Lethal attacks on humans had occurred in both areas, and villagers reported several near attacks by oras living there. We eventually captured and marked all the oras at Poreng, and it was immediately obvious that only one of the animals in that area acted oddly aggressive. This was 34W, a male approximately 2 m long. Its behavior was so unusual that the villagers soon began calling it *ora gila* (the crazy monitor). When met on a game trail this individual would just as often approach an intruder with mouth open and tail bowed as it would run off into the underbrush. It followed exactly the footprints of our children and those of villagers for hundreds of meters on the beach, suggesting it was not frightened of human scent. In fact, it often entered our camp and once almost stepped into our shelter, though we were clearly visible inside. It often came to our Poreng camp shortly after our arrival, perhaps because it was drawn by the scent of our campfire (a pattern frequently noticed by villagers as well). On one trip to Poreng it drove our expedition members out of their tent, stuck its head into a knapsack, removed a shirt, and tore it to shreds. Other pieces of clothing and things we handled were similarly manipulated or carried off. There was no question in our minds, or in the minds of the villagers, that if given the opportunity this animal might attack, whether provoked or not.

Many individual behavioral traits of large oras are related to locomotion, and in some cases these peculiarities allowed us to recognize the spoor of certain individuals (discussed further below).

Behavior Inventory

Greenberg (1977) states, "the literature of reptile behavior is sorely deficient in complete, objective behavior inventories. The importance of the ethogram, long known to students of animal behavior and emphasized by Tinbergen (1969) and Beer (1963), is too often neglected." In addition to Greenberg's very complete ethogram of *Sceloporus cyanogenys*, particularly noteworthy are the studies of Brattstrom (1971) in attempting to index the postures of *Amphibolurus barbatus*, the work in social behavior by Bogert (1959), Ruibal (1961), Heath (1965), Carpenter (1967), and DeWitt (1971), and on thermoregulatory behavior by Stamps and Barlow (1973). Because of a need for additional detailed behavior descriptions, and particularly because nearly all the work to date has been on iguanid and agamid lizards, the behavioral units of the ora as noted in the field are described here in physical and functional terms, following the outline provided by Greenberg (1977). Letters in the following diagrammatic inventory refer to Figure 6–2.

(1) Body-down: Venter resting on substrate, legs near body (L).
(2) Resting: Eyes closed (N).
(3) Ventral adpression: Entire ventral surface of body pressed to substrate, rear legs (and sometimes front legs) spread and pressed (N).
(4) Posterior adpression: Posterior portion of venter pressed to substrate, rear legs spread and pressed (I).
(5) Head-down: Head resting on substrate (N).
(6) Head-up—low: Head parallel with substrate (O).
(7) Head-up—high—parallel: Head held above body axis, parallel to substrate (J, L).
(8) Head-up—high—vertical: Head held above body axis, tipped toward the vertical, neck is clearly stretched (K).
(9) Head-cocked: Head rotated and tilted horizontally at the neck (E, I).
(10) Head-tipped—down: Head tipped down from neck axis (A–C, G, H, M).
(11) Neck-sigmoid curve: Neck in S-shaped curve.
(12) Neck-arched: Neck arched vertically (H).
(13) Neck expansion: Hyoid apparatus pulled down and forward, increasing apparent depth of neck (A–H, M).
(14) Roach-up: Median dorsal skin of neck and nuchal area raised (A, B, D, H, M).

(15) Head-scrape: Side of head scraped against substrate.
(16) Anterior body-up—low: Elbows bent at angle greater than 45° (M, O).
(17) Anterior body-up—high: Elbows bent at angle less than 45°, sometimes nearly straightened (I, J).
(18) Oblique stance: Anterior part of body greatly elevated by utilizing environmental surface irregularities.
(19) Body-up—low: Elbows and knees bent, body raised very slightly above substrate (M, O).
(20) Body-up—high: Elbows and knees maximally straightened (G).
(21) Tail-up—partial: Tail slightly elevated above substrate, but touching at tip (O).
(22) Tail-up—full: Tail straight, completely raised above substrate (G).
(23) Tail-arched: Tail raised above substrate in curving arc.
(24) Tail-to-side: Tail bent to one side (L).
(25) Tail-blow: Stimulus struck with tail (M).
(26) Tail tip-quiver: Tail tip quivers rapidly.
(27) Tail-lashing: Tail lashed from one side to another (H).
(28) Defecation: Tail lifted at base, distal portion limp.
(29) Regurgitation: Gastric pellets of undigestible material after large meals, or stomach contents regurgitated when under stress.
(30) Oviposition: Tail lifted at base, distal portion limp, rear legs elevate vent area (not seen, but undoubtedly present).
(31) Cloacal-scrape: Cloacal area dragged or scraped against substrate as ora moves forward.
(32) Pawing: Movement of loose substrate with front foot, shallow displacement of surface material.
(33) Leg-holding: Front foot used to hold food or other animal down or still.
(34) Mouth-holding: Holding (pressing) prey toward substrate with mouth.
(35) Digging: Considerable displacement of surface material with front foot.
(36) Scratch: Part of body scratched with one of feet.
(37) Nudge: Object pushed with snout.
(38) Food grab: Food object abruptly picked up in or grasped by the mouth.
(39) Food cutting: Flesh cut into sections with teeth by jerking backward with entire body, or by sawing motion of jaws on first one side and then the other.
(40) Food shake: Prey or flesh slung from side to side, sometimes pounded against the ground.

(41) Swallow: Inertial manipulation of food into gullet.
(42) Mouth-scrape: Lips and side of mouth scraped against substrate.
(43) Lip-licking: Tongue wipes lips.
(44) Tongue-touch: Tongue briefly touched to surface of substrate, water, or other animal.
(45) Licking: Rapidly repeated tongue-touch.
(46) Drinking: Nose placed in water, gular pumping movements.
(47) Mouth-open: Lips and teeth parted, but interior of mouth not visible (E).
(48) Gaping: Mouth wide open, interior visible (F).
(49) Salivation, foaming: Usually with mouth open, when overheated, or when frightened (Lederer, 1942).
(50) Epiglottal projection: Epiglottis erected and pushed forward.
(51) Negative perpendicular orientation: Animal faces away from stimulus, but maximizes exposure of its dorsal surface by postural adjustments.
(52) Positive parallel orientation: Animal faces stimulus and minimizes exposure of body to it by postural adjustments.
(53) Lateral orientation *(Schragstellung)*: Sagittal plane made to face stimulus point by postural adjustments (M).
(54) Face-off: Two oras in mutual lateral orientation.
(55) Walking: Body slightly elevated, moving forward relatively slowly (O).
(56) Running: Tail up, very fast movement.
(57) Bipedal position: Standing still, or walking forward slowly (K).
(58) Jumping: Jumping between elevated positions, or up or down.
(59) Climbing: Climbing up or down on trees and rocks, utilizing legs, body, and tail.
(60) Swimming: Feet held at side and body and tail strongly undulated.
(61) Rubbing: Slow, forward movement with venter pressed onto substrate (I).
(62) Sagittal expansion: Lateral compression of the body, increasing the profile of the sagittal plane (M).
(63) Side-nudge: Pushing against the side of another ora with the nose.
(64) Forearm-grip: Grasping another ora with forelimbs.
(65) Head-over-shoulder-face-off: Mutual placement of heads over one another's shoulders.
(66) Ritual-walk: Stiff-legged, stately, slow-marching walk (G).
(67) Head-up-body-down-alert: Body pressed to substrate, but head tipped up ("receptive") (L).
(68) Lunge: Rapid movement of the body toward prey or antagonist.
(69) Bite: Gripping with teeth.

(70) Mounting: Mating straddle of male on back of female.
(71) Tail-tuck: Tail of male tucked under tail of female and vents brought into apposition.
(72) Copulation: Insertion of hemipenis and pelvic thrusting while vents are in apposition.
(73) Shiver: Trembling of female before and during time eggs are laid.

Behavioral Complexes

Rubbing.—One of the most unusual habits of smaller oras is that of rubbing their bodies, particularly the belly, on the ground near or in carrion. Most oras rub themselves in hair or particularly intestinal (not stomach) contents. On October 1, I noticed "Speckles," a 1 m specimen, use its front claws to pull wads of "slipping" hair from a rotten goat (=*pawing*). It then rubbed its belly and sides in the hair pulled off the carrion.

The rubbing movements are often accompanied by *lateral expansion* and twisting of the body as the ora presses itself down onto the area in a manner similar to that of captive snakes bothered with skin problems. Considerable force is often applied to the scented surface area by that part of the body being drawn over it. The hind legs usually dangle posteriorly, while the front legs are used to pull the ora over the ground. At the same time the animal may rotate its body in such a way that a side, as well as the belly and chest, is brought in contact with the substrate. Sometimes this rotation is so great that the ora almost turns onto its back, as a dog might roll in an unusual scent. Occasionally the ora simply lies in the hair and intestinal contents, with legs stretched out to the sides, pressing its belly onto the intestinal contents by bending its back in a concave arc.

Rubbing was observed in males and females, and all were about 1 m in total length (102–121 cm). That none of the large oras was ever observed to exhibit this behavior is undoubtedly important. Though it is impossible to be sure, *rubbing* in intestinal contents and hair may be a method by which young oras mask their own scent. Because of the feeding habits of large oras, I assume that this behavior is important in protecting the young from being cannibalized by the adults, since the latter usually do not eat intestinal contents and hair. The scent may also be an important mechanism for relocating the quickly disappearing carrion after the young have been chased off by larger, agonistic specimens. The ora would reorient quickly by following its own scent trail back to the carrion. The entire pattern obviously needs additional study.

Basking behavior.—Basking refers to a group of distinctive postures consistently associated with heliothermic or thermophilic activity. In oras it is characteristic only in the early morning, more so in smaller than in larger specimens. The oras generally adopt a *negative perpendicular orientation* to the sun, often facilitated by land slope, rocks, and other irregularities. There is usually no *lateral*

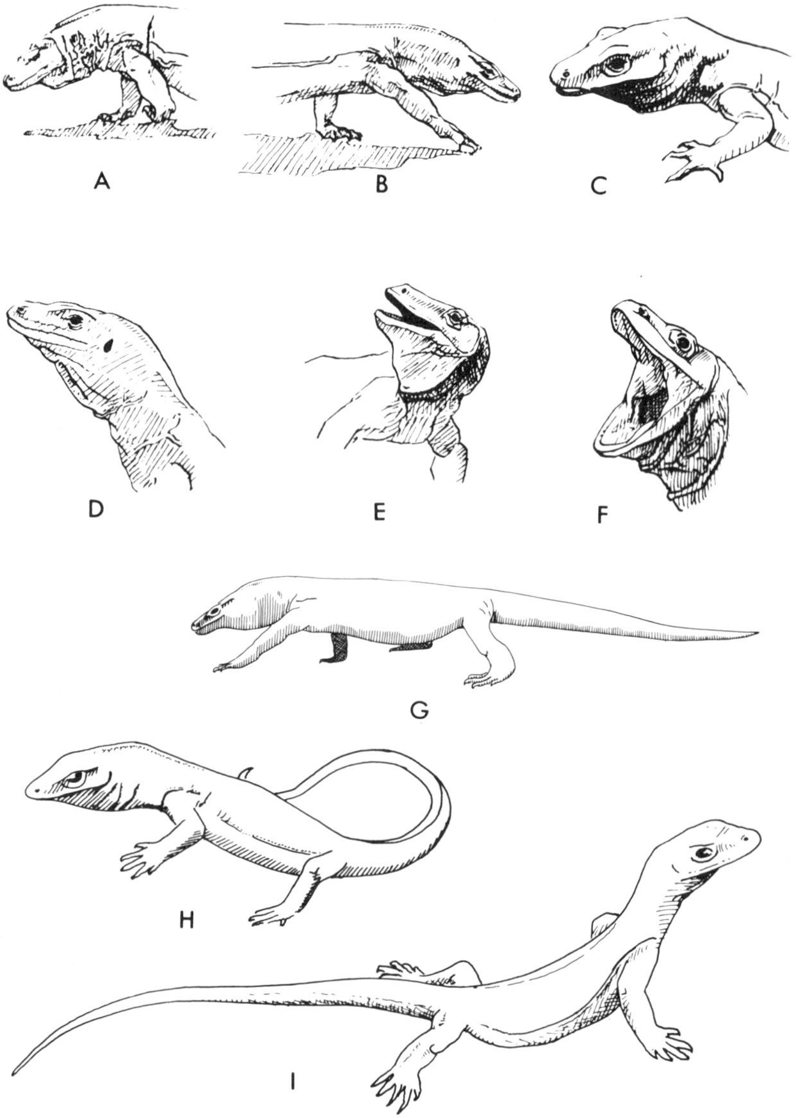

Fig. 6–2. Typical postures and movements of *V. komodoensis* (all from photos). A: adult, showing stress as part of threat: head down, roach raised, hyoid expanded. B: subadult, showing stress as part of appeasement display. C: hatchling, showing stress as part of threat display. D: adult, alert head position showing slight stress. E: hatchling, showing teeth and stronger stress. F: adult, showing high level threat displays. G: juvenile appeasement display. H:

hatchling, tail hooking with stress reaction shown in roach, hyoid, and head position. I: hatchling, plowing. J: adult, alert. K: bipedal stance. L: adult, resting and alert. M: adult, showing stress reaction while delivering tail blow to its left. N: basking position. O: adult, normal walking position, no stress (compare tail tip position with that in G).

expansion, as is common in other lizards, except as the belly is normally flattened when it is rested on the substrate. *Ventral adpression* is the most common posture assumed. It is essentially identical to that reported in other lizards (Brattstrom, 1971; Greenberg, 1977). Small oras frequently bask in the trees, adults always on the ground.

Thermal comfort postures.—Thermal comfort postures are consistently associated with high ambient temperatures and function to relieve thermal stress. *Toes-* and *feet-up postures* were not noted in oras in the field. The normal behavioral complex included *body-up, head-up—high, anterior-body-up—high, neck expansion, tail-up—partial, gaping,* and *epiglottal projection.* All of these are associated either with avoidance of hot substrate or with increasing air passage through the animal. *Anterior-body-up* postures are frequently affiliated with other behavioral complexes, such as the alert position or other strong feeding, social, or threat stimuli. In this connection it is often accompanied by *head-up—high,* with much scanning.

Tail postures are also related to both high temperatures and strong stimuli, as well as to walking and running. *Gaping* was common during breaks in a feeding sequence in the open savanna, when the individual had been compelled to accept high body temperatures. It is a common behavioral complex in overheated reptiles (see Greenberg, 1977, for review). *Gaping* is also associated with defensive reaction, but the mouth is often opened more widely in the latter, particularly if it is a strong response.

Alert behavior.—Alert behavior is equivalent to the "perching" behavioral complex found in many other lizards. In oras it is the adoption of a particular position by means of which the animal is better able to view the environment. However, it does *not* consume the major part of the ora's activity cycle, as it does in most other lizards. In oras alert behavior is distinctive in posture, but not site. The complex includes *anterior-body-up—high, tail-down, head-up—high—parallel.* This position allows the ora to peer over the top of the savanna grasses. It is not equivalent to what might be called a quiescent position, in which the head and body are both held at a lower angle, or rested on the substrate. When part of the foraging complex, there may be attendant *scanning, head-cocked,* and *licking* patterns. Sight of prey or carrion recognition usually leads to a change in posture, whereby the head and body are lowered into the *walking* pattern, accompanied by *licking.*

Foraging.—Searching for prey usually includes slow *walking,* with *head-up—low* (or *head-tipped-down*) and *licking* the substrate and air, alternating with *head-up—high—parallel, anterior-body-up—low* (or *high*) and *licking. Pawing* and *digging* may take place around nests or debris. Stalking consists of a slow, cautious approach to the prey, with frequent pauses. When within attack distance, the ora assumes a *body-up—moderate* position, with *head-tipped-down,* followed by a sudden *lunge* and *food-grab* and *bite*, sometimes with *leg-* or *mouth-holding*.

Hunting from ambush.—This complex (adults only) usually involves preparation of the ambush site, in which the ora *paws* at the substrate, clearing debris from an area approximately the size of its body. This alternates with *head-up—high—parallel, anterior-body-up—high*, and *tongue-touch*. Finally the ora settles in the cleaned pallet area in a *body-down* position, but with the legs to the side. The head is in a *head-down* or *head-up—low* position, usually the former. The ambush site is normally next to a game trail, and when prey moved along it, in front of the hidden ora, the lizard would suddenly *lunge, food grab,* and *bite*. *Food shake* and *food cutting* movements predominate. *Leg-* or *mouth-holding* are often used with smaller prey. *Tail-up—full* and *tail-lash* are both common and undoubtedly related to balance. *Holding* and *oblique stance* are sometimes included.

Feeding.—Feeding usually begins with the *food grab,* which is often quite violent. There are no chewing movements, but vigorous *food cutting* and *food shake* behavior is typical. In this way large pieces of flesh and bone are separated from the carcass and *swallowed* (Chap. 9). *Lip-licking* and *mouth-scrape* behavior eliminate debris (including foreign matter from the substrate as well as unwanted hair and intestinal contents).

Drinking.—Abdoessoeki (n.d.) was incorrect when he stated that oras drink by flicking their tongues in the water.[1] They drink by immersing the snout to the eyes and sucking up water in the same manner that snakes do. After several deep drafts, the head is raised and the water allowed to run down the throat.

Throughout the range of the ora, water is generally scarce for much of the year. There is no evidence that the animals migrate to available water sources during the dry season. Thus, most of the Komodo ora population manages without any free water from at least April to late December. However, when water is available, oras drink deep and long, usually from savanna hog wallows, particularly after eating a large meal (Chap. 12).

Aggressive display and defensive behavior.—Many workers have described defensive behavior in varanid lizards: Grijs (1899) and Faussek (1906) for *V. griseus*; Stirling (1912) for *V. giganteus*; Schneider (*in* Werner, 1900) for *V. rudicollis*; Cowles (1930) and Tofahr (1908) for *V. niloticus*; Hediger (1934) for *V. indicus*; Smith (1935) and Deraniyagala (1931) for *V. bengalensis*; Smith (1927) for *V. timorensis*; Murphy and Mitchell (1974) for *V. gilleni*; Honegger and Heusser (1969) for *V. salvator*; and Mertens (1942) as a review.

In oras, threat includes both the display and actual weaponry, such as the teeth

1. Mertens (1942) agrees with Hediger (1934) that the tongue is not used by adults in drinking, but the former reports that baby *V. niloticus* will lick up water drops. Captive *V. bengalensis* regularly use their tongues to obtain water from pools too small or shallow for immersion of their snouts.

in *gaping*, or the tail in *tail-lashing* and *tail-quivering*. These are usually combined with *lateral orientation* (*Schragstellung*), *sagittal expansion, head-cocked, head-tipped-down* (or *head-up—high—vertical*), *neck-arched—vertically, neck expansion, body-up—high, tail-arched,* or *tail-to-side* (Chap. 12) and often accompained by an auditory cue, such as a hiss. Slight foamy spittle is produced rarely under extreme stress. The *roach* may be raised. Further stimulus will usually elicit *tail-blow, lunge,* and *biting*. Physical restraint usually leads to *scratching, biting, tail-blow, defecation, cloacal-scrape,* and *regurgitation*. *Biting* may be accompanied by *food cutting* and *food shake* to produce a particularly serious wound.

Appeasement display.—This display in varanids consists of the following: *head-tipped—down* (rarely *head-up—low*), head often *cocked, neck-arched, neck expansion, body-up—high, sagittal expansion, tail-up—full,* often with *licking, lateral orientation* common, and usually *ritual walking* (Fig. 6-2). The mouth is closed and there is no hissing. It is important to recognize that no weaponry is shown. The behavioral complex is called an appeasement display (Auffenberg, 1976), because it is most often exhibited by subdominant individuals (Chap. 12). It apparently tends to reduce aggression and, if successful, leads to flight (*running*), with resulting loss of all parts of the display. It may be related to the assertion display (Carpenter, 1962a, b, 1967) of iguanids and agamids, but there are no head nods or push-ups.

Fighting behavior.—Fighting (between two adult males) is ritualized and nonritualized. Unlike many other lizard species, fighting in varanids, for whatever purpose, is often serious and leads to severe lacerations, or even death. Ritualized combat includes a *face-off,* sometimes of the *head-over-shoulder* type. One or both of the potential assailants may exhibit aggressive or even assertive displays. In some circumstances, a *bipedal position* may be assumed, followed by *biting* (Oesman, 1970) and probably *forearm-grip* (not seen, but assumed to occur on the basis of a ritualized fighting pattern in closely related species). Nonritualized fighting usually involves initial *tongue-touch* or *licking,* often followed by assertion or aggressive displays, at least on the part of one of the assailants. Suddenly thereafter, one of the combatants will *lunge* and *bite,* or the dominant member will *paw* the subdominant one across the back. The subdominant individual will usually *run* away, and the dominant one often will give *chase*, continually trying to *paw* the other. Submissive behavior usually includes *head-down,* often *cocked* in direction of aggressor, with *ventral adpression,* or at least *posterior adpression*. The tail is held tight against the substrate. The dominant individual may *mount* the subdominant one, at least momentarily.

Courtship.—Chapter 8 contains a complete description of courtship activities. The behavioral complex normally begins with the male *tongue-touching* or *licking* the female at the insertion of the hind limbs with the body, the area just anterior to the ear, or the side of the face. The male may *nudge* the female with his snout. The female may show an assertion display pattern, or may *run*, with the male following closely behind. The male may *bite* the neck and always *paws* the female on the back. The female may *tongue-touch* the male. *Mounting* is usually rapidly executed, followed by a *tail-tuck* and *copulation*. It is important to point out that the male only rarely shows overt aggressive behavior in the sense of exhibiting weaponry. When it occurs it involves *mouth-open* (no *gaping*) during the chase and before grasping the neck. Weak *neck-arch* and *neck expansion* may be exhibited by the male.

Locomotion

Oras are primarily quadrupedal and terrestrial. Though they readily climb into trees to feed, rest, or just sit and look around, they do so very slowly and deliberately. When danger threatens, larger oras descend and flee on the ground; the youngest specimens remain in the tree and find shelter in clumps of orchids, under loose bark, or in holes and crevices.

The largest oras spend all their waking hours on the ground, either lying, sitting, squatting, or walking around. In the typical stance, the one used almost exclusively during both walking and running, the soles of the feet are placed flat on the ground and the legs are thrust out to the side and then under the body in a rather stiff fashion. The body, head, and tail are undulated gently from side to side in rhythm with the four moving limbs. This quadrupedal gait is similar to that of other lizards, except that the motion is normally considerably slower. That is, the animal raises the two diagonally opposite feet and advances them while the other pair supports and propels the body forward. The head is normally swung slowly from side to side, with the tail swung oppositely, a manner identical to that described by Smith (1931) for *V. bengalensis*. When walking through dense vegetation the oras frequently move herbs and small brush aside with their heads, necks, and shoulders without breaking stride. Young oras tend to walk in a more stilt-legged fashion than older animals, approaching the stiff, formal gait of crocodilians.

When running, oras do not always move the body in vigorous and rapid lateral undulations as might be expected from watching them walk slowly. Rather, the body and tail are held fairly rigid. The hind feet are swung in a rather wide lateral arc, giving them the appearance of the movement of a double-ended kayak paddle. Zander's (1895) description of the body movements of swiftly running *V. griseus* is equally applicable to running oras. The rapidity of their stride is approximately equivalent to, and sounds somewhat like, that of a muffled machine gun.

Normal walking and searching is carried on at a speed of approximately 4.8 km/hr (our observations agree with those of Lederer, 1942). However, frightened oras (N = 10) run at clocked speeds of 4.1 to 5.6 m/sec (about 14–18.5

Table 6–1. Speeds of oras calculated on measured course

Ora number	Body length (cm)	Speed (m/sec)
1W	48.9	5.6
21W	52.7	4.5
43W	63.6	4.8
30W	65.6	4.2
18W	69.1	4.3
2W	70.2	4.3
10Y	75.7	4.1
3Y	76.1	5.2
28Y	81.0	5.3
19W	120.1	4.6

km/hr)(Table 6-1). There is no correlation with size ($r = 0.13$). Large oras sometimes resort to a slow trot that carries them over the ground at speeds of approximately 8–10 km/hr. We chased, on our motorcycles, an unidentified ora of about 2 m total length for a distance of slightly over 0.5 km. On the basis of odometer readings the ora maintained a speed of slightly over 14 km/hr for the entire course.

Oras move most rapidly (and at speeds probably greater than 18 km/hr) when frightened on rocky hillsides—particularly talus slopes composed of loose rocks. Here they may throw themselves downslope, propelling themselves with their feet and sliding as much as running down the steep inclines. Large oras may break shrubs of as much as 1.5 cm diameter in this downward flight.

Bipedalism.—Bipedal locomotion is reported for several *Varanus* species (e.g., Stirling, 1912). Though Lederer (1942) stated that an adult female ora could walk bipedally for short distances, we never observed bipedalism in these lizards during the study. Oras often assume a tripedal stance when food is suspended as a bait (Broughton, 1936; de Jong, 1937; Collins, 1956; King, 1968) (Fig. 6–2k), or when young oras raise the front part of their body off the ground to look over the grass. In this common stance of several monitor species (Mertens, 1942) including the ora, the tail is used as the third leg of a tripod. Bipedal adult oras that were evidently fighting have been described by Oesman (1970). Rapid bipedal locomotion has been described in early newspaper articles, but I doubt that it occurs, for it would confer little advantage in the dense vegetation and the precipitous hillsides of Komodo and would probably be undesirable in these situations.

Climbing.—Most monitor species climb well, often to heights of over 30 m

(Hagen, 1890; Green, 1909; Symons, 1912). Only *V. griseus* is a poor climber (Grijs, 1899). Some species, such as *V. prasinus* and *V. gilleni,* are highly adapted for an arboreal life, having a prehensile tail (Mertens, 1942), while others, such as *V. dumerilii, V. rudicollis,* and *V. grayi,* live almost exclusively in the trees (Müller, 1923; Barbour, 1926; Auffenberg, 1979b). Several species commonly spend the night in tree holes (Loveridge, 1923; Stebbins and Barwick, 1968). Although the adults of *V. exanthematicus* live on the ground, Grote (1911) stated that the young live in trees, in epiphytes, under the bark, and in tree holes. Burden (1928) and Kern (1968) both reported that the young of *V. komodoensis* are often found in trees, a fact corroborated by our own data. Oras are found in the trees throughout the year and almost all are less than 150 cm long ($\overline{X}$ total length, 105.6 cm) (Table 6-2). While they are capable of climbing to the very tops of the trees, oras spend most of their time at the lower levels ($\overline{X}$ height above the ground, 3.6 m). They may spend slightly over two days in one tree without descending and then often for only a few moments. The major reasons for arborealism seem to be foraging and resting, and sometimes escape from larger pursuing oras. Oras over 150 cm long seldom ascend trees because of their great weight. Small oras, from hatchling size to approximately 150 cm long, have no difficulty climbing vertical trunks as high as 10 m. They ascend the trees with ease, though rather slowly and deliberately, pulling themselves up largely by their claws, which larger oras dig deeply into the bark. Vines, branches, and irregularities in the trunk are all used for support. On the whole, their climbing seems to be extremely cautious, in sharp contrast to that of smaller arboreal lizards in other parts of the world. The reduced arboreality of adult oras is obviously correlated with their great bulk.

Except for oras in their first year, none jumps readily from branch to branch or actively runs in the tree. Each foot is placed carefully and firmly. Their feet are incapable of holding onto branches securely, and several times we saw oras misstep or slip, saving themselves from falling by one or two claws. Though larger animals sometime test a support of dubious quality by stepping on it gingerly with a front leg, oras appear to be poor judges of what branches can support their weight, and smaller dead limbs may break beneath larger specimens. Very small hatchling and juvenile oras are the most adept at climbing, easily turning upside down when moving down the trunk and sometimes jumping short distances. They occasionally fall, but were never observed to be injured. Most falls occur when oras attain too much momentum in descending the trunk. Once in the trees, smaller individuals, particularly, seem at ease, and the amount of climbing they do is not noticeably affected by an observer on the ground nearby. In several instances it was possible to climb within several feet of the lizards.

Oras rarely jump. Sometimes they leap from branch to branch or from branch to a hanging bait. However, when in the trees and frightened by an observer

Table 6–2. Tree utilization (37 observations)

Month	Length of ora (mm)	Height in tree (m)	Activity	Minimum total time in trees (hr)
January	110	6.2	?	1.0
	120	1.3	To take bait	1.2
February	150	1.6	Food search	0.5
	110	6.4	Resting	2.4
March	158	2.3	Resting	4.1
	120	3.3	Resting	1.2
April	45	5.0	Food, resting	2.0
	45	3.5	Resting	1.2
	50	3.5	Food search	0.5
	46	4.4	Resting	3.4
May	60	3.5	?	0.5
	110	2.2	?	0.5
	50	3.0	Food, resting	2.0
	70	4.5	Resting, sleeping	12.5
	147	4.0	Resting	0.5
June	55	3.0	Food search	0.5
	105	3.8	Resting, sleeping	52.0
	110	6.0	Resting	1.5
July	108	3.0	Resting, sleeping	22.0
	165	2.5	Food search	0.5
August	125	5.0	Food search	1.5
	65	2.5	Food search	2.5
	70	4.5	Food search	1.0
September	65	3.0	Food search	0.5
	72	3.5	Resting, food	21.5
	72	2.5	Food search	0.5
	75	1.5	Escape	0.8
	143	10.0	Escape	3.5
	170	5.5	To catch ora	0.2
October	168	3.0	To take bait	0.5
	126	1.5	Escape	0.5
November	85	2.5	Resting	1.0
	125	5.5	Resting	3.5
	130	1.8	Resting	0.5
	90	3.2	Resting	4.0
	131	3.0	To take bait	0.5
	143	2.5	Resting	0.5

climbing with them, or when someone is attempting to noose them, they often jump to the ground from as high as 6 m. Deraniyagala (1931) reported that *V. bengalensis* may leap from heights of up to 20 m without injury. Oras momentarily interrupt their fall just as they slide off the end of the branch. This is accomplished by turning the hind feet posteriorly as they come down out of the trees and dragging the claws through the terminal twigs and leaves.

Oras rarely jump while on the ground. However, by a combination of jumping and clambering, oras 1–1.5 m sometimes scrambled over the corrugated galvanized metal fencing of our activity pen, which was 1.5 m high.

Swimming.—Almost all monitor species are good swimmers. Hediger (1934) reported that *V. indicus* swims and dives as well as a highly specialized water animal. *Varanus niloticus* and *V. salvator* are particularly adept in the water. Smith (1932) reported specimens of *V. salvator* swimming from one island to another in the Gulf of Siam; Annandale (*in* Boulenger, 1903) reported the same in the Bight of Patani. In the Mergui Archipelago,where its normal habitat is mangrove forest, *V. dumerilii* often takes to the sea (Smith, 1931). *V. mertensi* from northern Australia inhabits similar environments. Even desert-adapted species, such as *V. varius* and *V. griseus,* swim well and enter the water voluntarily (Flower, 1933; Mertens, 1942). *Varanus niloticus* has been known to remain underwater for as long as 1 hour (Mertens, 1942). Waite (1929) believed that aquatic varanids can close and seal the narial openings to keep water from entering. The escape behavior of many monitors includes diving and swimming away underwater (Peters, 1870; Scherer, 1907; Smith, 1931).

Van Huut (1937) and DePrato (MS) stated that captive oras in their care were good swimmers. Burden (1928) reported that oras do not seek the sea to escape and are only mediocre swimmers when thrown into the bay (much of the distance they traveled was underwater). Pfeffer (1965) saw one in salt water on Rintja and another in a mangrove swamp there. During our study we saw one voluntarily enter the brackish water of a black mangrove swamp on Pulau Padar. At the mouth of Vai Nangebabu, Komodo, one plunged into a saltwater pool and swam approximately 9 m underwater to the opposite shore. We also saw them swimming toward a bait in the freshwater river of Nanga Loök (western Flores). One villager told me he saw an ora swimming along the coral-strewn bottom of a bar in very clear water 1.8 m deep near the foot of Gunung Vau, Komodo, and we photographed adults underwater in similar circumstances near Gunung Klinta. We were never able to determine what the oras were doing there. After regularly rising to the surface to breathe, they would then dive under again.

Perhaps the most remarkable, regularly occurring swimming feat of which I am aware is the movement of large individuals between Komodo and Nusa Mbarapu. The water gap, with a considerable tidal current, is about 450 m across the narrowest part. Two small islets lie in the channel, but near them tidal currents are sometime exceptionally strong, with many whirlpools. The roar of the running tide through this narrow gap can sometimes be heard for several hundred meters.

Nusa Mbarapu is only 1.4 km^2 in extent, too small to maintain a large predator permanently. The lizards have been seen swimming to the island, when fishermen are drying large quantities of fish on the beach or when many goats are being kept there (see Chap. 12 for further discussion). The oras swim to Nusa Mbarapu only when the scent is blown northward to Komodo. Presumably they must also be quite hungry to attempt the crossing. We hung a dead goat on the Nusa Mbarapu beach on one occasion and went back two days later, but found no indication that oras had crossed to the island. However, one goat produces little scent and during the 3-day period the wind was not steady and often blew in the opposite direction.

At least four people have seen oras swimming the water gap. All said the oras swam underwater for about 100 m, surfaced to breathe, and returned underwater.

During the last week of October, 1973, several different-sized oras were taken to the water gap and thrown overboard to be observed and photographed. It is clear that swimming ability is not correlated with size. All the oras tired fairly rapidly and none could be called a strong swimmer. When tired they tended to stay at the surface, rolling greatly and awkwardly in the swell, with weak lateral undulations. The low temperature of the water (about 26°C) may be an important factor in determining how long a period the oras can swim effectively.

When fresh (and warm?), oras swim reasonably well, with the propulsive force provided by the slowly undulating body and tail. The hind and front legs are held at the sides and are not used either at the surface or while diving. However, when the tail is restrained, or when they swim from the bottom toward the surface, the legs are sometimes used alternately, as in walking. The oras dive vertically only when they are threatened at the surface. When fresh they dive easily to a depth of at least 4 m, sometimes resting on the bottom.

Digging.—Excavation is a common activity and one in which oras are particularly effective (Lederer, 1942). They dig with their front feet, using one leg repeatedly for several strokes and then changing to the other. Even in deep holes, the hind limbs are not used to throw dirt out of the burrows. Rather, the lizard backs up slowly, continuing to scratch and throw dirt posteriorly with the front limbs. Digging may occur any time of the day, sometimes even during the night. In addition to excavating burrows and nest cavities, oras dig to locate megapode eggs and to find rodents, lizards, insects, and snakes. Loose bark is sometimes torn from the trees with the claws, and sticks, stones, and other debris are often pushed aside during pallet construction or when looking for food. When actively excavating, an adult ora will dig a hole almost 1 m deep in less than 1 hr in unconsolidated earth and extend a burrow at least 20 cm into the hard, baked clay of the arroyo walls in the same length of time. When unable to find a way out of the large stockade funnel traps, they usually tried (often successfully) digging under the walls.

Spoor

An ora moving through the forest or the savanna leaves behind numerous signs, which persist for several days or, occasionally, even months. Recognition of such signs is essential in the study of oras, because they are needed in tracking the animals and they aid in estimating relative population densities. Recent studies on *Varanus bengalensis* (Auffenberg, in press b) have shown that varanid spoor provide accurate data on which to estimate local population numbers. They are also invaluable in giving a rough idea of size and composition of populations, food habits, daily travel distances, and other aspects of behavior. The crude, elongate areas cleaned free of all surface debris are the most frequently seen and most lasting evidences of oras in their habitat. These are the areas where oras often lie day after day on spurs and along game trails, engaged in either thermoregulatory or digestive activities. Only the larger oras habitually use such forms. Sometimes the forms are constructed adjacent to game trails in anticipation of game movements, in which case they are used as staging areas to initiate attacks on passing deer and hogs. Each large ora apparently maintains four to five such forms as thermoregulatory resting areas (see Chap. 9). The actual number used for hunting purposes on the valley floor and on the slopes of the hills is unknown. The thermoregulatory forms, used intermittently for long periods (perhaps many years), remain visible throughout the rainy season. I suspect that the evidence for many of these lasts for several years, even without repeated visits by the large oras. However, hunting forms are used much less, and under most circumstances their evidence lasts for little more than a month or two. The major difference between these two types is their size. The thermoregulatory, or resting, forms on mountain spurs are often more than 1 m wide and approximately 2–3 m long. Hunting forms are approximately 0.5 m wide, and the area scraped bare of vegetation is usually about 1.5 m long.

There are essentially two types of visible trails: actual tracks of the lizard's feet, and the imprint of the dragged tail on sand, mud, or volcanic ash. Such trails are often seen in areas denuded of vegetation. They are most effective in establishing population numbers or individual size when they are produced on the extremely fine-grained volcanic ash that irregularly falls on the islands in this area. In these ashfalls every scale on the underside of particularly the larger animals' feet can actually be counted. The foot length can be accurately determined and compared with a plot of foot length against total length of known animals. In combination, these measurements, plus the distinctive way in which the tail was swung from side to side, provide a spoor of sufficient individuality to identify many of the large lizards of our study area.[2]

The Senses

Sight.—Much of the oras' behavior suggests they are largely scent-oriented. Sight is used mainly in situations needing no scent confirmation or in which scent confirmation is unattainable, such as identification at great distances. My observations agree with those of Piazzini (1960), who stated that the animals are quick to spot any slight movement either on the horizon or in the thickets through which they are moving. However, their capacity to do so does not seem to be superior to that of a man who is reasonably conscious of the surrounding forest. Werner (1904) claimed a visual perception radius of about 2 m, and Mertens (1942) about 50 m for larger monitors. Lederer (1942) reported that captive *V. komodoensis* could recognize individual keepers in good light from a 5–6 m distance. Food could be seen at greater distances.

Hearing.—Though the anatomy of the internal ear of varanids is poorly known, the literature suggests they have moderate acoustical capacity (Miller, 1969). Unfortunately, very few experimental data are available regarding the level of this ability, and opinions based on field observations are not in agreement. Thus Waite (1927) stated that *V. gouldi* is deaf and based this conclusion on the observation that the animal failed to react to shots fired nearby. Von Scherer (1903) stated that newly caught *V. niloticus* reacted to certain sounds, but that after a short time in captivity they failed to react to the same sounds.

Most reports concerning wild oras also suggest that they are incapable of hearing sounds (e.g., Burden, 1928; Broughton, 1936; Collins, 1956; Piazzini, 1960). Pfeffer (1959) believed they may become deaf with age. Shotgun blasts, firecrackers, rattling plates, and shouting have all been tried, apparently with no noticeable effect on the lizards. However, de Jong (1937) believed that his field observations showed oras capable of hearing sounds. There were several instances in which the animals could not have seen de Jong as they moved through high *Imperata* stands. Everytime de Jong shouted the oras stopped and began to move again only after he remained silent. In addition, he reported that natives of Flores are convinced oras can hear the sounds of hunters and their dogs as they drive deer over the savannas.

Procter (1929) believed a captive ora in the London Zoo was capable of hearing because, although the ora could not see anyone, it moved toward its cage door (apparently in anticipation of food) whenever the keeper turned the key in the lock. These and other observations convinced Procter (in a letter to Burden,

2. Both Pfeffer (1959) and Hoogerwerf (1954a) implied there was considerable difference in the spoor and locomotor patterns of *V. komodoensis* and *V. salvator,* but I disagree.

1928) that *V. komodoensis*, "in common with many of the larger and more powerful species, do not always bother to pay attention." Lederer (1942) reported that the captive ora "Bubchen" would come when persons familiar to her called her name. In her early years at the zoo the noise of loud vehicles and low flying aircraft disturbed her greatly. Pfeffer (1959) reported that oras in the Surabaja Zoo responded to the voice of their keeper.

Very little sensory physiological work has been completed on reptiles. Recent auditory experiments have shown that some reptiles formerly considered deaf are capable of hearing certain sounds produced at the level of common forest noises (Campbell and Evans, 1967). The fact that these animals fail to respond to these sounds is not so much an indication of their inability to do so, but rather of their disinterest. This seems perfectly clear to us in our observations.

Oras of different sizes clearly responded to the noise of flies suddenly roused from carrion. The oras were often lying in the shade nearby, paying no attention at that moment to the carrion from which they might have been feeding just a few minutes previously. However, when we, or other oras, disturbed the carrion and the flies left the carcass in a loud swarm, the resting oras usually turned their heads to look at the carrion.

On several occasions those of us watching several feeding oras from behind a blind noticed that some members of a feeding group would suddenly raise their heads at the sound of our motorcycle engine starting at the Base Camp; the villagers talking as they passed unseen on trails near the bait sites; or certain sharp noises we produced from behind the blind, such as striking two pieces of wood together, clapping our hands, shouting shrilly, whistling, hissing, playing a harmonica, and quickly turning up the volume on a radio. However, in every instance the response was short lived. Normally it included only the slight adjustment of the individual's head raised into an attentive position, often with slight eye movement. Occasionally they would survey the surrounding area by turning their heads from side to side before they resumed feeding. Only rarely could we elicit a second response to the same sound. Only hissing provided a flight reaction, and then usually only in the smaller oras. However, at the sight of a distant moving motorcycle or person the oras usually lowered their heads and quietly moved into the brush.

Oras make very few sounds themselves. Hissing is common in an aggressive context (Chap. 12), particularly as part of a defensive display. Sneezing is loud and frequent during feeding—usually to dislodge fly larvae from the nasal passages. The sound produced by a male scratching its claws on the rough dorsal scales of the female during courtship (Chap. 8) can sometimes be heard for many meters.

Olfactory.—Burghardt (1970) has published a review of chemical perception in reptiles. His conclusions indicate that chemical signals are of great importance in snakes and lizards, so many of their responses seem almost completely controlled by chemical cues. This is certainly true of the ora, for it is largely scent-oriented. Its olfactory efficiency is considered in detail in Chapters 9–12. In this species scent apparently plays an important role in territorial marking (Chap. 12). The ora also locates much of its prey by olfaction and quickly sensed our presence behind the blind. When downwind they often walked to the blind and placed their snouts against the small camera ports, exploring the air in the cavity behind by waving their tongues in all directions. They often approached the blind with their tongues flicking constantly, and on occasion would walk to the open back of the blind. touching the tips of their tongues to our equipment and clothing (Chap. 12).

In general, visual inspection alone seldom convinces an ora that an unfamiliar object is edible. In almost every instance food was eaten only after it had been touched by the tongue.

7

Demography

None of the usual reptilian secondary sex characters is developed in Komodo dragons. In fact, monitors in general are difficult to sex. This, plus the strict protection given *Varanus komodoensis*, placed severe handicaps on our investigation of several important demographic factors, because we obviously could not kill specimens and make internal studies, as is usual in unprotected forms. Injuries, deficiencies, and diseases could just as easily have been placed in the chapter on individuals, but they are discussed here because of their demographic importance.

Sex Ratio

Two methods were initially utilized to determine sex ratio: cloacal probing and caudal injection of sterile water to evert the hemipenes. The latter was quickly found to be completely impractical. Darevsky and Kadarsan (1964) had previously used cloacal probing on 22 Komodo monitors, concluding that the sex ratio was about 6 males to 1 female. However, we abandoned this technique because we were unable to consistently duplicate results on the same specimen. Fortunately, another method came to light about the same time. This involved a qualitative difference in the precloacal scales that was later found to have a sexual basis. In males some of these scales were arranged in two clusters, resembling a crude rosette (Chap. 2). Proof for this relationship was provided by: (1) the dissection of 6 preserved hatchling oras in the Jogjakarta Zoo, 2 larger specimens recently killed by other monitors, and 2 subadults killed by me under a permit; and (2) a comparison of the skins of museum specimens whose sex had previously been determined by dissection. I found the results completely consistent; that is, males possess the rosette, females do not. Unfortunately, no other easier method has yet come to light, and to my knowledge it is impossible to sex these animals from any distance.

Table 7–1. Male : female ratios of various *V. komodoensis* populations

Sample	Males	Females	Ratio
Dunn (1927a)	14	3	4.6 : 1
Darevsky and Kadarsan (1964)	22	4	5.5 : 1
All live specimens, this study	38	11	3.4 : 1
Preserved siblings, Jogjakarta Zoo	8	2	4.0 : 1
Mature only (170 cm +), this study	18	6	3.0 : 1
Total, all dissected specimens	50	15	3.3 : 1

Dunn's (1927a) conclusion of approximately 4.5 males to each female was based on the dissection of 17 specimens brought from Komodo by the Burden Expedition. With the exception of 3 dissected specimens, Darevsky and Kadarsan's 5.5:1 ratio was based on specimens probed in the field. During our field study of 49 oras, we obtained a ratio of 3.4 males to 1 female (Table 7–1) by determining the presence or absence of precloacal rosettes. Our ratio compares favorably with the ratio of 3.3:1 obtained when 65 dissected and sexed specimens from previous studies are pooled (17 of Dunn, 3 of Darevsky and Kadarsan, 13 in the Bogor Museum, 16 in zoological gardens, and 8 of this study). I believe it is reasonably accurate, at least for the Komodo population.

The data available on reptiles suggest that most have a 1:1 sex ratio. However, several workers (e.g., Tinkle, 1961) have reported asymmetric ratios in turtles, though geographic variation occurs in this ratio (Tinkle, 1961; Auffenberg and Weaver, 1969). Gibbons (1970) concluded that the data available to him did not substantiate a generally stated trend of female dominance in turtle sex ratios. What little information is in the literature on sex ratios of lizard populations suggests that, among adults, females almost always outnumber males (Blair, 1960; Hirth, 1963; Harris, 1964: and Tinkle, 1967). However, in adult varanids a higher number of males is common (Pianka, 1968, 1969, 1971; King and Green, 1979).

Darevsky and Kadarsan (1964) suggested that the high ratio of males to females in ora populations is a possible mechanism for reducing population levels in a marginal environment but failed to explain how this would be operative. Both Anderson (1961) and Williams (1966) concluded there are no empirical data in which sex ratio can be shown to be a population-regulation mechanism. Williams (1966) stated that, "sex ratio has significance in organic adaptation only in a population that is out of equilibrium . . . ." Cycling populations are almost continuously in this situation and so should represent an area in which to investigate the role of selection in establishing sex ratios. It is not known if ora populations are in oscillation or not.

King and Green (1979) suggested that male varanids may have larger activity ranges than females, making them more liable to capture, but no data are available. Auffenberg (1979a) has shown that male *V. bengalensis* are more active than females in captivity. Unfortunately, no data are available on these aspects in *V. komodoensis*.

In some cyclic populations, such as mice, certain genotypes produce asymmetrical sex ratios in favor of males. Myers and Krebs (1971) believed that if this genotype is at a selective advantage at high densities, there should be an increase in favor of males and a decrease in potential for population increase (the Darevsky and Kadarsan hypothesis); and with low differential male survival, excess male production will occur during the positive phase of the population cycle. As the decline begins, the "female" allele gains a selective advantage over the "male" allele. Myers and Krebs concluded that while changing sex ratios are probably not the principal driving factor in population cycles, variation and selection can lead to sex ratio changes which may have demographic consequences.

In most animal species in which one sex is numerically favored as adults, including the human species, the ratio of fetal or newly hatched young is usually 1:1. For example, in tigers the ratio of newborn is 1:1, though the adult ratio may vary from 2 to 5:1 in favor of females. However, information on primary sex ratios, especially in reptiles, is practically nonexistent. In the 10 hatchling oras available of the originally larger Jogjakarta clutch (Chap. 8), the male to female ratio is 4:1. In a small group of 6 sibling embryos in the Bogor Museum studied by de Jong (1946), 2 were rotten, and the male to female ratio of the remaining 4 was 3:1. Two additional embryos in the same museum from other clutches were males, making a total embryo male to female ratio of 5:1. Though the sample is small, it is clear that male oras predominate, not only in adult ratios, but in the primary sex ratios as well. This is apparently rare in vertebrates, and the few primary sex ratios available for other reptiles show equal sexual representation. Thus, there is no evidence of post-hatching sexual selection in the differential sex ratio among wild oras. Spieth (1974) suggested, on the basis of a theoretical model, that only a small bias is required to shift the primary sex ratio, and that skewed sex ratios are not necessarily anomalies.

In conclusion, data from several sources mitigate against Darevsky and Kadarsan's theory that the asymmetric sex ratio in oras is a special mechanism to prevent overpopulation. There is no evidence of sexual selection, and asymmetrical ratios are apparently common in many varanid species. Furthermore, it is not clear how such a mechanism would operate. Comparative numbers of subadult and adult classes indicate that overpopulation of oras does occur. This overproduction is obviously controlled by factors independent of sex ratio. Furthermore, the disproportionate total number of each of the sexes is offset by the ratio of *breeding* males to females that is close to 1:1, due to pair bond behavior (see below). Thus, the sex ratio favoring the ora males may serve to insure that every female has a mate as well as to assure high fecundity. This has already been suggested by Cuellar (1968) for bisexual forms of *Cnemidophorus*, in which the sex ratios are also skewed in favor of the males.

Population Density

Two related methods were used to determine ora density. In both, carrion was offered as bait, with individuals either trapped, marked with plastic tags and painted with large numbers, or sprayed with paint in distinctive patterns from behind a blind. Several types of traps were used (Chap. 1), most of which were located near the base camp (Fig. 1–9).

Though the traps drew many oras during the year, it was not possible to sample all parts of the study area by this method. The second system involved spraying the animals with paint from behind a blind. Five baiting sites were selected (Fig. 1–11) at which conditions for observing and spraying the lizards were ideal, with emphasis placed on area coverage and habitat:

> The *Base Camp* observation blind was approximately 300 m north of the beach, at the common edge of a small *Zizyphus* savanna, extensive monsoon forests, and an old garden plot overgrown with *Hypoestes* brush. It drew oras to it from much of the monsoon forest in the lower valley of Kali Besar, including Gunung Rankin and hills 19 and 23.
>
> *Kalo* was at the base of Ntodo Kalo, in the *Zizyphus* savanna approximately 300 m north of the beach. Bait in this area drew oras from the *Zizyphus* savannas on the east flank of Nggokodjarung, the entire south flank of the high ridge containing Ntodo Adjung, and the *Borassus* savanna on the west flank of Ntodo Klea.
>
> The *Nanenepi* site was located 100 m from the beach, on the edge of a mangrove swamp near the mouth of Kali Besar and Vai Nggulung. A very tangled area of thickets and dense vegetation lay between Nanenepi and the Trap 1 baiting area—the "overflow area" of the two major streams of the area. *Zizyphus* savanna surrounded the mangrove swamps and overflow thickets to the north.
>
> Vai *Nggulung* is a stream rising high on the central ridge below Gunung

> Satilibo, and running down a broad valley to Nanenepi. The bait site was approximately 3 km inland from the beach and drew oras from the small *Zizyphus* savannas on either side of the streambed, as well as from the extensive *Tamarindus* forests in the valley and on the slopes of the surrounding hills. That oras were drawn to it from the central east-west ridge of the island is certain.
>
> Kali *Inaloah* is a small tributary of Kali Besar, approximately 2.5 km inland from the beach, and almost due north of the Loho Kalo baiting locality, but on the north side of the high ridge separating Loho Liang from Loho Kalo. The approximate boundary of the bait area was the east-west central island ridge to the north, Gunung Adjung to the east, and Gunung Nikipolak to the west. Most of the entire area was monsoon forest, including some small and somewhat isolated *Zizyphus* savannas.

In each case, an average size adult goat was killed and hung from the branch of a tree at the bait site for two days. At the same time, a portable observation blind was erected at the site. At 0800 of the third day at least two observers were stationed in the blind and remained there until approximately 1200; the next shift lasted from 1400 to 1700, making a total of 7 hours/day. All oras coming to the bait during those 7 hours, for a period of 3 days, were counted and marked. A trombone type insecticide sprayer, slightly modified so that it would spray paint thinned with turpentine, was used for marking. The observation blind was generally placed within 1.5 to 3 m of the bait. This meant that the lizards were usually quite close, and it was relatively easy to spray them in a selected pattern on the body, tail, or legs. A rough sketch of the lizard was made, showing the paint pattern applied, and the animals were given code names reflecting these patterns and colors, such as "Speckles," "Three spot," "White face," "Red tail," "Blue left leg," etc. All lizards that came to a particular bait site were sprayed the same color: Kalo bait site (green), the Base Camp (white), Nanenepi (red), Nggulung (yellow), and Inaloah (blue). Total length was estimated for all these specimens, often when the lizard was against a horizontally placed 2 m board marked off in alternating blue and white bands 30 cm wide. The combination of color, paint pattern, and size made it relatively easy to identify individuals from afar, or months later in totally different parts of the study area. When all lizards seen in the study area were sprayed, total density was established. Population turnover rates and movement were easily obtained. Some of the painted oras were eventually trapped and given permanent tags. The lizards were easily sprayed from the blind, and there is no evidence that the fresh paint sprayed on them in any way changed their normal activity patterns or behavior towards one another.

In a study area of approximately 10.3 km^2, 67 lizards were marked in this manner, and 50 individuals were trapped, permanently tagged, and had large

Table 7–2. Number of scavenging oras in vicinity of all baiting stations on Komodo

Bait sites	Avg. number oras per day	Estimated avg. number oras per km^2
Vai Nggulung	9.7	16.8
Kali Inaloah	6.3	13.1
Loho Kalo	3.0	9.3
Nanenepi	4.7	11.2
Base Camp	7.6	14.3

numbers painted on their sides. Thus a total of a least 117 individuals were known to have inhabited the area during our study.

The average number of lizards/day at each of these stations is shown in Table 7–2. Density estimates were calculated by means of Jolly's stochastic method (Jolly, 1965; Southwood, 1966). The Lincoln Index estimate was not used, because the population of the study area is an open one, with immigration and emigration an important feature of population dynamics.

The results of a Jolly analysis on the data obtained from Trap I, Loho Liang, Komodo, is provided in Table 7–3, and practically all of the individuals within the "drawing area" of the scent-baited trap had been captured after 8 months (Fig. 7–1). Average density for the 12 specific baiting periods analyzed is 21.8 ($\pm$3.1). On the basis of the estimated range over which carrion scent is effective in attracting oras to sites in this valley, the population estimate is 14.3/km^2. However, this does not include oras in their first year of life (up to about 1 m) that rarely visit carrion. Nonscavenging young of the year are recruited into the population from approximately May to January, but no data on hatching success are available. Blair (1960) and others have shown that nest failure in lizards is often as high as 75 percent. If this estimate is used as a base of calculation in the ora, the *maximum* number of juveniles recruited into the population/km^2 is approximately 3, providing a size-inclusive estimate averaging less than 20 individuals/km^2 on Komodo (probably close to 17). Mortality in young specimens during the first few months is undoubtedly very high and is responsible for further significant reduction of population densities.

When the numbers of scavenging oras/day at each of the baiting stations are compared (Table 7–2), it is apparent that the density estimates for each of these areas are not equal, undoubtedly reflecting food and shelter resource availability. Loho Kalo, with an estimated average density of 9.3 scavenging age animals/km^2, has the lowest population of the areas sampled, and this is probably

Table 7–3. Base Camp area mark and recapture analysis

Bait period i	Proportion of recaptures $\bar{\alpha}_i$	Number marked wild oras $\hat{M}_i$	Total population P_i	Standard error of P_i
1	—	0	—	—
2	0.703	10	13.3	3.6
3	0.645	13	20.0	4.1
4	0.487	17	34.7	4.9
5	0.452	20	44.4	7.1
6	0.273	22	80.2	10.6
7	0.502	22	44.4	4.8
8	0.730	23	31.5	3.5
9	0.514	24	47.1	4.8
10	0.825	24	29.8	3.6
11	0.800	26	32.5	4.1
12	1.000	26	26.0	1.2

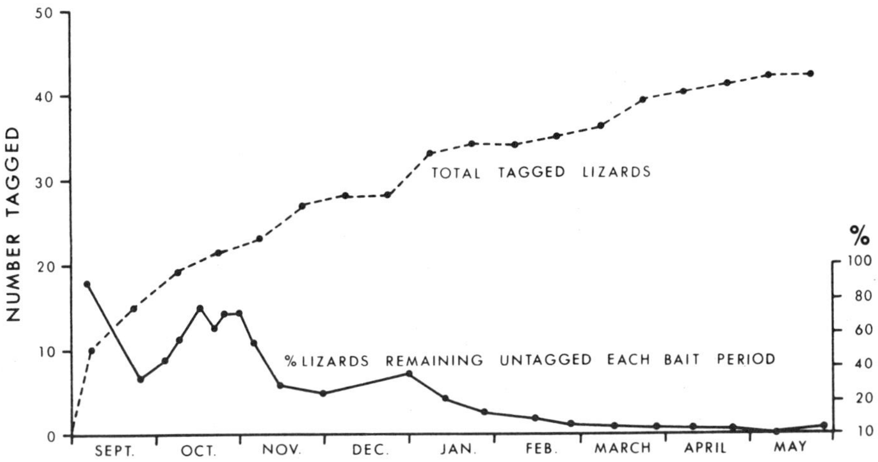

Fig. 7–1. Diagram of tagging program in Loho Liang, Komodo.

correlated with the low wild boar and deer densities and the generally more open habitats represented there (Chap. 13). Jungle fowl and megapode birds also have low densities. At Nanenepi (approximately 11 scavenging age oras/km^2) densities of both oras and wild boar are higher than at Loho Kalo and cover is extensive. The highest densities were in mixed habitats of monsoon forest interspersed with *Zizyphus* savannas. Of these, the highest density of scavenging-age oras (16.8/km^2) is at Vai Nggulung (almost twice that of the poorest habitat).

Unfortunately, there are no comparable data from any of the other islands on which the ora is found. The only available information is based on crude estimates of the number of individual oras seen at carrion baits by past expeditions. De Jong's conclusion (1937) of an increase in ora density on Flores and Komodo between his 1926 and 1937 field trips was based on the number of oras seen per entire trip on each island and possible seasonal differences. While Hoogerwerf (1954a) admitted the faultiness of such comparisons, he drew his own conclusions from the same data. The only clear facts regarding the populations on islands, other than Komodo, is that Komodo and Rintja have higher densities than either Padar or Flores (de Jong 1937; Hoogerwerf, 1954a; and our own observations on Padar and Flores). Whether Rintja actually possesses a higher population density than Komodo, as suggested by Hoogerwerf (1954a), remains to be demonstrated.

Estimates of total ora numbers throughout the entire range suffer from the same lack of significant data (Table 7–4); for the most part, they are simply unsubstantiated guesses. Pfeffer (1959) stated that it is difficult to establish the total number of oras in existence, because the number of young is almost completely unknown, and visits to the islands had been too brief. The present study provides reasonably sound data for several habitats on Komodo. Presuming the density is no higher on the remaining islands than an average of 17/km^2 on Komodo and as low as 6.4/km^2 on Padar and parts of Flores, the maximum total standing crop of scavenging-age oras is estimated at 5713, the estimated number of nonscavenging hatchlings is 1500. The total of 7213 oras is higher than was estimated previously (Table 7–4).

Ora Dispersion

It has been shown above that oras are not randomly dispersed throughout their habitat. Their dispersion with respect to extrinsic factors, such as food and shelter, is largely due to the intrinsic response and behavioral patterns of the individuals.

As pointed out elsewhere, though the scavenging areas of large and small oras overlap completely, their predation territories do not. The hunting techniques of the very large and very small oras are different and are dependent on the type of prey pursued and the habitat in which it is found. The geographical and ecologi-

Table 7–4. Estimates of ora population sizes

Source	Komodo	Padar	Rintja	Flores	Total
Van Heurn (1932)	—	—	—	—	"Several hundred to several thousand"
Lallemant (1929)	—	—	—	—	1000
Hoogerwerf (1954a)	—	—	—	—	1000
Pfeffer (1959)	400–500	100	400–500	500	1500–2000
Piazzini (1960)	1500	—	1000	—	2500
This study	2348	60[a]	792	2448	5713 (+ 1500[b])

a. Plus an estimated 40 on Gili Mota and 25 on Oewada Sami.
b. Yearly hatchling crop of 1500 estimated as follows: 400 mature females × 15 eggs/avg. nest = 25 percent nesting success.

cal dispersion of individual oras, including the component size groups of the population, is thus largely the result of prey-dependent elements. However, ecological factors become very important at the periphery of the range, as in Flores, where population density undoubtedly becomes increasingly unstable at high elevations (Chap. 3) and at the extreme eastern limit of the range—even in lowland areas. Though physical conditions of the environment are important, food is the ultimate extrinsic factor determining density of the entire population, as well as dispersion of size groups.

In oras the most effective intrinsic convention to bring about dispersion within equal-size classes is a high level of intraspecific aggression (Chap. 12). This behavior apparently limits the number of hunting territories occupied by both large and small lizards in a certain area. Such aggression, particularly when cannibalism is involved (as it is in the ora), probably imposes a ceiling on population density as well as on food resources. In its action it is completely density-dependent, allowing the habitat to fill amid mounting rivalry to a maximum density, after which further intrusion is apparently repelled. In whatever fashion their claim is established, once it occurs, hunting must take place without undue interference from rivals in the same size class.

Biomass

The little available information on the biomass of reptiles refers to small, chiefly insectivorous lizards. The data presented by Fitch (1954) of approximately 2200 g/acre for the prairie skink (*Eumeces obsoleta*) in the central United States agree reasonably well with Tinkle's (1967) calculation of 2000 g/acre in the rainbow agama (*Agama agama*) in tropical Africa (based on Harris, 1964). The agama is much larger than the skink, and this fact is indicated by the comparative densities

(*Agama*, 36 g/acre, *Eumeces*, 50–100 g/acre). Jenssen (1970) reported that the clouded anole (*Anolis nebulosus*) in Mexico had a biomass of approximately 230–314 g/acre. Tinkle (1967) estimated an average yearly biomass of 67 g/acre in the fringe-footed lizard (*Uta stansburiana*) in arid habitats in western North America.

For Komodo average ora density is 0.026 animals/ha. The mean body length for Komodo oras is about 55 cm, and the average weight for specimens of this size is 6000 g. Thus ora biomass on Komodo is calculated as 33,590 g/km^2 or about 335.9 g/ha. On the island of Padar, where densities are estimated as only 6.4/km^2, and where average ora weight is about 5000 g, the biomass is estimated as 23,160 g/km^2, or 231.6 g/ha. The biomass in the western half of Rintja is approximately the same as it is on Komodo, but in the eastern half it is equivalent to that of Padar. Ora densities on Flores are probably roughly equivalent to those of Komodo in only a few localities, such as the area around Loök-Nggoer, Mangarrai District. Over most of the remaining Flores range, ora populations are extremely low.

It is obvious that the biomass of ora populations is not significantly different from the biomass of other lizards studied. However, as far as can be determined, the biomass seems to be higher in mesic forest lizards, whether temperate or tropical, than it is in those from drier habitats.

Mortality

We found no direct evidence of interspecific predation, though intraspecific predation is quite common. Pfeffer (1959) cited a native report that a female dug into an ora nest just as the eggs were hatching, and that she attempted to eat the young and had to be driven away. They also kill and eat one another at later life stages. Ora eggs are also eaten by captive adult males (experiences in Surabaja, Jogjakarta, and Jakarta zoos, personal communication). Adult females also eat them (Lederer, 1942; et al.), although apparently not as often as do males. Most intraspecific aggression occurs when small feeding aggregations are formed. Much of the antagonism is directed toward the smaller members of the group; when unable to escape, they are sometimes killed and eaten (Chap. 9). Adults may also be eaten, but there is no direct evidence that other oras killed them. On the basis of scat analysis, the level of instances of intraspecific predation in oras is as high as 8 percent in some parts of Komodo.

Though no evidence is available, it is certain that on Komodo dogs, wild boar, civet cats, several species of raptors, and probably some snakes eat the young whenever they find them; all are known to feed on other lizard species. *Varanus salvator* may eat young *V. komodoensis* in western Flores, where the two species are sympatric. On Rintja and Flores, the crab-eating macaque is a potential predator of the young. It is also quite likely that many nests are destroyed each year by wild boars.

On the basis of present evidence, it seems highly likely that the greatest predation pressure on oras at all stages of their life is generated from within their own species. Is the high intraspecific rate of ora predation related to extreme hunger, as postulated by Pfeffer (1959), or is it an important mechanism related to population control?

We must assume that the average breeding season produces sufficiently large numbers of oras to replace losses incurred during the year. However, the population level of young oras may be lowered directly by parents actually killing the offspring (kronism), or by the offspring killing each other. Kronism is generally practiced by male birds and mammals (Wynne-Edwards, 1962), and there are at least some data showing this is also true of oras. When competition for food is severe, the eating of young by adult oras is an efficient system of post-hatching recruitment control. Prolicide (eating of eggs) is probably density-dependent, and, to my knowledge, is unknown in any other lizard species studied, though it probably occurs in at least some crocodilians (Cott, 1960).

Most deaths resulting from activities unassociated with intraspecific fighting probably occur as a result of attacking prey species. Under certain circumstances almost any but the smaller prey are potentially dangerous. Rats, dogs, civet cats, and some snakes can easily kill oras smaller than 1 m total length. According to Mertens (1942) *Varanus* species are not immune to the venom of poisonous snakes. He did not agree with Spatz (1931), who claimed that *V. griseus* was immune to the bite of the desert viper (*Aspis cerastes*). Deer, and particularly kerabau (water buffalo) and wild boar, are potentially dangerous to even the largest of oras. Domestic kerabau are known to have gored large attacking oras on Flores. Wild hogs of several species are dangerous adversaries, capable of killing even the largest mammalian carnivores. Examination and interpretation of the scar patterns of oras suggest that many of those in the shoulder areas, particularly the puncture wounds, are produced by wild hogs (Chap. 6). The size distribution of oras with this wound type makes it clear that only the largest specimens dare attack a medium to large wild hog.

Oras are undoubtedly sometimes suffocated by burrow collapse. On 2 March 1970 the head of a live 2 m ora was seen protruding from the soil and rocks of a collapsed burrow. Though it struggled to free itself, it was able to move only its head; the limbs, body, and tail were all completely covered. Furthermore, several large, angular rocks had fallen in such a way that they were "keyed" into one another. Without our assistance the ora would not have been able to extricate itself. On 30 April 1970 a 1.2 m ora was found in a well-like depression 1.6 m deep and 1.8 m across. It was completely emaciated, very weak, and would have died in the hole had not we lifted it out.

Surface water is rarely available on Komodo, Rintja, Padar, and even many parts of western Flores for much of each year, and a few oras undoubtedly die as a result. Some cannibalism may stem from the death of the younger specimens which are particularly susceptible to desiccation. Starvation is also a potentially important factor. Oras commonly fail to secure food for many weeks, particularly during the long dry season. Starvation death may be frequent, because many are thin and seem to lead a marginal existence. Darevsky and Kadarsan (1964) reported seeing an old, hungry, emaciated ora of about 2.8 m seeking dead molluscs that washed up on the beach. They believe many die of starvation.

August and September, 1969, was a very critical time for ora 8W (2.89 m), when its life hung from a slender thread indeed. It weighed 36 kilos and was so emaciated it could hardly walk, (19W, slightly shorter, weighed 48 kilos and was not fat). It was completely unable to run, and even walking seemed difficult. During this period 8W spent more time than usual hunting, further depleting its already low energy reserves. It was obviously too weak to have successfully attacked any medium-sized prey, and it spent much time searching for jungle fowl and other smaller animals. It came to our bait several times but was easily driven away by much smaller specimens. I am convinced that if it had not eaten approximately one-half of a dead wild dog we offered, it would have shortly died. The one good meal apparently tipped the scales. Soon thereafter its droppings contained hooves and hair of a wild pig it had eaten. By November it had gained approximately 10 kg, and by January it weighed 50 kilos and appeared to be in good health.

Disease

While disease must play a role in reptile population density, little information is available. In general, reptiles are resistant to many bacterial diseases. Though fungal and bacterial skin infections are common in most mammals of Komodo, oras show no evidence of them. Ora bites on deer (and humans) often become badly infected (Chap. 10), whereas we saw no evidence of sepsis on other oras.

Parasitemia is not particularly high in oras (Chap. 12). Intestinal, gastric, pulmonary, and rectal endoparasites are all known, though there is no evidence that wild specimens are affected by them. However, amoebiasis (*Entamoeba* sp.) is common in captive oras. Amoebic, combined with bacterial, infections are often widespread, frequently with hepatic and cloacal foci. Causes of death are liver necrosis, acute hemorrhagic proctitis, and/or severe colitis and enteritis (Gray et al., 1966). However, analysis of 38 ora droppings from Loho Liang showed that amoebic infection is probably uncommon in the wild, and it is therefore not considered a major controlling factor in nature.

Longevity

In general, most large reptiles have relatively long life spans, with the longest-lived species of tortoises having a potential of over 100 years. The only reported evidence of aging in any oras is ossification of bronchial cartilage in a large captive male (Gray et al., 1966).

It is generally recognized that the death rate of most animals is relatively high in the very young age groups; middle life is often characterized by a sustained period of low mortality, succeeded by a sharply rising rate in old age. On the basis of population structure (Chap. 2), it is suspected, though unproven, that the second and third parts of this pattern are characteristic of oras. The ora apparently maintains a steady high level of mortality, even in adult life. Probably few, if any, individuals become senile, at least at the present time when human predation is high. Thus, each year-class undergoes a roughly exponential decline in numbers as it grows older, and the remnant, if any, surviving to ultimate senility is undoubtedly so small that we have only the vaguest notion of how long the potential life span of an ora may be. Longevity records for captive individuals of other species of *Varanus* are: *V. griseus* about 9.5 years (Flower, 1933); *V. flavescens* 5 years, 8 months (Flower, 1933); *V. varius* and *V. gouldi* almost 7 years (Flower, 1933); and *V. niloticus* 15 years (Flower, 1937).

Several oras have had a captive life of 20 years. Darevsky (1963) stated that a specimen lived in the Surabaja Zoo for that length of time, and he believed a life span of 50 years is possible for the ora. Lederer (1942) concluded that the female captive in his care at the Frankfurt Zoo was 7–8 years old when received, and lived 16 years, 8 months, and 21 days in captivity (Mertens, 1946b), making it about 24 years of age when it was killed during a World War II bombing raid. A specimen in the Berlin Aquarium lived in captivity for 16 years, 7 months, 6 days. However, it was already 2.49 m long when received. Assuming a minimum age of 10 years before it attained this length, it would have had a total life span of about 26 years. Both Darevsky (1963) and Pfeffer (1959) estimated that longevity in the wild is about 50 years, and I agree.

The potential use of layering in the cortical bones as the basis of age estimates of reptiles is well known (Griffith, 1961). However, the utilization of such banding for age determinations is complicated by several factors. The major problem is that two types of bands occur in cortica bones of reptiles; those produced as a result of remodeling, and laminar bone produced as a result of growth (Enlow, 1969). Suzuki (1963) and others have shown that the laminar structure of bony tissue in reptiles reflects periodic, seasonally related waves of activity, indicating feeding periodicity. While several workers have stressed a direct correlation between age and "annual" rings in compact reptile bone (e.g., St. Girons, 1965), others (Suzuki, 1963; Gibbons, 1977) have questioned the validity of such a correlation since it is known that remodeling processes often obscure, modify, or destroy original stratification.

However, the great climatic seasonality in the Lesser Sunda Islands suggested that if ora bones could be found that were not subject to much remodeling, an age estimate of larger specimens might be possible. Several bones that were useful in similar studies on other reptiles were investigated to determine their potential usefulness in studies of the ora, but most were found to be wanting, mainly due to extensive remodeling. In ora skeletons the growth rings on the pre- and postzygapophyses, glenoid cavity of the centrum, and in transverse sections of the osteoderms could be counted and potentially useful. The vertebrae of 14 specimens of *V. komodoensis* were examined (UF 28219–20, 29000 [2], AMNH 37879 [3], 37908–13, 74606), representing individuals varying in total length from approximately 1.5 to 3.0 m.

Prezygapophysial growth rings varied considerably in size and completeness. However, when only the large complete rings were counted, they varied from 6 to 13. Within the entire vertebral column of a single individual, the number of major rings never varied more than ± 2. Ring number is correlated with centrum length and thus is potentially age-dependent. The counts are clearly not the total number of major rings produced throughout current total life span, but represent only those rings not yet remodeled. However, the counts do suggest that at least some oras with a total length of slightly over 2 m are at least 11 years of age. If growth estimates by Darevsky and Kadarsan (1964) and this study are correct, these specimens are then at least 18 years of age. If we use the same reasoning, the largest specimens in the sample are at least 20 years old, thus agreeing with the data on captive specimens.

The number of rings on the glenoid cavity of the centrum has been used most extensively in fishes (particularly in species lacking scales) to estimate age. The advantage of this total when compared to that obtained from the prezygapophyses is that there is less possibility that remodeling has obscured the rings laid down during the smallest size stages of the specimens examined. These data show a correlation with centrum length. The number of rings varies from 6 to 17, with the maximum somewhat closer than the prezygapophysial counts to the estimated age-size estimate on the basis of captive specimens.

Osteoderms from the dorsal body region of three specimens of different sizes were sectioned to determine whether growth rings are present, whether the number of rings is correlated with centrum length, and whether the number of rings is equal to that in the centrum; on the basis of osteoderm development, remodeling of the central portion of osteoderms is less likely than in other elements (Enlow, 1969). The number of growth rings (Fig. 7–2) was identical to that found in centra of the same specimens from which the osteoderms were taken, suggesting that osteoderm ring number may be a practical measure of age estimate. However, it has not yet been possible to evaluate this technique in terms of a larger sample size or, more important, to obtain osteoderms from individuals of known age. Because skin samples are easily obtained, osteodermal ring number offers potential in determining the age of wild oras without destroying any individuals.

Great longevity generally corresponds with large size in reptiles (Gibbons, 1977), and such a correlation in the ora is not unexpected. Wilbur et al. (1974)

theorized that a long life expectancy in adults would be associated with those species that have evolved effective means of defense against predators, usually by large size. However, there are no predators of adult oras and large size may be related to the increased efficiency of feeding on large prey, a factor not considered by Gibbons (1977) in his excellent review of the Wilbur model.

In addition, long life tends to stabilize ora population numbers, allowing for an admixture of individuals representing widely disparate age classes. Thus the

Fig. 7–2. Section of dorsal osteoderm from subadult ora (UF 28000), showing tendency of bone to form rings, in this case thought to represent four major (?annual) growth periods.

adults often interact with the young, in this case regulating at least some population recruitment through kronism. Long life also tends to be associated with permanent, extensive, and stable environments of a type characterized by that of the Komodo monitor. In this potentially long-lived species, a steady mortality drain year after year throughout a long middle life leaves only a minute remnant to fill out the full potential life span and to die of senility. The population structure of other large long-lived predatory reptiles, such as crocodiles, suggests the same pattern. However, population structure in giant land tortoises (at least in insular situations) suggests a pattern highly skewed toward older individuals, with death rate due to senility relatively high in comparison to mortality rates throughout all the life stages. However, what little information is available for the ora suggests that senile death rate is very low, and, by itself, is probably a minor factor in population turnover.

Growth

Only two publications report data on growth of captive oras and very little information is available on wild specimens. The most complete data are provided by Lederer (1942), who included measurements for 15 years of growth in the captive female in his care (Fig. 7–3). Through its entire captive life this specimen grew at an average rate of 3.6 cm/yr. However, at a total length of approximately 210 cm, growth rate slowed from 5.6 cm/yr for the 7-year period prior to this time to 1.3 cm/yr for the 6 following years, suggesting that this specimen was nearing the lower level of maximum size reached by females. During this time the length increased 30 percent while the weight increased 106 percent.

Additional growth data for a pair of adults are available from the Tokyo Ueno

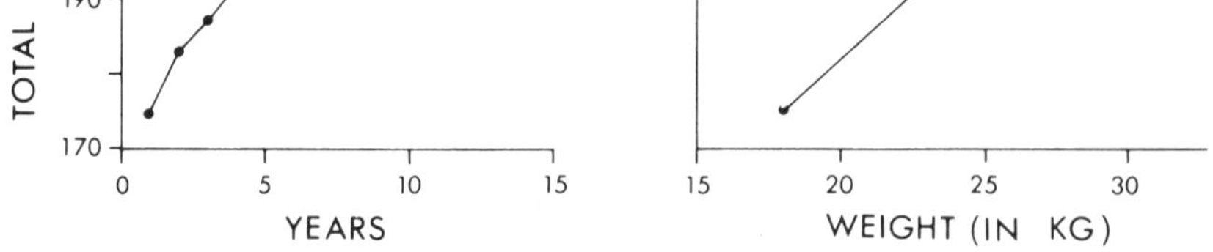

Fig. 7–3. Growth and weight changes in "Bubchen," Berlin Zoo (after Lederer, 1942).

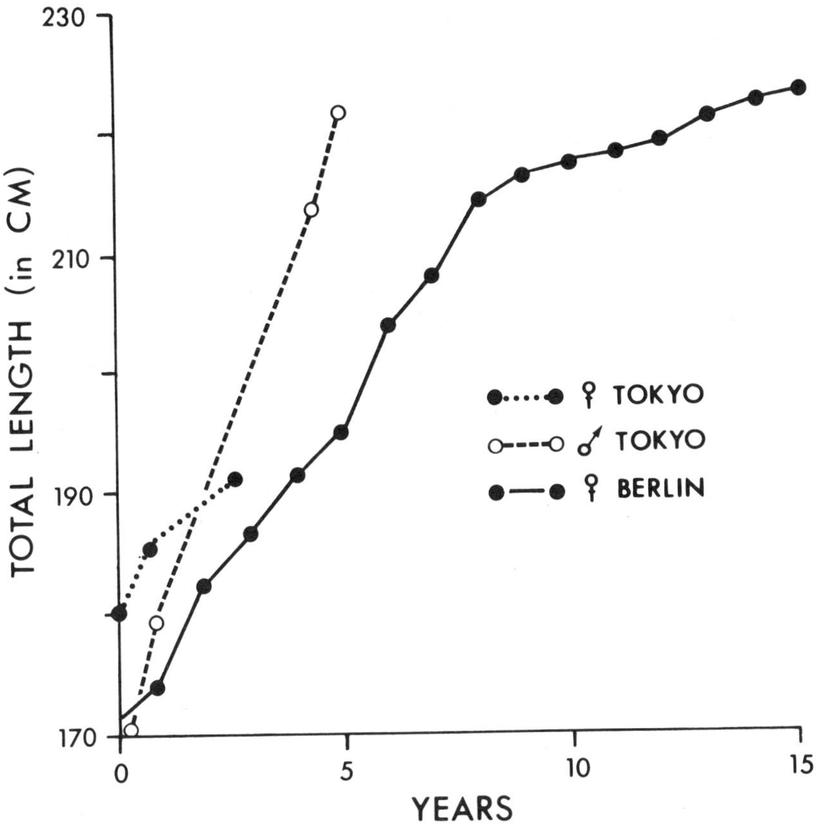

Fig. 7–4. Growth in three captive oras, showing the faster male growth rate. Tokyo data after Hisada (in a letter).

Park Zoo (Hisada, personal communication). The growth records for the female cover too short a period to determine overall pattern. However, the male grew 12 cm/yr for the first 4 years of captivity (Fig. 7–4), a rate almost twice as great as the highest annual rate of the female. That growth continues as maximum size is approached, but at a decreased rate, as shown by data (F. W. King, personal communication) from the New York Zoological Society specimen, "Jago" (total length 2.68 m). This large monitor grew 2.6 cm in 4 years, with an overall rate of about 0.8 cm/yr (0.3 percent of total length/yr).

Busono (n.d.) reported growth in a pair of *V. komodoensis* maintained in the Gembira Loka Zoo (Jogjakarta). The two specimens, obtained from the Surabaja Zoo in 1960, were each almost exactly 1 m total length on their arrival. Six years later the male had attained a total length of 2.5 m and the female 1.75 m. During this period the average growth rate of the male was twice that of the female: male 25 cm/yr, female 12.5 cm/yr. This supports the suggestion by Hoogerwerf (1948) that males and females grow at different rates[1] and accounts for the disparity in average size of adult males and females.

Busono (n.d.) also provided growth data for hatchlings during the first few months of their life in captivity. On June 3, 1968, 29 young hatched in the zoo; total length varied from 25.3 to 35.7 cm, $\bar{X}$ 30.1 cm ($\bar{X}$ weight, 80.3 g). Approximately 3 months later (September 2) the average length of the 5 specimens measured was 55.2 cm ($\bar{X}$ weight, 300 g). Thus, during this short period length increased by 63 percent and weight by 370 percent.

De Jong (1948) reported on 25 young hatched in the Jakarta Zoo in April, 1942. The largest of these hatchlings was 20 cm total length, 2 months later it was 35–40 cm, and by the end of October (5 months) it was 50 cm. At the end of December (7 months) the length of 10 measured specimens was said to range from 45 to 60 cm. In June, 1943 (14 months), "several" ranged from 75 to 80 cm. In May, 1945 (3 yrs), one specimen was 115–120 cm; by 1946 (4 yrs) it was 150 cm. De Jong stated on July 25, 1947, that four of the specimens "were of the same size as the original parents." These specimens grew at a slower rate than those at the Jogjakarta Zoo. Hoogerwerf (1948) recognized this slow growth rate at the time and concluded it was abnormal, due to very low feeding levels during the war years.

On the basis of present captive specimen data, the approximate life growth pattern (for females) is: ± 60 percent increase in total length during the first few months (to ± 50–80 cm, 50 g), ± 50 percent increase for the remaining 10 months (to ± 1 m), ± 25 percent increase over the next 3 years, or $\bar{X}$ 8.1 percent/yr (to ± 170 cm, 17 kg), ± 2 percent, or $\bar{X}$ 0.3 percent/yr increase over the next 7 years (to ± nearly 250 cm, 44 kg). Growth rates in males are probably higher.

Very few data are available for growth in the wild. Darevsky and Stusak (1968) reported a 27 cm increase in total length during a period of 3 years in a specimen originally 125 cm total length (= 9 cm/yr, or 7.5 percent increase/yr). Darevsky and Auffenberg (1973) provided data for growth in two specimens, including one whose body length had increased by 65 cm in 7 years; 46 percent total increase, or 6.6 percent/yr. This is within the range of growth expected for this size class (± 130 cm initial total length).

The only other data for growth derive from this study. While more than 49 Loho Liang individuals were marked and carefully measured, only 13 were subsequently captured and remeasured after a minimum of 5 months. Of these, only 9 showed any growth (Fig. 7–5). This varied from 1.5 to 16.8 cm in absolute measurements, but from 1.4 to 11.3 percent/yr increase over initial length when corrected for 12 months (assuming equivalent growth rates for the remainder of

1. The same pattern has been noted with captive *V. bengalensis*.

the year). The average annual percent increase for individuals >200 cm is 0.3, agreeing with data on large captives. However, the maximum of 6.9 percent/yr is far greater than any other data available for this size class. The average annual percentage increase for specimens 150–199 cm is 4.7; for specimens 150–170 cm it is 4.3; for specimens 171–215 cm it is 5.0. The difference between the last two percentages is not statistically significant. However, the overall value of 4.7 percent is within the growth range expected for specimens of this size class on the basis of captives.

Figure 7–6 is a summary of the growth data for the smallest size classes of oras, supporting my conclusions that during the first year after hatching wild specimens nearly double in length and, at that time, join the main body of the population within which there is no, or hardly any, indication of year classes beyond the young of the year group.

Komodo monitors exhibit considerable change in form from hatchling to adult stages; the juveniles are slender, with long necks, limbs, and tails, while the adults are stocky with shorter appendages (Fig. 7-7; Pl. 2; see Chap. 2 for details).

Each ora on Komodo tends to shed its skin in patches over a period of about 6 months of the year (Table 7–5). Sloughing usually begins in September with the

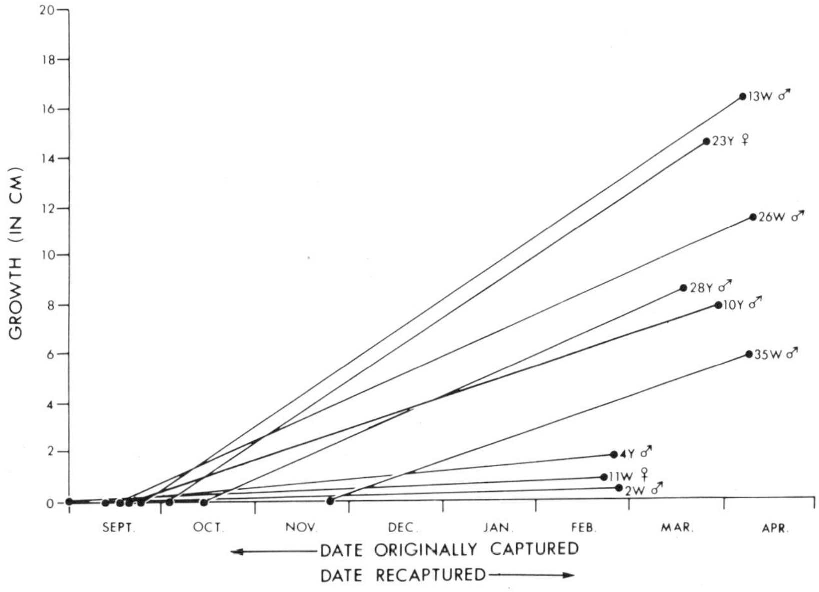

Fig. 7–5. Growth of selected free-ranging oras, Loho Liang, Komodo, 1969–70.

skin of the entire venter and around the insertion of the legs with the body. The shedding pattern continues over the lateral surfaces of the body and tail, as well as the limbs. By January and February only the skin along the dorsal surface of the tail and body and on the head usually remains unshed. The process is normally completed by the end of February or March. Sometimes the entire remaining portion of skin is sloughed over a short period—particularly during the monsoon season. The annual shedding cycle sometimes begins immediately after the

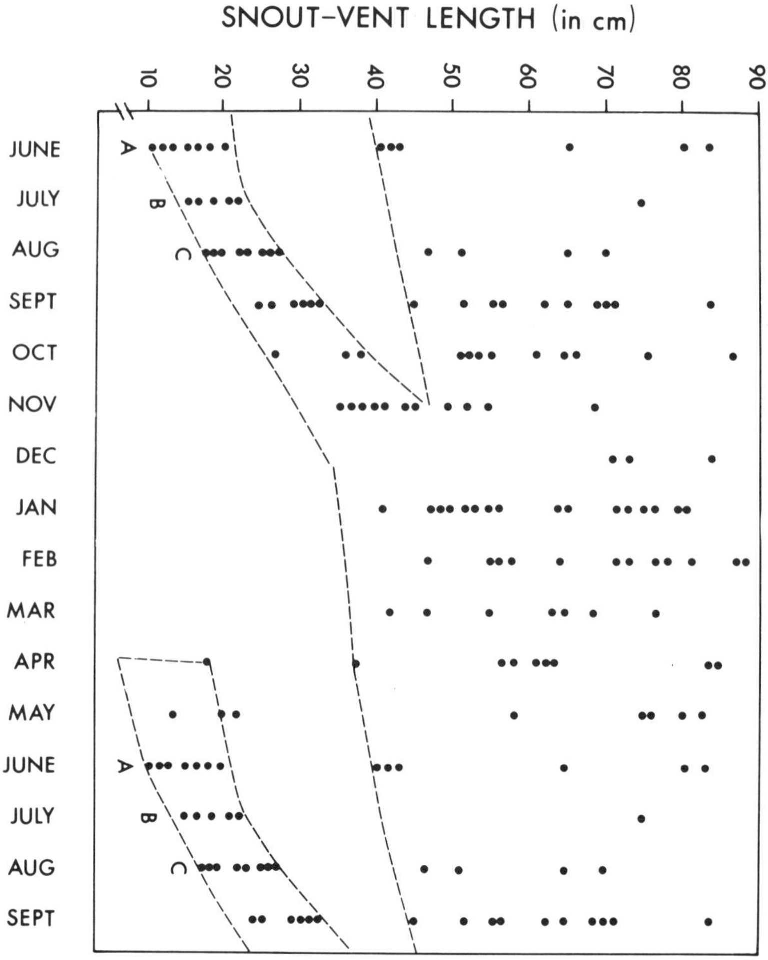

Fig. 7–6. Age classes are not clearly defined, and hatchlings enter the main population group after a period of about one year. A, B, and C represent data from captive young (after Busono, no date).

previous one has been completed, but usually several months pass before the new skin begins to shed. Skin is being shed somewhere on the body in the majority of specimens during the wet season (September–March, see Table 7–6). Complete skin shedding occurs only during the period of heaviest rainfall and is undoubtedly associated with the opportunity to soak in the water-filled game wallows.

Maturity

For *V. komodoensis* very little is known of the size and age at which sexual maturity and the cyclic changes in gonads or fat bodies occur. This is primarily due to the necessary protection afforded the species by the Indonesian government. However, data from dissections of a few specimens that had died in captivity or were legally sacrificed in the field are available. Darevsky and Kadarsan (1964) dissected two males of 157 and 238 cm total length on Komodo on August 3 and 5. In both specimens the epididymides were in a loose, wrinkled, resting stage. No mature spermatozoa were seen; the presence of completely formed ducts involved in the initial stages of spermatogenesis suggested both were sexually mature. In a 171 cm female dissected August 26, the oviducts were empty, but white oocytes as large as 8–10 mm in diameter were found in the ovaries. There were 15 well-defined corpora lutea, clearly proving this was a sexually mature animal.

Fig. 7–7. Changes in proportion in the ora shown from hatchling (*below*) to large adult (*above*). Note the increased massiveness of the body and, particularly, of the tail.

Table 7–5. Season shedding (X) of 12 oras on Komodo

Specimen	Aug.	Sept.	Oct.	Nov.	Dec.	Jan.	Feb.	Mar.	Apr.	May
19W		X	X	X	X	X	X	X		
11W		X	X	?	X	X	X	X		
2W						X	X			
10W		X	?	X	X	X				
8W		X	X	X		X				
22W		X			X					X
13W		X	X	X	X	X	X	X		
20W				X	X					
26W		X	X	?	X	X		X		
14W						X				
10Y		X	X		X	X		X	X	
6W	X	X	X	X	X	X				

Table 7–6. Percentage of specimens (N = 46) with partial and complete skin sheddings per month

Month	Partial sheddings	Complete sheddings
July	0.0	0.0
August	0.2	0.0
September	10.2	0.0
October	14.3	0.0
November	9.2	0.0
December	14.3	9.4
January	19.9	58.3
February	12.0	25.0
March	10.9	8.3
April	4.0	5.0
May	5.0	0.0

On June 29 I sacrificed a single male and female on Flores for parasite and reproductive studies.[2] The reproductive organs were fixed in Bouin's solution. The testes were dehydrated in Dioxane and imbedded in paraffin. Sections were made 10 μ thick and stained with Delafield's hematoxylin and eosin.

The female (UF 29001) was 152.5 cm in total length. No corpora lutea were found in the ovaries, but there were 8 enlarged ovarian eggs on one side and 9 on

2. Under a special permit issued by the Indonesian Wildlife Office in Bogor.

the other. Of these the largest were 5–7 mm in diameter. Thus this individual had not yet laid any eggs and was probably in its first season of sexual maturity. The total length of the male (UF 29000) was 155.5 cm. The epididymis was slightly convoluted and somewhat turgid. There were, however, no sperm in the testes, suggesting the specimen was not yet mature.

Females in captivity are known to have laid eggs at minimum total lengths of 150 cm (Bunsono, n.d.) and 148 cm (Galstaun, 1973). Maximum lengths at which eggs are produced are 190 cm (Brongersma, 1932) and 205 cm (Galstaun, 1973). The smallest captive sexually mature males reported are 191 cm (Galstaun, 1973) and about 2 m (Busono, n.d.). Thus, data on both field and captive specimens suggest that males and females become mature at about 150 cm and probably retain the capability of producing eggs to the largest lengths attained. Very few data are available on age of sexually mature individuals. Senility is probably an important factor, but information is lacking. The age of onset of sexual maturity is, however, roughly established. Darevsky and Kadarsan (1964) estimate this as approximately 5 years for both males and females of the Komodo population based on the presumed growth rates in the wild (though founded on very few data). That this estimate is probably fairly accurate is shown by data obtained from captive individuals. The breeding pair at the Jogjakarta Zoo were acquired in 1960, when they were 100 cm and 150 cm in total length. The smaller female was undoubtedly in her second year, and the male was probably in his third. The first eggs were laid 3 years later, so that the female was probably mature at an age of 4–5 years. However, fertile eggs were not definitely proven until 1966, so that by this time both specimens were clearly mature. The data derived from the pair in the Jakarta Zoo suggest a similar age for onset of maturity. Galstaun (1973) estimated the ages of the male and female in 1967 (146 cm and 118 cm, respectively) as 4 years. Eggs were not laid until 1969, when the animals were probably 6 years of age.

8

Reproduction

Local ora population recruitment takes place through both immigration and reproduction. The former is particularly important in view of the high number of nonresidents. However, considering the entire insular population, reproduction is most important. Recruitment through nonresidents is discussed in Chapter 7. In this chapter the reproductive process, from courtship through hatching, is described and discussed.

Courtship

There is seldom any difficulty in recognizing courtship, because in most animal species the males display to the females in some fashion or other immediately before mating; they may dance, posture, call, or otherwise behave peculiarly and conspicuously. Often the females join in—sometimes behaving in the same way, sometimes reacting to male displays with demonstrations of their own (Bastock, 1967). However, a strict definition of courtship that includes display is not applicable to the Komodo monitor. This species lacks the distinct visual display signals that often characterize pre-mating behavior in other lizard species. However, a broader definition, such as "courtship is a heterosexual communication system leading to copulation" (Morris, 1956), would include the precopulatory informational exchange between male and female monitors. This is the definition used here.

Since the early studies of Noble and Bradley (1933), courtship and mating have been described and studied in many lizard families. In general, the findings indicate that behavioral patterns differ more among families than among genera and species of any one family. The courtship behavior of many lizard species is often stereotyped and is sometimes confused with the less stereotyped and more spectacular fighting reaction. The challenge, or fight reaction, of lizards generally consists of lateral body compression, extension of dewlap or hyoid apparatus, arching of back, lowering of head, assertion pushups, and head bobbing. In courtship the head is often lowered and the back arched as in the fighting response, but the body is generally not laterally compressed, and the dewlap is less extended. After approaching the female, the male may touch her side or snout or grasp the skin of her neck with his jaws. He then places his body over hers, with his tail twisted beneath hers until the cloacal openings meet; hemipenal insertion occurs and coition is completed (see, e.g., Carpenter, 1961a, b, 1962a, b; Hunsaker, 1962; and Clarke, 1965 for reviews).

Very few data are available on courtship and mating in varanid lizards. Falk (1921) reported pairing in *V. exanthematicus* and Werner (1893) in *V. griseus*, but important details were not given. The most complete analysis is that of *V. bengalensis* by Auffenberg (in press a). A photograph of a breeding pair of *V. komodoensis* is provided in Stadt (1939), but there are no accompanying remarks. Tanzer and van Heurn (1938) simply stated that the captive male ora "in the course of copulation . . . lies over the female, in the usual manner." DePrato (MS) provides a good description (see below) in unpublished form. During the present study ora sexual behavior was observed on 10 different occasions, and these data, plus those available on breeding in zoological gardens, constitute our knowledge of courtship and mating in this species (Auffenberg, 1978).

Dates of courtship and mating in zoological gardens and in the field are provided in Table 8–1. The dates for courtship extend from January 19 through October 1, or nearly the entire year. However, dates for successful coition have a narrower overall variation, extending from June 28 to early October, or about 3 months. The first wider distribution is apparently typical for tropical lizard species, or for those living in areas with long, favorable reproduction periods (Harris, 1964; Rand, 1967). Lengthy courtship periods are often associated with multiple egg clutches during a long reproductive season (Tinkle, 1969). In oras the restriction of actual copulation to a small part of the year leads to a strongly peaked cyclic annual pattern. The frequent sexual behavior of at least certain adult male and female oras during most of the rest of the year raises the question of the purpose of this unconsummated courtship activity. Though the relationship between courtship activity and successful breeding is not clear, it will be shown below that in the ora the repetition of courtship behavior may be related to pair bond formation.

On the basis of the size of young observed in the field and the dates of eggs laid in Indonesian zoos by recently captured females, de Jong (1944) concluded that in the field oras breed in June and July. Darevsky and Kadarsan (1964) and Maleyev and Darevsky (1963) believed that breeding takes place on Komodo at the end of July or early August. This was based on the dissection of a single female on August 26; the oviducts were empty, but there were white oocytes in the ovaries, some as large as 8–10 mm in diameter. The presence of well-defined corpora lutea with distinct cicatrices suggested to them that the eggs were depos-

Table 8–1. Courtship and mating data for *V. komodoensis*

Data from this study			
Date	Place	Courtship partners	Others in group
July 12, 1971	Flores	no numbers	None
Sept. 10, 1969	Komodo	19W (♂) + 11W (♀)	3W(♂)
Oct. 1, 1969	Komodo	19W (♂) + 23W (♀)	5W, 7W, 13W, 22W (all ♂)
Oct. 1, 1969	Komodo	"Whiteface" (♂) + 23W (♀)	None
Oct. 1, 1969	Komodo	"Whiteface" (♂) + 19W (♂)! (see text)	23W(♀)
Jan. 19, 1970	Komodo	19W (♂) + "Blue Right" (♀)	"Blue Ear," "Blue Spot" (both ♂) 23W (♀)
March 2, 1970	Komodo	19W (♂) + 23W (♀) (3 times)	44W (♀)
March 2, 1970	Komodo	"F" (♂) + 23W (♀)	13W, 8W (both ♂)
March 27, 1970	Komodo	19W(♂) + 23W (♀)	None

Data from other studies			
Date	Place	Remarks	Authority
July 4–10, 1937	Jakarta Zoo	"Male in course of copulation . . . lies over female in the usual manner."	Tanzer and van Heurn (1938)
June 28, 1938	Jakarta Zoo	"Mating attempts made first in water, then on land."	de Jong (1937)
July 15–first week of Aug., 1968	Jogjakarta Zoo		Busono (in letter)
End Sept. and first week of Oct., 1969	Jogjakarta Zoo		Busono (in letter)

ited less than a month before. Oesman (1967) reported that a pair of oras in the Jogjakarta Zoological Garden bred intermittently from July 15 to the first week in August, and again from the end of September to the first week in October. Tanzer and van Heurn (1938) and de Jong (1937) listed breeding dates of July 4 through 10 and June 28, respectively. This total time of 3 months corresponds roughly with the breeding periods known in several species of monitors (Fitch, 1970), although the months during which breeding actually takes place may be different.

That courtship and mating often take place in the presence of small aggregations of oras is very significant; in most other lizard species the mating pair is usually alone. The high correlation between sexual activity and aggregations is due to the fact that courtship and breeding often occur near carrion. The number of individuals in an aggregation, in addition to the courting pair, varied from 0 to 4 (Table 8–1), and were of both sexes and of various sizes (Table 8–2). These data make it clear that the dominant male of the study area (19W) courted 23W most consistently, even though other large females were occasionally present (October 1, 1969; March 2, 1970).

All courtship observed between 19W and 23W occurred within the common, shared portion of their activity ranges, which overlapped one another broadly (Chap. 5). The sharing of activity ranges between a male and one or more females is of course very common among vertebrates and widespread in lizards (Rand, 1967). The same oras also utilized the same thermoregulatory sites and burrows, though never at the same time.

Table 8–2. Size distribution of ora groups including a courting pair

Courting pair			Others in group	
Male		Female	Male	Female
19W (247.5 cm)	×	11W (190.5 cm)	3W (189.7 cm)	0
19W (247.5 cm)	×	23W (235.9 cm)	5W (167.9 cm)	22W (225.2 cm)
			7W (131.1 cm)	
			13W (160.7 cm)	
19W (247.5 cm)	×	23W (235.9 cm)	13W (160.7 cm)	44W (196.6 cm)
			8W (252.6 cm)	

In addition to 23W, 19W courted only two other lizards out of the 10 courtship sequences observed on Komodo. Only two other males, Whiteface and F, courted 23W. Both of these males were in the study area for relatively short periods of time and were not permanent residents, as were 19W and 23W.

Both Whiteface and F were clearly young animals and seemed inexperienced when their behavior was compared to that of 19W. In his one attempt to court and mount 23W, Whiteface was severely bitten by the larger female. The frequency and smoothness of courtship between 19W and 23W, as well as their infrequent sexual behavior with other individuals and the confusion often attending these attempts, lead me to several important conclusions. First, 19W and 23W probably recognized one another as individuals. Rand (1967) came to the same conclusion in his behavioral studies of the iguanid *Anolis lineatopus*. Partly on this basis, Rand and others (e.g., Fitch, 1956; Milstead, 1961) have suggested the possibility of pair-bonding in some species of iguanid lizards. The present information suggests that this is also possible in *V. komodoensis*. The frequency of unconsummated courtship in coactions between 19W and 23W and other oras, as well as the virtual nonexistence of precopulatory displays, are here interpreted as a behavioral complex serving to maintain a pair bond throughout the year. Pair-bonding would be extremely important in organisms that are not only capable of killing one another, but that are often cannibalistic. Such pair-bonding is a function of "territorial" display well established in bird literature.

Homosexual courtship display and even mounting attempts were witnessed on several occasions. All the attempts were by young, perhaps relatively inexperienced individuals, such as Whiteface. These courting and mounting attempts were completely ignored by large males, such as 19W. Rand (1967) found similar homosexual behavior on the part of male *Anolis lineatopus* toward young males of the same size as mature females. No homosexual mating was witnessed in any ora coactions. The only difference in homo- and heterosexual courtship behavior was that males always ran away, whereas females being courted ran away only when the male was unacceptable.

The following journal excerpts describe two courtship sequences mainly between 19W and 23W. Observations and photographs were made from blinds located 2–5 m from the carrion and lizard aggregations.

October 1, 1970

1721. . . 19W (♂) steps out of brush from behind (the smaller) Whiteface (♂), who is on savanna edge at carrion. 19W presents full threat display to Whiteface, who runs off. 23W (♀) is already feeding at carrion. With its tongue tips 19W touches 23W over her entire back (Fig. 8–1A), then scratches her back and then her side with the claws of the right front foot (Fig. 8–1C). The ratchet-like noise made by the claws being pulled over the scales can be heard for at least 30 m. When 19W tries to mount

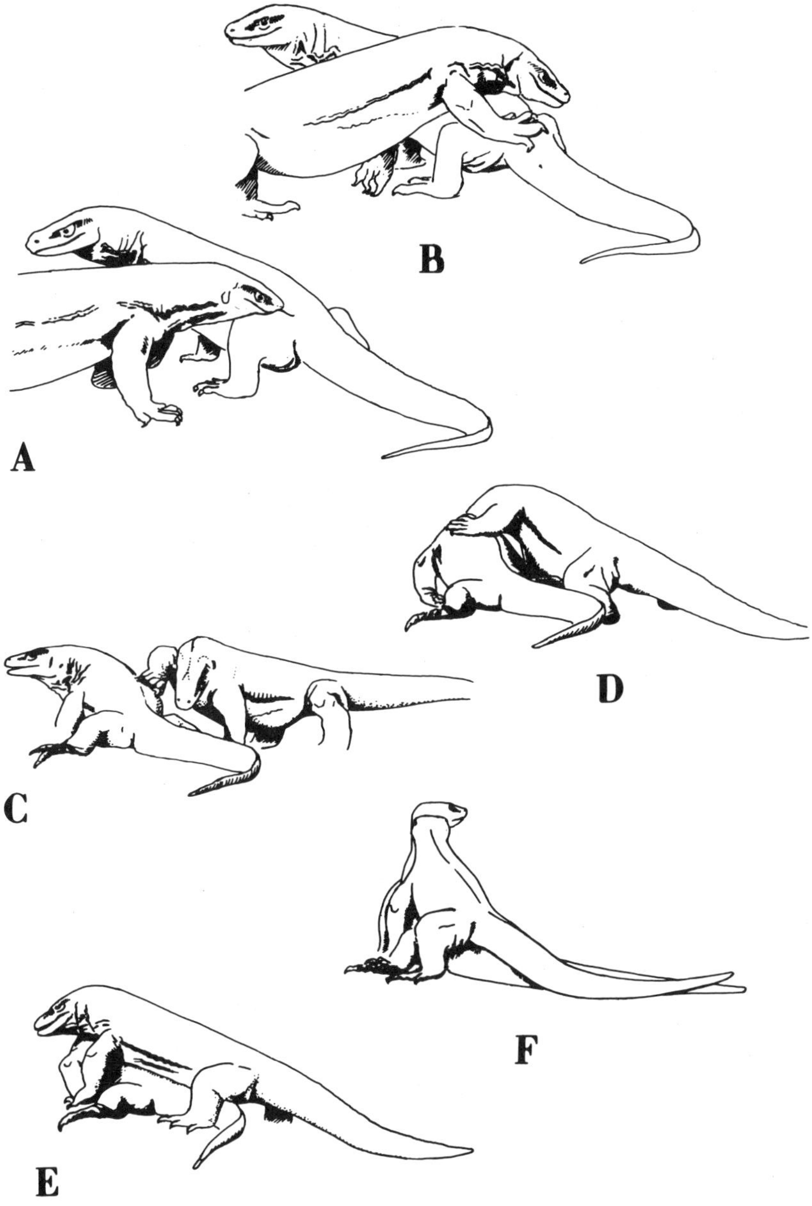

Fig. 8–1. Elements of courtship in *V. komodoensis* (from photographs; male 19W, female 23W). A: male tongues and presses snout into insertion of hind limb and body. B: male crawls onto and over female. C: male scratches female on sides and back. D: male tongues area around tympanum. E: male quickly mounts female. F: male fully mounted on female.

23W, the latter produces a threat display, hissing loudly. 19W mounts quickly (Fig. 8–1E) but slides off her back after a few seconds. When 19W again scratches her back, 23W raises herself on all four legs and arches her tail. 19W again draws his claws across her back and then mounts 23W very quickly. As 23W slowly moves toward carrion she presents a weak threat display, with throat inflation, slightly lowered head and tail held higher above the ground than usual. At first 19W follows only a few feet behind, but then rushes forward, causing her to do the same. 19W suddenly stops, turns and moves toward carrion, causing 7W (♂) to move away from the carrion on which it had been feeding. All lizards eventually move to shade of brush nearby.

1732. . . 23W again approaches carrion and is eating by the time 19W approaches carrion while presenting stress display to 23W. 23W trots off, returning stress display to 19W several times in succession. She then licks 19W on his back. While both are lying on the ground, looking at one another, 19W lunges forward and begins to mount 23W, but immediately slides off. Both return to carrion and feed. Whiteface moves to carrion and all three feed peacefully.

1744. . . 19W attempts a stress display with head and neck of dead feral dog in its mouth. Immediately disgorging the dog's head, it licks 23W over entire back and then attempts to mount her, but slips off. He then scratches her back, after which 23W runs 15 m into the brush, presenting stress display the entire time he chases her. They suddenly stop in the shade and sit looking at one another, with 19W periodically showing stress. 19W suddenly mounts 23W, but she slips out from under him and runs 8 m away. 19W is in close pursuit, trying to scratch her with one front leg as he runs behind and slightly to one side of her. He again threatens as he approaches her and continues as they lick one another on the side of the head. 23W moves away with 19W following about 3.7 m behind. 23W returns to carrion with 19W still following her. 19W threatens 23W and she moves from carrion, with 19W still following her; she begins to run faster; both show stress displays as they run into brush together.

1753. . . 23W again approaches carrion with 19W following and threatening her as she circles behind him to obtain the flesh. In so doing, she nudges and licks 19W on the back several times, while 19W continues to threaten. 19W then turns, walks toward her and licks and rubs his chin on her back, the sides of her head and neck, and then the head again. He suddenly threatens her as he walks away and returns in an arc. 23W lowers head and watches 19W out of the corner of her eye. 19W scratches her back with the claws of one of his front feet, while lying almost parallel and partly on top of her (Fig. 8–1D). 19W nudges and rubs the sides of her face and then the sides of her body with his snout (Fig. 8–1A), starts to scratch her back and suddenly mounts her.

1802. . . 19W slides off 23W as she moves toward carrion. When 19W starts to follow her, she stops and hisses loudly several times; 19W moves to shade of brush; 23W also moves into brush and into deeper forest beyond ecotone. Shortly thereafter the remaining lizards (19W, 7W, and 13W) move into forest at 1805 hr.

March 2, 1970

1043: 23W approaches carrion and begins to eat.

1055: 19W moves out of brush, approaches 23W, attempts to bite her back, and is unsuccessful in mounting her. 23W presents appeasement display to 19W and then moves to shade. 19W eats carrion.

1105: 19W, still eating, threatens 23W as she approaches carrion. 23W retreats to shade; 19W still eating. 23W approaches carrion and they eat side by side. 19W stops feeding and scratches 23W on her side and her face with the claws of his right front foot. 23W moves in circle and returns with weak appeasement display. They again eat side by side. 23W moves into shade, then back to carrion where they again feed side by side.

1115: 19W suddenly mounts 23W, slides off and retires to brush. 23W continues to eat. 19W presents appeasement display to 23W several times as he moves to carrion, though he quickly returns to shade.

1120: With no prior movement on the part of either individual, 19W dashes out of shade and chases 23W, who runs 15 m into brush. He returns and begins to feed, suddenly spins around and chases 23W for 8 m. When she stops 19W threatens her and she runs away, stopping only when 19W retreats to return to feed.

1140: 19W feeding and 23W again approaching carrion.

1145: 19W retires to shade as 23W comes close to carrion, but is then chased by 19W. She is quickly mounted, though 19W slides off in a few seconds. As she rests with belly and tail on the ground and head in an attentive pose, he rubs her entire back with the sides of his face. 19W again mounts her, but immediately slides off. He scratches her lower back with the claws of his right front foot, then stops and moves more anteriorly, again scratching her with his right foot (Fig. 8–1C). He then mounts her completely, with his front legs clasped around her body just behind her front legs, with his hind legs on either side of the base of her tail (Fig. 8–1F). 23W remains completely passive. After a few moments he slides off and scratches her over her hip and back, then suddenly mounts her completely again. After a few seconds he again slides off, nudging and licking her side at the insertion of the hindlimb with her body, and rubbing the sides of his face in the same area. He then scratches her back anterior to the hips. She remains very still while he again mounts her completely, but slides off and then rubs his face on her side just before momentarily mounting her again. When he slides off after several seconds, she moves a few feet forward. He again scratches her back with his right front foot. She walks forward and he follows her into the brush, where she begins to run. He follows very closely at the same pace,

1215: 19W trying to place his right front foot on her back.

1314: 19W returns to carrion.

1408: 13W presents appeasement display to 19W as the former approaches from downwind.

1425: 13W runs away as 23W approaches carrion from downwind. 19W immediately walks to 23W and mounts her without any preliminaries. She presents a stress reaction and moves a few feet as he slides off her back and feeds on the carrion.

1452: 23W returns to carrion and both eat together. 19W seems to be trying to crawl underneath 23W, forcing his head, neck, and anterior part of the body between her body and the ground surface.

1502: 19W scratches 23W on the back so hard with his left front foot that he pulls her to him. He clambers over her back and they both feed on the carcass. He scratches her on the side of the face with his right foot, nudges her in front of her hip with his snout, tries to stick his head underneath her belly again, scratches her on the side of her face with his right foot, licks the side of her face a number of times and then both eat together.

Successful copulation between these two individuals during late June to early October was preceded by identical courtship behavior. However, on one of the frequent mountings, the female raised the base of her tail after the male repeatedly touched this area with his hind foot. The male then followed the female's movement with a rapid tail twist, placing his vent in contact with hers. This was followed by a short period of feeble thrusting. They remained in copula approximately 12 min. Only once have I seen the male bite the dorsal surface of the female's neck, as reported by Honegger and Heusser (1969) for *V. salvator.*

DePrato's (MS) description of courtship in two captive specimens in the National Zoological Park (Washington, D.C.) includes some of the pertinent behaviors observed here, as well as tongue licking by both specimens, male threats, and the *female* pressing her head under the body of the male. Kern's (1968) description of part of the courtship pattern in two adult oras on Komodo included raking the claws over the back. His explanation that this represented one individual punishing the other for having eaten the bait is completely untenable. He stated both specimens were males, but provided no proof.

Courtship and territoriality.—Many workers feel there may be some significance in the fact that courting animals often hold territories at mating time, and that territorial and mating behavior often coincide. Early in the breeding season the males of such species usually establish themselves in separate localities, which they defend against other males of the same species. They often proclaim their occupancy by displaying themselves conspicuously, particularly by patrolling the boundaries of the defended area. Females rarely establish territories, but they usually visit the males in theirs, where courtship and breeding occur. Such territorial behavior conveys many advantages for the breeding animal. However, the Komodo monitor establishes no territory of this type. What territory is defended is vague, seems to change constantly, and is most often associated with an area around carrion. Thus it is a typical feeding territory.

As pointed out above, courtship in the ora normally takes place around a carcass. Thus the place where it occurs is not correlated with boundaries of a specific male's territory in the usual sense. However, the behavior of a group of oras around a carcass is clearly related to a social hierarchy associated with a feeding territory (Chap. 5).

The feeding territorial behavior of a dominant male ora, such as 19W, may complicate the process of pairing and mating by introducing an element of hostility. As oras are not obviously sexually dimorphic, the male's first reaction to a female is usually aggressive, because in the usual territorial context she is an intruder into an ill-defined territory around the carcass, and she possesses essentially all the morphologic characteristics of males of his own species. Hence, large male oras may be ambivalent in their approaches to females. The females may often be threatened and, theoretically, even attacked. Either obviously interferes with successful courtship.

The meeting of two or more lizards around a carcass is obviously a conflict situation, resulting in particularly the smaller individuals behaving in a characteristic way: they may be restless; they change rapidly from one activity to another; and their movements may be incomplete or distorted. In the case of courtship, this means that male oras may on one occasion court a female intently and on another ignore her completely, as is shown above to be a typical pattern between 19W and 23W. The variation in his sexually directed responses and their duration and intensity are all probably reflections of the conflict produced.

In monitors, as in many fishes and some birds, courtship is very short and mating takes place almost immediately. In this case there is evidence of three separate reactions to females: attack, avoidance, and mating. Attacks of males upon females are common at the moment of, and immediately after, the encounter and less common a short time later. The data on courtship among oras not sharing the same activity range agree with those of almost all the investigations in birds and mammals, bolstering the conclusion that attack behavior in these individuals predominates in the early stages of courtship, but is superseded by mating attempts, although such attempts seem generally unsuccessful in "unacquainted" individuals.

In these monitors neck expansion display occurs only in the aggressive phase of courtship and is apparently associated with conflict situations (stress reaction). Inflation displays are thus not necessarily cues related to dominance as they are in most iguanids and agamids. Although normally two opposing reactions cannot be equally strong for more than a brief, intermediate period, in this case we question whether attack and mating reactions are actually in opposition. Similar situations in which the male appears to attack the female while mating with her have been described in feral house cats (Leyhausen, 1956) and in some fishes (Lorenz, 1966). The simultaneous neck or shoulder biting and copulation by males of many lizard species may fall into the same category, although the biting may be an initial immobilization technique, as it is in some land tortoises (Auffenberg, 1964).

Many observations suggest that females of most reptile species are initially coy, but become responsive when stimulated. This is very typical in ora courtship, as illustrated in the sequences provided above. Such behavior on the part of the females usually prevents male oras from mating with them. As in tortoises (Auffenberg, 1964; Weaver, 1970), the behavior of the female ora is apparently very carefully regulated to avoid driving the male away. However, on at least one occasion (October 1) 23W allowed Whiteface, a young male, to mount her, but she then threatened him. He slid off and she attacked and bit him so severely that he apparently left the entire study area, for we never saw him again.

In fact, most female acts during the early phases of courtship, or in unsuccessful sequences, are agonistic and may include actual weaponry use. Continued association with particularly large, powerful males who are able to restrain aggressive females reduces female agonistic behavior in at least *V. bengalensis* (Auffenberg, in press a). The large size of *Varanus komodoensis*, plus the aggressive nature of the females and their capability to inflict serious wounds or even kill smaller specimens, have undoubtedly contributed to the development of an important reproductive behavioral strategy. This strategy stresses courtship throughout the year among the mature male and female pairs who share the greater part of the same foraging area. As a result, familiarity is assured among individuals potentially dangerous to one another, thereby reducing stress during the actual breeding period itself, and long-term pair bonds develop.

The probability for courtship success is thus enhanced by what amounts to an almost year-round courtship with otherwise very aggressive and powerful females. The same strategy in its basic format is seen in *Varanus bengalensis* (Auffenberg, in press a). Though it is not known what role, if any, pair-bonding plays in this species in the wild, it is obvious that a long association (several months in captivity) effected through repeated courtship by the male is usually necessary for successful mating.

The entire social system of *Varanus komodoensis* (and perhaps all species of the genus) seems designed to select the most powerful mates for ultimate breeding. I base this on the fact that visual displays intimating social rank are of much less importance in varanids than in agamids or iguanids. On the other hand, warning and actual weaponry use are common components of even adolescent interactions (Auffenberg and Ganci, in press). Unlike lizards in other families, female varanids are much more apt to engage in agonistic displays and use their teeth and tails more often than do males of the same species. This means that males interested in copulation must be able to so completely restrain the females that they remain uninjured after breeding.

Cues used in courtship.—Visual display is not as significant a factor in the courtship of oras as it is in agamid and iguanid lizards. There are no changes in speed or relative timing of movements, such as the rhythmic repetition of head bobbing often involved in the extremely effective visual displays of other lizards. However, courtship in Komodo monitors does include many tactile cues. There is much rubbing the chin on and pushing the snout to the sides and top of the body and neck of the female.

Important chemical cues in reptile courtship are reviewed in Burghardt (1970). In oras the tongue tips are frequently touched to the insertion of the hind legs and the body, the temporal area, and the sides of the snout (Fig. 8–2). Males

Fig. 8–2. Shaded areas are those frequently licked and touched with the snout during courtship.

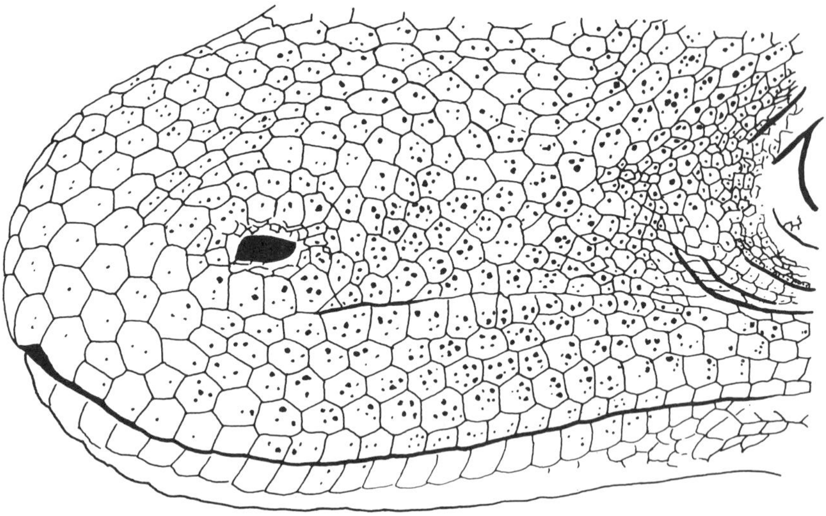

Fig. 8–3. Snout of adult female (from photograph), showing distribution of pustules and their concentration between the eye and nostril.

lick females more than the reverse. Mudrack (1969) described similar tonguing of the female's shoulders, neck, and the back of the head by male *V. storri*, and Auffenberg (in press a) analyzed tonguing by courting male *V. bengalensis*.

While it is not clear what information is being obtained from these areas, or what structures may produce the pheromones, the skin of at least the side of the snout exhibits noticeable modifications. While most head scales have a single, slightly swollen pustular scale organ, the scales between the eye and nostril have 4 to 6 (Fig. 8–3). Their function is unknown, but they may be associated with important chemical and/or tactile cues.

Hard scratches administered to the female's back by the male's claws are obviously important tactile cues that may serve to quiet the female. Neck and shoulder biting may serve to immobilize the female as well as to anchor the male and place him slightly posterior, where cloacal correspondence is more likely. Included in Table 12–7 are cues important in courtship behavior.

Mounting behavior.—Mounting occurred several times in each courtship sequence and was always accomplished rather suddenly, giving the appearance of being a contrived surprise, as in the copulation of many land turtles. The first mounting was usually the most vigorous, becoming less intense with each successive attempt. When the intensity decreased, the females often took the opportunity to dash into the nearest cover, very rarely onto the open savanna. The fact that the female ran away often seemed to stimulate the male to follow her again.

Quite often the male tried to place one foot on her back as she ran away—probably an abortive attempt to scratch her back.

It is perhaps significant that almost all the mountings witnessed were initiated by the male approaching the female at an angle from behind and slightly to one side of one of her hind legs. Approaches were made from either side. Several times the males mounted the females directly from behind, moving straight up the tail and back. At no time did the males attempt to force their tails under those of the females.

There is almost no information on how much time normally elapses between successful copulation and egg laying. A female in the Amsterdam Zoo laid fertile eggs 1 year after having been received, without any apparent copulation with a male in the same enclosure (Brongersma, 1932). If the above account is true, sperm storage by the female is extremely likely. De Jong (1944) estimated that breeding in the wild took place in July, and the eggs were laid several weeks later in August. This is corroborated by the only data available; that provided by Tanzer and van Heurn (1938), who reported captive breeding from July 4 to 10, with egg laying occurring 2 weeks later on July 24.

Nesting

Very little data are available on the nests of any monitor species. Several are known to lay their eggs in termitaria—*V. flavescens* Theobald, 1868 (= *V. dumerilli* ? Mertens, 1942), *V. niloticus* Cowles, 1928, 1930, and *V. monitor* = *bengalensis* Smith, 1935. The eggs are usually deposited in tree holes, stumps, and particularly holes excavated in the ground (Mertens, 1942). At least two species are said to place leaves, rubbish, or even manure in the burrows in which the eggs are laid—*V. monitor* = *bengalensis* Smith, 1935, and *V. exanthematicus* Rose, 1950. Much remains to be learned regarding the nesting habits of almost all species, and *V. komodoensis* is no exception. The complete lack of information on this species led Lallemant (1929) to suggest they might produce living young; this has since been shown to be incorrect. Eggs have been observed both in the field and in captivity.

Most eggs deposited in zoological gardens are simply laid on the floor of the cages. However, a few instances of relatively natural nesting sequences are known. Unfortunately, opportunities to obtain considerably more information have been allowed to slip by. Thus, though many eggs have been deposited in the Surabaja Zoological Gardens, Oesman (1967) simply stated that the "females laid their eggs in sand." His statement that the females bury the eggs in the sand after the shells harden in the sun is clearly incorrect. Schoenmakers (1935) stated that one of the oras in the Surabaja Zoo excavated its nest under the roots of some living shrubs in the outdoor pen. Tanzer and van Heurn (1938) reported that a nest in the Surabaja Zoo was about 45 cm deep, dug in a humus hillock. Fifteen eggs were in an upper layer and 10 below them. The most accurate description is by Brongersma (1932), who described the digging of a nest hole by an ora female in the Amsterdam Zoo. At about 1500 hr, the female began to excavate a hole by throwing sand over her back with, alternately, her left and right front feet. The nest hole, dug in the sand within several hours, was so deep that it caved in and the female had to be pulled out by her tail. During the night a second hole was dug, but apparently no eggs were deposited in it. Busono (n.d.) stated that in August, 1968, the female in the Jogjakarta Zoo "dug into the sandy ground with her forelegs. She then burrowed herself in the ground and came out after about an hour. . . ." In a discussion with Dr. Busono, he extended and clarified some of his brief published remarks. Because his observations are the only details available on nest construction, I include his verbal description in its entirety (edited):

> On the 26th of August, I observed the female, which was put in the pen when it was first built, digging into the beach sand. The area selected was in a spot free of grass and in front of the female's burrow. The excavation seemed to be no deeper than 0.5 m, and while the animal was completely under the earth, the ground at the surface was seen to heave slightly. The earth heaved up by her apparently U-shaped movement through the soil caved in when she reappeared at the surface in approximately one hour. I presume she laid some eggs in the hole. The same behavior was repeated three or four times in the same general area until the end of the first week of September.

Apparently each excursion under the surface was in the same general area, but roughly U-shaped and in a direction different from previous subsurface movement. Evidently additional eggs were laid on these occasions. That this is not unusual is suggested by Tanzer and van Heurn (1938), who reported that the same captive female constructed two adjacent nests, each with egg clutches. Whether multiple clutches in the same area are simply common, normal, or an artifact of captivity remains unknown. The multiple nesting activity in the same sandy area in Jogjakarta eventually produced a low mound of loose soil. The collapse of the burrow and absence of a noticeable mouth, as in this nest, is also mentioned by Tanzer and van Heurn (1938). Unfortunately, this constitutes our total knowledge of nest construction in captivity.

Extremely little data are available on nests in the field. All the local villagers I talked to claimed complete ignorance of ora nests, although it was generally conceded that the eggs were deposited underground. A story current on Komodo is that a former expedition to that island discovered ora egg shells in a large burrow (4 m long). However, H. Oesman, who was familiar with ora eggs, claimed the pieces were hard and fragile, like bird or gecko eggs. Pfeffer (1956) reported a more plausible native tale of a plugged nest hole on Rintja that was 30–40 cm deep. On July 11, 1971, we discovered an abortive ora nest on Flores. Nine eggs were found on the surface along a game trail 80 m above the valley floor on the slope of an open, grassy hill (Galo Muri) about 1 km west of Djarek, Mangarrai District. Very close by a hole was dug about 37 cm deep into the sandy clay. Though the eggs (UF 37031) were slightly shriveled, they appeared normal in size, texture, and color.

Father Erwin Schmutz, SVD (in a letter), provided the best information on a natural nest. Because of the importance of his find, I visited the nest on July 12, 1971. The site was about 150 m above the Nanga Loök Valley floor, just south of the village of Nggoer, Mangarrai District, Flores, on the north slope of a high hill known as Todong Tali. The immediate area was a boulder-strewn, ungrazed slope at an angle of about 45°, covered mainly with *Themeda gigantea* (Fig. 8–4). The nest was located between two boulders in a slot 65 × 45 × 45 cm, originally filled with a lateritic gravelly sand comprised of nodules up to 2.5 cm in diameter. When originally found, the eggs were spilled out in front of, and slightly downslope from the hole. All of the shells (representing 13–15 eggs) were neatly sliced in several areas, suggesting that the clutch hatched successfully. However, their presence outside the nest cavity, from which all the loose soil had been excavated either prior to or at the time of hatching, is a problem discussed further below.

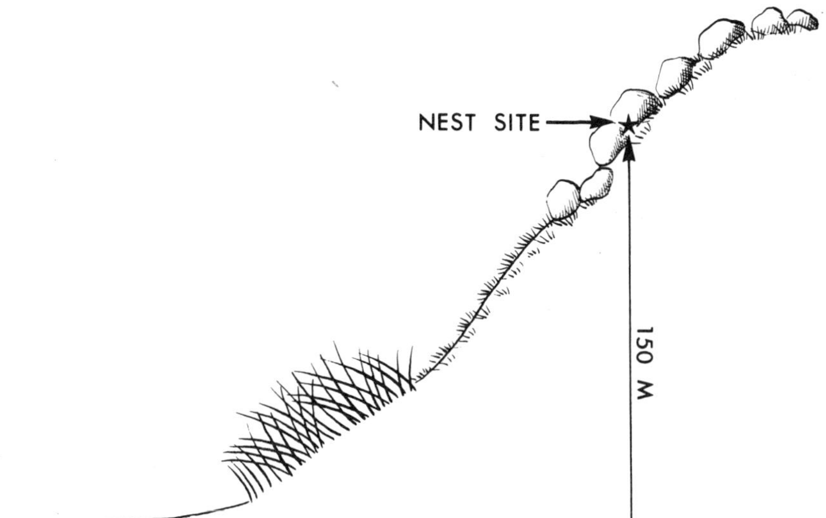

Fig. 8–4. Nest site location (diagrammatic), near Nggoer, Mangarrai District, Flores (Nanga Loök valley).

Hilmi Oesman (personal communication) suggested that certain clusters of ora holes on hillsides (see Chap. 5) may be used for nesting. To test this hypothesis, we excavated a number of burrows in the Ntodo Kalo cluster, but we found no evidence of the usually highly resistant eggshells. However, I have recently learned that several female *V. bengalensis* may use the same area for nesting. Thus it is still possible that the hillside hole clusters on Komodo and Flores may be nest sites.

Eggs

Laying behavior.—Very little information is available on behavior of the female during egg deposition. Oesman (1967) stated that the female's belly appears larger than usual before the eggs are laid. Brongersma (1932) reported that for as much as 15 minutes before each egg is laid the female's head "shivers," her respiration rate increases, the skin of her back moves back and forth, and her hyoid area is inflated. As far as can be determined from his publication, the average rate of egg deposition is about 1 hr, though at least 2 eggs are laid at one time. Lederer's (1942) report is essentially the same. The captive female was restless, with an increased breathing rate (no specifics) noticeable as much as one day before laying the first eggs; her head vibrated or shook immediately before the eggs were laid. Whenever an egg was deposited the female stood on all 4 legs and lifted her tail slightly.

While these data are useful, in neither of these reports were the eggs described as being deposited in a nest cavity. Therefore the described behavior details may be inaccurate for a natural field situation.

Seasonality.—Though much remains to be learned regarding the time of year and the seasonal conditions under which varanids lay their eggs, certain generalizations can be made. Species with wide geographic ranges, such as *V. bengalensis* and *V. salvator*, also lay their eggs over a broad time period. The reproductive periods of some, such as those that deposit their eggs in termitaria (e.g., *V. niloticus*), occur during the wet season, when the termite hills are wet and easily dug. Others, such as *V. komodoensis*, lay their eggs during the dry season. Egg-laying dates for this species are provided in Table 8–3. Unfortunately, most of these dates were obtained from captive animals. However, except for the Frankfurt specimens, most data were obtained shortly after the animals were captured. Only two dates are based on observations in the field: July 24, based on our discovery of the aborted nest on Flores; and "end of July through early August," based on the examination of the reproductive tract of a specimen killed on August 26 (Darevsky and Kadarsan, 1964).

Dates for egg laying by captive oras are commonly available, though they probably reflect far greater seasonal variation than occurs in nature. De Jong (1937) believed the oras held in captivity in Java lay their eggs as much as 2

Table 8–3. Laying periods and egg complement per nest in *V. komodoensis*

Date	Number of eggs	Place[a]	Authority
Jan. 20–22	11?	(Washington)	DePrato (MS)
March 13–17	3	(Frankfurt)	Lederer (1942)
April 15	2?	(Jakarta)	de Jong (1941)
April 27–29	5	(Frankfurt)	Lederer (1942)
April 24–30	14	(Surabaja)	This study
May 13–22	7	(Washington)	DePrato (MS)
June 1	29	(Jogjakarta)	Busono (n.d.)
June 29	16	Surabaja	Schoenmakers (1935)
July 24	9	Flores	This study
"End of July–early August"	15	Komodo	Darevsky and Kadarsan (1964)
Aug. 6–13	30	(Tokyo)	Hisada (1966)
Aug. 13	15[b]	(Surabaja)	Tanzer and van Heurn (1938)
Aug. 15–22	22	(Jakarta)	Galstaun (1973)
Aug. 18–31	4	(Jakarta)	Galstaun (1973)
Aug. 21–25	2	(Jakarta)	Galstaun (1973)
Aug. 26	?	(Jogjakarta)	Busono (n.d.)
Aug. 28	23	(Surabaja)	This study
Sept. 1–7	22	(Surabaja)	This study
Sept. 9–30	3	(Washington)	DePrato (MS)
Sept. 4–8	12	(Amsterdam)	Brongersma (1932)
"September"	40[c]	(Jogjakarta)	Oesman (1967)
Oct. 19–26	13	(Washington)	DePrato (MS)
"Mid-November"	24	(Jogjakarta)	Oesman (1967)
Dec. 19–29	27	(Washington)	DePrato (MS)
?	13–15	Flores	Schmutz (in letter)
?	10	(Surabaja)	Tanzer and van Heurn (1938)
?	26	?	Pfeffer (1959)

a. Parentheses indicate captive specimens.

b. Plus 2 on surface.

c. Obviously several clutches.

months earlier than they do within the natural range of the species, but the data available for captives suggest laying dates are simply more variable, ranging throughout the entire year.

In the Bogor Museum a single, unnumbered female specimen that had died in the Jakarta Zoo on May 21, 1956, contained a number of ovarian follicles, the largest being 2.2 cm greatest diameter. They would not have been fully developed until about July or early August.

Dates on which eggs are known to have been laid both in the field and in captivity are given in Table 8–3. The data make it clear that the majority are laid in August and September, even in captive situations. The dates furthest removed from these (December to April) tend to apply to long-held captives in zoological gardens in the Temperate Zone (Washington and Frankfurt).

Egg complement.—Varanids tend to lay large numbers of eggs (Table 8–4). Though Fitch (1970) stated they are probably the most prolific of all lizards, it is clear that size is extremely important. For example, varanid species, such as *V. caudolineatus*, that are about the same small size as most other lizards have proportionately much smaller egg clutches (3–6) than the larger forms. Data on the number of eggs deposited by *V. komodoensis* are given in Table 8–5. The number varies from 1 to 30, though both extremes are unusual. The mean number of eggs in 17 clutches is 18.7.

It will be shown below that *V. komodoensis* lay one or more eggs several times during several, successive days. Thus, some important questions relating to varanid egg clutch comparisons are: What constitutes a clutch? Is it the number of eggs per nest? per day? In this study a single clutch refers to the total number of eggs laid by a female in the same place over a period of a week or so; that is, it represents the entire seasonal complement. Not all the published information corresponds to this scheme. For example, Pfeffer (1959) stated that 25–26 eggs are laid by large females, but it is not clear if he means one clutch or several. In the Jogjakarta Zoo at least 27 eggs were laid by one female in one nest site, for this many hatchlings were found later. The nest burrow was first excavated on August 26, but later dug into several more times by the same female until the first week in September. Therefore, not all the eggs may have been laid at one time, though this is most likely because the young all appeared at the same time.

Hilmi Oesman, of the Surabaja Zoo, provided me with what are probably the best data available at present. In a 6-day period from April 24 to April 30, 14 eggs were found scattered on the surface of the pen; on August 28, 23 more eggs were found; and from September 1 to September 7, 22 more eggs. Thus, one female dropped 59 eggs in a period of approximately 4 months, in three clutches, during two blocks of time (April and near the end of August), with the clutch number varying from 14 to 23.

Additional data on egg clutch and deposition time are presented in Table 8–6. Average clutch size is 18.5 eggs, ranging from 1 to 30 (Table 8–7). Range in

Table 8–4. Egg complement and egg size in varanid species

Species	Number of eggs	Size (mm)	Weight (g)	Place	Authority
V. salvator	15–25	92–100 × 35–38		Ceylon	Deraniyagala (1931)
V. salvator	15–30			Burma	Smith (1931)
V. salvator	10	75–80 × 29–41		Hainan	Schmidt (1927)
V. salvator	7	70.5–74.5 × 37–38.5	50.8–52.4	Sumatra	Meer-Mohr (1930)
V. salvator	?	70 × 38		Java	Karawier (1905)
V. salvator	6	75 × 38		captivity	Lederer (1942)
V. salvator	14	65–71 × 34–37	46–55	captivity	Kratzer (1973)
V. varius	12	63.5 × 38.1		Australia	McCoy (1878)
V. varius	"to 20"			Australia	Worrell (1963)
V. bengalensis	24			Ceylon	Deraniyagala (1957)
V. bengalensis	25–30	45 × 25		Burma	Smith (1931)
V. bengalensis	19–29			W.Pakistan	Minton (1966)
V. caudolineatus	3–6			Australia	Pianka (1970)
V. griseus	10–20			Africa	Thilenius (1897)
V. griseus	15–20			W. Pakistan	Smith (1935)

Table 8–5. Egg clutch size and deposition interval in *V. komodoensis*

Clutch size	Interval (days)	Authority
30	7	Hisada (1966)
28	1?	This study
27	11	DePrato (MS)
22	7	This study
22	8	Galstaun (1973)
19	10	DePrato (MS)
14	6	This study
12	5	Brongersma (1932)
11	3	DePrato (MS)
9	1	This study
7	10	DePrato (MS)
5	3	Lederer (1942)
4	14	Galstaun (1973)
3	3	Lederer (1942)
2	9	DePrato (MS)
2	5	Galstaun (1973)
1	1	DePrato (MS)

deposition time is from 1 to 14 days, and there is no correlation between clutch size and total deposition time. The number of days between layings within a clutch varies from 1 to 8 ($\bar{X} = 3.0$, $n = 38$). The number of eggs laid per day varies from 1 to 20 ($\bar{X} = 3.5$, $n = 38$). Oesman's (1967) report of 40 eggs laid in the Surabaja Zoo obviously represents several clutches.

Individual females rarely lay two clutches in a single year; most females lay only one, with a 12-month lapse before the next one. Some captive females have a clutch every year, others do not. Tanzer and van Heurn (1932) reported that a female in the Surabaja Zoo laid eggs in 1931, 1933, and 1937; a Jogjakarta female laid eggs in 1963, 1965, 1966, and 1968. The number of eggs laid by captive females appears to be greater in earlier years than in later ones. For example, the average number of eggs from 3 captive females is 16.5 for the first year, 7.0 for the second year, and 2.3 for the third year, based on data in Table 8–6. It is obvious that the apparent health of a female, based on growth and

Table 8–6. Number of eggs laid within a single clutch

Authority	Days on which eggs were laid														Date
	1	2	3	4	5	6	7	8	9	10	11	12	13	14	
DePrato (MS)	9?		2												Jan. 1967
DePrato (MS)	1				3					20	3				Dec. 1968
DePrato (MS)	2			3						2					May 1970
DePrato (MS)	2		4			7				6					Oct. 1970
DePrato (MS)	1								1					1	Sept. 1971
Brongersma (1932)	8	1			2										Sept. 1931
Galstaun (1973)	1		1	2	8	4	5	1							Aug. 1969
Galstaun (1973)	1							1					1	1	Aug. 1971
Galstaun (1973)	1				1										Aug. 1972
Lederer (1942)	1	2	2												April 1940
Lederer (1942)	1		1		1										March 1941

weight gain or maintenance, does not reflect an ability to produce eggs. There is much we do not yet understand about the influence of diet on reptilian reproduction.

Morphology.—The eggs of all varanids have a soft, smooth leathery shell. Magnification of the shell surface of *V. komodoensis* reveals anastomosing wrinkles (Fig. 8–5). While the eggs of some species, such as *V. exanthematicus* and rarely *V. salvator*, adhere to each other in one or more masses (Lederer, 1942), those of the ora are separate. Though Oesman (1967) reported the shells harden in the sun, all other workers, including myself, have found them to remain flexible but turgid throughout incubation.

Sizes and weights of 69 *V. komodoensis* eggs are provided in Table 8–7. Misshapen, shriveled eggs are common under captive conditions (Brongersma, 1932; DePrato, MS). Mean and overall range in length, diameter, and weight are 86.5 mm (59–115), 55.5 mm (42–66), and 124.7 g (55–210), respectively.

Incubation and hatching.—Most varanid eggs tend to have long incubation periods, when compared with those of other lizards. In *V. niloticus* the eggs normally hatch after 8–10 months (Cowles, 1930); *V. griseus* up to 10 months (Thilenius, 1897); *V. salvator* from 2.5 to 10 months (Kratzer, 1973); and others, such as *V. bengalensis* and *V. varius* (Smith, 1931; Deraniyagala, 1957; and Worrell, 1963) have short to intermediate lengths of 2.5–5.5 months. The incubation period of the Komodo monitor is usually 8–8.5 months.

As pointed out above, most ora eggs are laid from July to early September. On Komodo we observed hatchlings on six occasions, all in April and May. Two of the sightings (May 13 and May 24) were on hillsides near hole clusters at Ntodo

Table 8–7. Characteristics of eggs (N = 69) of *V. komodoensis*

Month	Place	Number of eggs	Length (mm)		Diameter (mm)		Weight (g)	
			$\overline{X}$	Range	$\overline{X}$	Range	$\overline{X}$	Range
April	Surabaja	13	78.5	73–85	45.9	42–52	76.4	72–81
August	Jakarta	8	93.2	74–101	57.3	45–66	157.8	80–210
September	Amsterdam	8	94.1	90–100	55.0	50–62.5	158.4	109–203
April	Frankfurt	3	105.7	98–115	57.7	50–58	130.0	130
August	Surabaja	1	92.0		60.0		176.0	
August	Tokyo	30	83.0	59–102	52.0	44–66	121.0	55–144
January	Washington	6	77.3	75–82	45.0	44–47	73.3	69–82
Overall means and ranges			86.5	59–115	55.5	42–66	124.7	55–210

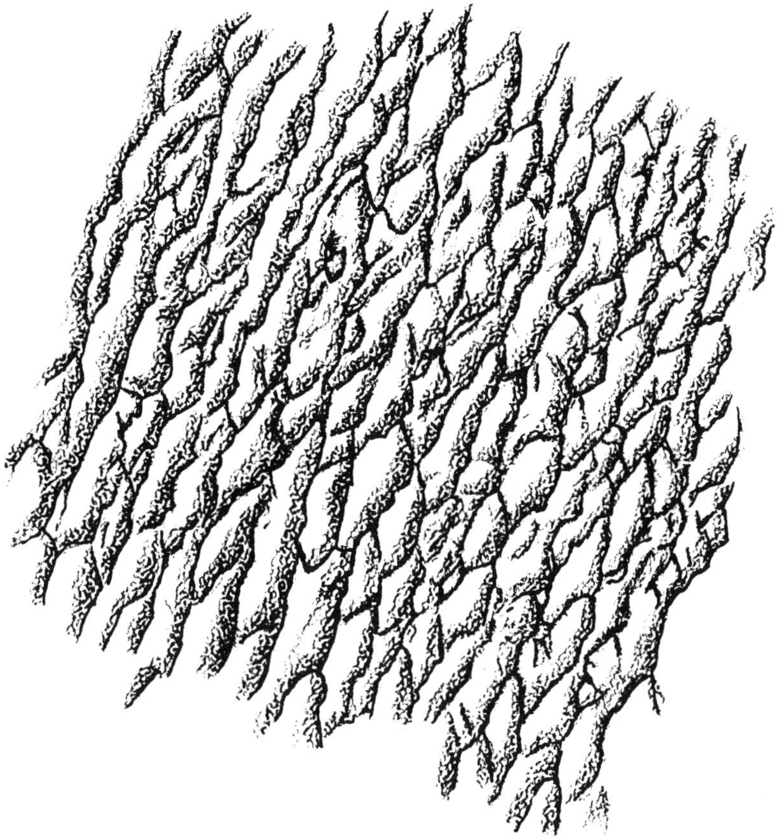

Fig. 8–5. Egg shell surface.

Kalo and Gunung Kubur (see Chap. 5); one (May 24) was in the overgrown garden area north of the base camp; and three (April 10, 11, and 15) were on the beach near the base camp. Of these, the individual(s) seen on April 10 and 11 was clearly several weeks old. Kern (1968) photographed what appeared to be a newly hatched ora along the same beach, also in the month of April. The specimen we found on April 15 was only a few days old, with an unhealed umbilical scar. De Jong (1944) reported that two hatchlings appeared in the Jakarta Zoo on April 15, and Busono (n.d.) stated that the 29 young in the Jogjakarta Zoo were first seen on June 1. Therefore, the hatching dates range from the last week of March to the beginning of June, a period of approximately 10 weeks.

There is some evidence that female varanids, at least *V. griseus*, return to their nest holes after the original deposition. This behavior has not been reported for any other species. There are, however, some anecdotes and observations suggesting that this is at least a possibility in *V. komodoensis*. Lallemant (1929) was reporting local folklore when he stated that females sometimes protect the nest when they return to it. Pfeffer (1959) reiterated the tale of a Komodan who claimed to have seen a female (?) "near" an ora nest, attempting to catch (perhaps to kill and eat) the agile young as they left the nest cavity. After driving off the female (?), the villager examined the nest and eggs. His description of both were in complete agreement with published observations, and Pfeffer was convinced he had at least seen the nest. That the rest of the villager's story is also possible is suggested by literature reports, as well as by our own observations. That oras will kill and eat their own kind, particularly young individuals, is a fact (see Chap. 11). Many observations of captives indicate that the adults will also eat eggs. Both males and females have been reported to eat *V. komodoensis* eggs, though apparently only those lying on the surface. Males are reported to have eaten 1 out of a 12-egg clutch (Brongersma, 1932), 20 out of a 24-egg clutch (Oesman, 1967), and 4 out of 24 (Galstaun, 1973). Females have also been reported to eat ora eggs—usually their own. Thus DePrato (MS) reports 2 eaten of 4 on one day and 3 of 6 on another day. Lederer (1942) mentioned that "Bubchen" ate the only egg she laid on one day, and Galstaun (1973) reported all 5 eggs laid in one day were eaten by the female. In almost every instance these females consumed only a part of the daily complement, and never all or even a large part of the clutch.

There is a slight possibility that an ora (the maternal individual?) may have been at the nest site on the boulder-strewn hillside when the young in that nest hatched. The eggs were all slit in the usual fashion, showing that the young had all hatched normally. Yet, all the shells had been dug out of the nest and scattered nearby, and none appeared to be torn or chewed.

No one has yet reported on the manner in which hatchling oras escape from their eggs or the nest, though there is no reason to believe it would be different from the process described by Kratzer (1973) for *V. salvator*. Slit eggshells of *V. komodoensis* are illustrated in Figure 8–6, showing that the hatchling apparently makes several attempts to open the egg, that these attempts are made at both ends, and that the slits tend to be horizontal rather than vertical.

For hatchling lengths and weights see Chapter 2.

Reproductive Strategy

The paucity of data on the reproductive biology of *V. komodoensis* allows for little discussion of adaptive reproductive strategy. When compared with data on many other lizard species provided by Tinkle, Wilbur, and Tilley (1970), the ora is a very late-maturing form (5–7 years), with maturity occurring at about 700 mm SVL. It normally produces one brood/year, comprised of an average of 18.5 eggs/clutch, deposited in the same nest area over a period of several days. The

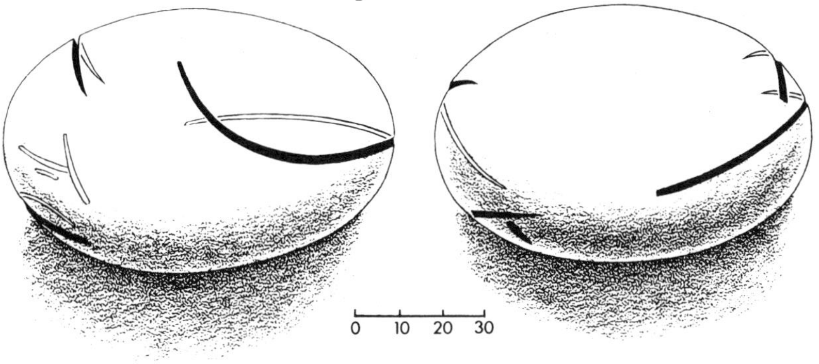

Fig. 8–6. Eggs of *V. komodoensis,* showing horizontal slitting concentrated at egg ends.

correlation of large size and advanced age at maturity, highly seasonal breeding, and large clutch size is completely consistent with Tinkle's (1969) hypothesis that lizard species of large size and exhibiting delayed maturity are expected to be single-brooded and have long life expectancies. In the wild most ora eggs hatch just after the short wet season ends (April–May) at a time when the density of potential prey (insects, etc.) is highest. This timing, plus the long incubation period of the eggs (8–8.5 months), requires that breeding take place (July–August) long before the wet season begins (December–January). Low population densities, coupled with large activity ranges and long residence, have apparently led to pair formation in most (all?) large, mature residents.

9

The Ora as a Scavenger

The published reports of almost every recent expedition that has studied, photographed, or captured Komodo monitors have emphasized the idea that *Varanus komodoensis* is a scavenger. It is true that oras regularly feed on rotten carcasses used as bait. The constancy and relative speed with which they come to rotten flesh has suggested to some zoologists that large oras are entirely scavengers (King, 1968; Oesman, 1970). Piazzini (1960) even asserted that oras are incapable of killing an animal as large as a chicken.

The plenitude of carcasses, however, is undocumented. Notwithstanding Darevsky's statement (1963) that the island abounds in carcasses, Hoogerwerf (1954a) found only one mummified deer in several trips to the island. More importantly we saw no mummified animals in our year there. The villagers see them rarely; most have never seen any. Hompes (1952) reported many mummified kerabau remains on Komodo during the exceptionally dry period of 1951 and said that wild boar and deer were also dying at the same time. Extensive falls of volcanic ash may destroy large areas important for grazing and browsing (Chap. 2). However, drought and volcanism are rare and do not cause enough deaths annually to supply carcasses to sustain ora populations.

Most medium to large monitor species feed on both live and dead prey (Mertens, 1942), and the degree to which one or the other is ingested apparently depends on the species of monitor, the animal's age and hunger level, and the opportunities presented. This is particularly true of the Komodo monitor. For example, small oras (40–100 cm in total length) are largely arboreal predators (Chap. 10). They start scavenging when they are about 1 m long (approximately 1 year old). Of 351 oras observed scavenging, only 8 were less than 1 m in total length. No hatchling or near hatchling-size individuals were ever seen at a carcass.

In addition to ora size, prey condition is also a factor in scavenging. When oras were offered a choice of a live, a freshly killed, or a decayed goat (N = 12), the fresh carcass was always chosen first. Pfeffer (1959) reported the same result from a single similar test on Rintja. When offered live and dead goats simultaneously (N = 8), the latter were always chosen. De Jong (1937) obtained a similar result in a single test conducted near Mburak, Flores.

Data provided below make it clear that an individual ora may be a successful predator one moment and an effective scavenger the next—which is a matter of opportunity, choice, or necessity. Where carcasses are not abundant even full-grown oras become pure predators of a very resourceful and successful type, contrary to what some authors have suggested. Thus the ora falls into a carnivore category known as the scavenger-predator type, which includes many large vertebrates, such as some species of felines, hyenas, jackals, and raptorial birds.

Biotic Interactions and Scavenging

Carrion associates.—The ora probably shares less carrion with other organisms in its range than does any other large vertebrate in the world. Carrion competitors include only a few birds, a suid, several species of beetles, flies, and feral dogs (Table 9–1). Except for the last, none is an important competitor (Chap. 11).

Though there are no true carrion birds in the Lesser Sunda Islands, three species occasionally feed on carcasses on Komodo: the brahminy kite (*Haliaster indus*), the jungle crow (*Corvus macrorhynchos*), and the white-fronted sea eagle (*Haliaetus leucogaster*). Throughout their ranges all three species reach maximum abundance in coastal areas.

Though we never saw a sea eagle over carrion (a pair lived near the Loho Liang base camp), Hoogerwerf (1954a) reported one at a carcass on Komodo. Pfeffer (1959) stated that it is a carrion associate on Rintja, where it is apparently much more common than on Komodo.

The brahminy kite is not common on either Komodo or Flores. Each large valley seems to possess only one or two pairs. Though we found it over carrion only once, villagers stated that it was more common than our observations indicated. Hoogerwerf (1955) also reported it as a carrion associate on Komodo. On the north coast we found evidence suggesting that an ora caught and ate a brahminy kite, probably as it fed on a fresh water buffalo carcass nearby.

Jungle crows are often found at carcasses of all ages, even new kills. Hoogerwerf (1954a) reported dozens of these birds feeding on carrion in Loho Liang. Normally there were no crows at our baits, but when they did appear we saw only one or two at a time, usually in a tree over both the carrion and the feeding lizards. Generally they would come to the ground only after the oras either had finished the carcass or had retired. The crows would then feed on small bits of flesh from the ground and from the carcass.

Ali (1964) and Schaller (1967) reported that the jungle crow is a common associate of vultures at carrion and that its movements often lead to the discovery

Table 9–1. Carrion competitors of *V. komodoensis*

Competitor	Incidence and Habits
Feral dogs	
(*Canis familiaris*)	Common, major feeding at night
Wild boar	
(*Sus scrofa*)	Rare, major feeding at night
Jungle crows	
(*Corvus macrorhynchos*)	Regular, diurnal
Brahminy kites	
(*Haliaster indus*)	Rare, diurnal
Carrion beetles	
(*Trox* sp.)	Common, diurnal, old carcasses
(Sp. B)	Regular, mainly nocturnal, found in fresh carcasses
Fly larvae	
(*Sarcophaga* sp.)	Regular, abundant in old carcasses
(*Calliphora* sp.)	Regular, abundant in old carcasses
(*Eristalis* sp.)	Regular, abundant in old carcasses

of tiger and leopard kills. We also located a very recently killed deer by tracing the calls of several particularly noisy crows while they sat in the trees above the still bleeding animal. Though apparently able to locate kills quickly, the role of both the crows and the kites as carrion competitors of the ora is negligible, since they occupy at least a tertiary position in the scavenger hierarchy of Komodo.

Wild boar (*Sus scrofa*) often fed on carrion bait at night, if the oras left part of a carcass or kill uneaten. However, they are not important competitors for carrion during the day because large oras sometimes attack them. On September 5 a large boar approached the bait to within a few feet of several oras feeding on it. Suddenly one of the larger lizards lunged at the boar and then dashed after it into the brush. In other instances boar came to the edge of the forest near the bait but would not step onto the savanna where the carcass lay, even though no oras were nearby.

Other than oras, feral dogs probably eat the largest number of both fresh kills and carrion on Komodo and Flores. Unlike boar, these animals are much more active and aggressive in their attempts to take carrion from the oras, or to drive them away. The latter behavior was particularly common. Normally the dogs sat several meters from the closest feeding lizards, harassing the oras from time to time by barking and snapping at them. Larger oras usually paid little attention, though on several occasions the largest would rush at the dogs. Sometimes the dogs would chase small oras from the carcass and then quickly return to the bait before the oras could. On November 21 we found a dog and several oras at a fresh deer kill. As we approached, the oras feeding on the carcass ran off. At that instant a dog, previously sitting at the side, rushed in and tried to drag off the body but was unable to do so. In most instances the dogs became particularly excited and active when the oras were dragging the carcass, or when it was almost completely eaten. It was only under these circumstances that they snapped at the large oras. However, their proximity sometimes led to large oras lunging at dogs attempting to snatch carrion and run away with it. In no instance was there any indication of cooperation among the dogs to drive away the oras. The largest number of dogs seen at carrion with oras was seven.

When bait or carrion is left overnight the dogs are extremely likely to eat most or all of it (also reported by Hoogerwerf, 1955). They often entered our traps at night. As with boar, nighttime is the period of most important direct competition with oras for available carrion. Flores villagers told me that dogs will feed on dead oras.

Two species of scarabid beetles feed on the flesh of dead animals and are thus in competition with the ora. One of these (genus and species unidentified) is mainly nocturnal and thus important only if the flesh remains after nightfall. They were found only from late December to March. On January 26 we had hung a freshly killed goat near our base camp and at 2130 hr I heard a hissing noise near the bait. My flashlight revealed the carcass, most of it covered by these beetles. By the following morning they had consumed approximately 5 kg of flesh and vanished. Hilmi Oesman (in a letter) reported that during one of his expeditions to Komodo, hordes of these insects consumed almost all the flesh of a medium-sized goat in two nights.

The second carrion beetle (*Trox* sp.) is found only at rotten flesh, never at fresh kills. Unlike the first species, it fed mainly during the day, though often concealed under the carcass and under nearby debris during the hottest part of the day. We sometimes found them on fresh ora droppings that contained hair, hooves, or bone. They and the larvae of several species of flies were probably the most important insect competitors of the ora.

The ora carrion competitors do not include the usually common carrion beetles of the families Silphidae (*Silpha, Necrophorus*) and Staphylinidae (*Creophilus*). I presume their absence is due to the extremely high scavenger efficiency of the oras, which simply leaves little of the carrion for any competitor that depends on rotted flesh for its existence.

Ora interactions.—As mentioned previously, oras often congregate at carrion and kill sites. The average number of individuals per day at a carrion bait in Loho Liang, Komodo, is 7.0 (range 1–17). In the case of a fresh kill, the intestinal contents slung out of the viscera apparently serve to attract other lizards to the vicinity. The aggregations of oras at such feeding sites are rather small (Table

5–2) and last for no more than 1 or 2 days, by which time the prey is usually completely eaten. However, in the somewhat unnatural situation of baits hung from a rope in a tree, aggregations might last 3 or 4 days, depending on the "life" of the bait animal (if not eaten by lizards it either decomposes, is consumed by dipteran larvae, or becomes mummified in the sun).

These small, casual ora aggregations are not really social, because there is very little true coordinated behavior of any sort. However, it is important to note that cooperative and disoperative factors are present in feeding aggregations. The latter is best illustrated by the fact that the largest individual at the bait usually takes the "best" parts, the best position around the bait, and most of the food. This individual determines (in the case of a natural, untethered bait) in which direction the carcass will be moved, when it will be moved, and under what conditions it will be moved. This disoperative factor is obviously dependent on rank order, which is based largely on size, although there is some evidence that prior experience may also be important (Chap. 12). The hierarchy determines not only the feeding sequence but the amount eaten by smaller individuals as well. Young oras often remain at carrion long after the larger specimens have left, picking up the few scraps remaining or dropped during the main feeding period. In general, the entire hierarchical system is established and reinforced largely in these feeding aggregations. The agonistic behavior associated with hierarchy formation is discussed more fully in Chapter 12.

No true cooperation was seen in ora aggregations, though a form of protocooperation (see Allee, 1951) sometimes occurred when large lizards were unable to swallow particularly large parts of the carcass because ligaments and flesh connected it to the body. After swallowing as much as possible, they would pull or jerk backward to break the connections. At these times smaller individuals would rush to the carcass, tearing the flesh from that part immediately in front of the jaws of the larger specimen.[1] By removing all the flesh in that area and cutting many of the tendons holding the pieces together at that joint, the carcass was easily separated by the larger ora, who had continued to pull at the piece. Thus larger oras obtained their share of the pieces that were more difficult to separate from the carcass, and at the same time it was possible for the smaller specimens to obtain more meat than would ordinarily be the case.

Techniques and Adaptations

Orientation to carrion.—Most scavengers find their prey by scent. Unfortunately, we have not yet established any clear link between the sensation of smell and the substance that causes it (the olfacient). A truly enormous gap exists between our sophisticated knowledge of optics and acoustics and our limited

1. Captive *V. bengalensis* behave similarly, many smaller individuals being more interested and intent on obtaining what another has than picking available food off the substrate.

knowledge of olfaction. Every sound and color can be accurately linked to a specific frequency of vibration or wavelength, but no one knows with certainty what distinguishes the various olfacients from each other (von Buddenbrock, 1958). Furthermore, most of the properties of olfacients have so far been explored only from the point of view of the human sense of smell. We still know nothing about those that affect the olfactory sense of most animals. We usually assume that animals as different as humans, oras, and insects react to the same smells, though perhaps differently. This view is supported by at least some data. For example, von Frisch (1967) showed that bees and humans respond to the same olfacients and that bees can be conditioned to react to any substance we can smell, but not to those which are odorless to us. Rensch and Eisentraut (1927) conducted a series of experiments suggesting that, except for bitter, lizard taste sensations were similar to those of humans. Both Lederer (1942) and DePrato (MS) mentioned that captive oras are attracted to the scent of freshly polished shoes, and Broughton (1936) stated that oras followed trails of linseed oil. Though humans are largely sight-guided and oras are apparently largely scent-guided, it is clear that both can and do react to at least some of the same scents. Most obvious is that both react to the scent of carrion.

Snakes and monitors are reputed to possess extremely sensitive olfactory mechanisms. In order for the lizards to find a carcass by olfaction the latter must emit a powerful odor over large areas of several square kilometers. During the first day after death, the volatile oils of decomposition are not produced in sufficient amount to attract any but the very closest lizards. However, after 48 hours the decomposition process is sufficiently accelerated that considerable scent is produced, and lizards from a large area will usually move toward the carcass. The direction, steadiness, and strength of the wind determine both the number of oras at the carrion and the direction from which they approach it (Table 9–2 and Fig. 9–1). Everything else being equal, the scent of a larger carcass will be stronger and be sustained for a longer period of time, tending to attract more individuals to it. The greatest proven distance from which we have attracted lizards to a bait is approximately 11 km, from the central ridge of the island (Ntodo Poreng area) to trap 1 near the beach. I am sure that under most circumstances on Komodo this distance would not be greatly exceeded, because a single deer carcass produces only a finite quantity of volatile decomposition products, and the source disappears before the diffusion gradient can pass over a great distance.

In perfectly still air the diffusion of volatile particles from the carcass establishes a concentration gradient having concentric spheres of equal intensity. But the gradient is fragile and is easily modified and deflected by complex wind currents in the mountainous area the oras inhabit. We placed carrion high on hills 19 and 23 in Loho Liang four times and found that a total of only 3 oras came to them. I presume oras have more difficulty locating a bait placed on a hill than one

Table 9–2. Number of oras and wind velocity at base camp site

Number of oras	Wind velocity (kph)
18	0–1.9
29	2.0–3.9
47	4.0–5.9
10	6.0–7.9
6	8.0–9.9
3	10.0–11.9
1	12.0–13.9

in the valley because the scent is carried out of their range. Thus, the position of the carcass, its size, and particularly the wind characteristics are important interacting factors determining which of the oras in an area are able to detect the scent.

Both topographical and vegetational features modify wind patterns, and it follows that some locations are more favorable than others as sentinel areas from which to detect windborne odors. Because of the generally prevailing onshore, daytime winds, these areas tend to be the higher, exposed knolls and ridges above forested valley floors. As pointed out previously (Chap. 5), these locations are prone to be utilized only by the largest individuals, which often pass much of each day at specific thermoregulatory sites, which are located so that odors produced by a rotting carcass on the valley floor are quickly carried to the oras by prevailing currents. On windy days even I could smell our carrion bait from about 1 km as I stood on either the upper slopes of hill 23 or on Ntodo Rankang.

Some organisms possess special mechanisms that aid in sampling the surrounding medium in the absence of natural currents, such as the sniffing of many mammals, most familiar in dogs. As far as is known, monitors do not utilize a "sampling current" of this type but depend entirely on wind currents. When such currents are present, the lizards orient toward the source simply by moving upwind, that is, directly against the wind current carrying the stimulant.

As humans, we are accustomed to making a distinction between taste and smell. In reptiles, however, there is probably little difference between the two. Normally the sense of smell gives us some indication of the chemical properties of a substance placed some distance away, whereas taste is operative only when the substance is in our mouths. However, lizards and snakes thrust out their moist tongue to pick up the particles constituting the stimulus (scent) and to carry them into the mouth (taste). The principal sensory receptors, the paired Jacobson's

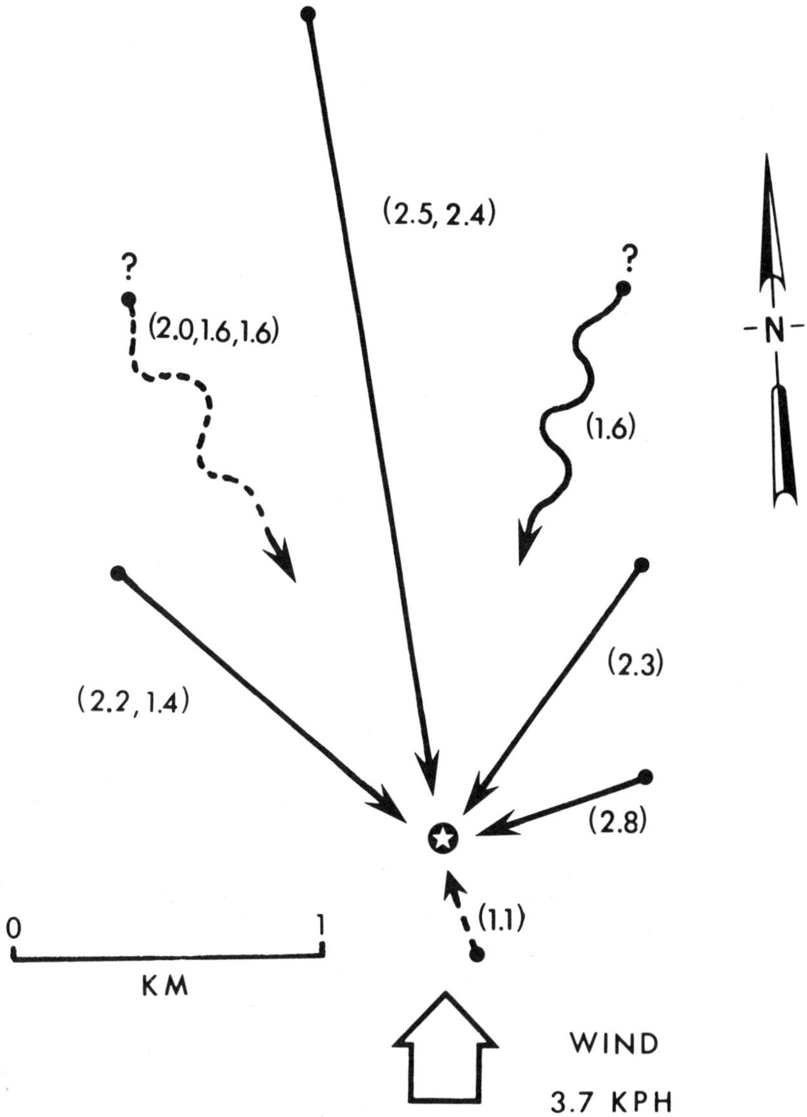

Fig. 9–1. Direction of origin, distance, and size (in m) of resident (solid lines) and transient oras (dashed lines) that came to carrion on Nov. 21–23, 1969, Loho Liang, Komodo. Wind direction and velocity shown. Wavy lines represent individuals that came from unknown distances and directions.

organs, are located in the upper portion of the palate of the mouth.[2] Both Broman (1920) and Parsons (1970) suggested that these vomeronasal organs in tetrapods probably represent the primitive nasal organ of fishes and are chemoreceptors adapted primarily for functioning in an aquatic environment. The main olfactory organ of tetrapods is believed to be a more recent development and one adapted to chemoreception in an aerial environment. There is no general agreement on whether the vomeronasal organs of reptiles are normally filled with water, though their surfaces are certainly moist. Unlike the situation in fishes, the chemical susbstances reacted to are not carried into the organs by water; rather, the divided tongue in monitors and snakes becomes the transporting mechanism in at least certain circumstances (see Burghardt, 1970).

Each tip of the tongue receives slightly different information by means of particles adhering to its surface, and the particles from each tip are apparently transferred to the right and left members of the Jacobson's organ, respectively. It is clear that if the intensity gradient of the chemical stimulus is sufficiently steep so that there can be an appreciable difference of intensities at the divergent tongue tips, the ora can turn and move toward the source of stimulation. However, steep odor gradients normally occur only in the vicinity of the carcass. Generally comparisons must be made between two points farther apart than the tongue tips. It is under these circumstances that the ora swings its head from side to side or moves in an undulating path, with comparisons made between the two tips at the ends of the swings. The accuracy with which some reptiles trail prey is notable. Thus Brock (1977) reported that rattlesnakes are able to differentiate between the trails of envenomated rats and the trails of other rats laid down over them.

The accuracy with which oras follow a scent trail was tested in the field a number of times. In most instances a goat carcass was dragged across the surface of the ground for approximately 300 m, the wind direction noted, and the exact path of a trailing lizard plotted in respect to the goat drag on the basis of the ora's tail in the powdery volcanic ash. The results show that oras are able to follow such drags extremely accurately over long distances. On May 17, a large, nearly 3 m lizard (19W) followed such a drag for approximately 2 km and would, we believe, have done so for much greater distances if we had extended the drag. The extent and direction of the deviations from the drag trail are largely determined by wind direction and velocity and are usually less than 0.7 m, or approximately the width of the large lizards (Fig. 9–2). In any event, the lizards could hardly have lost the trails.

When a rotten goat was used to produce the trail, the lizards followed the bait in the correct direction a significant number of times ($N = 21$ trials, $\chi^2 = 0.02$).

2. Kahmann (1939) believed that the nose of the varanid is as important in searching for an olfacient as are Jacobson's organs, but this interpretation is not accepted by all workers (see Burghardt [1970] for a review of chemical perception in reptiles).

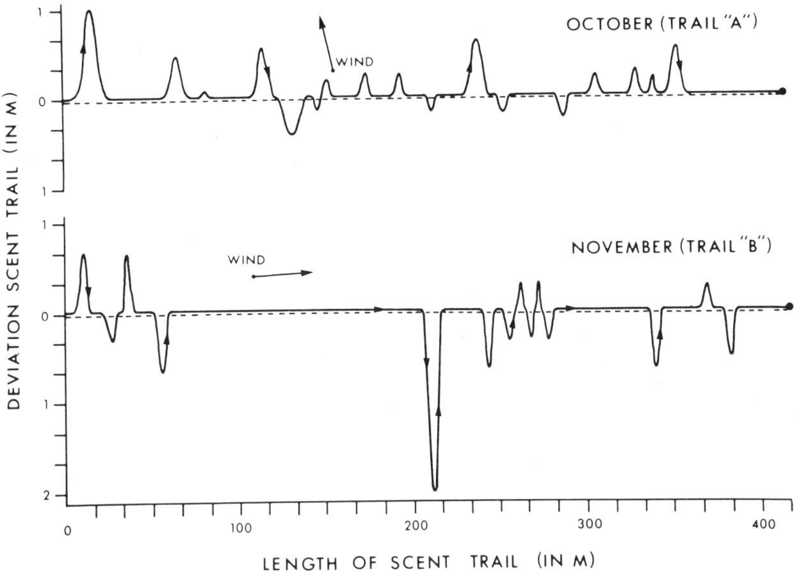

Fig. 9–2. Ora trailing deviations from scent trail produced by dragging dead goat over the ground (dashed line). In reality, the trail was very sinuous; here it is straightened to simplify graphics and interpretation.

Presumably they were able to determine the direction of the scent gradient quite easily, much as a hound might in following a deer scent. However, with recently killed goats the lizards often followed the drag in the wrong direction (N = 13 trials, $\chi^2 = <0.05$). The ability to follow a drag trail accurately is extremely important to the lizards when other oras or feral dogs have already located and started to drag a carcass away from the center of the diffusion gradient established in the air. Young oras may use certain of the carrion scents to mark a trail when they are chased from a carcass so that it can be followed back to the food.

However, most carrion is not found by the oras locating a drag, but by casually walking a rather convoluted path along game trails, trails of their own making, and other natural pathways such as dry riverbeds. The general aspect of the early part of the search pattern is such that directional changes occur irregularly, so that the oras move toward the carrion in a series of largely nondirectional convolutions. It is, of course, not clear whether these early turnings are truly random, or if the monitor is turning toward greater scent concentrations due to directional wind changes. If movement changes are completely random, there is no reason why the majority of oras moving toward a bait should not sometimes turn and walk in the wrong direction. Clear trails laid in the volcanic ash showed that this was more common the greater the distance between the ora and the bait (Fig.

9–3). Thus I conclude that more rapid directional changes are likely to occur when the monitor is moving into a weaker scent field than when it is moving into a stronger one (Table 9–3). This results in the general drift-like movement toward the bait that seems so obvious to an observer in the field but for which it is normally so difficult to obtain conclusive data. Therefore the movement of oras toward a carcass represents a typical case of klino-kinesis, similar to that described by Fraenkel and Gunn (1940) for other lower vertebrates.

Accidentally and rarely ingested food.—Almost all the food ingested by oras is either animal flesh or animal products. Very little vegetation is found in the gut or in droppings. Most of that found in either originates from the stomach contents of ungulate prey animals (though the stomach and intestines are usually deliberately emptied of their contents before being eaten). In addition, much vegetation is sometimes ingested during feeding—particularly if the flesh becomes covered with fallen leaves or dry grass. Parry (1932) reported Assamese native tales that *V. salvator* (or *V. bengalensis*?) feeds on melons, cucumbers, and rice. Sundanese villagers insist that oras eat only flesh. On several occasions we deliberately offered the animals a variety of vegetables and fruit. Boiled rice was eaten several times by small- to medium-sized oras. Papaya, breadfruit, mango, and various citrus peelings were often carefully investigated but never eaten.

Soil is frequently ingested accidentally by oras while they are feeding. Many of the droppings, particularly those lacking animal remains, are often composed

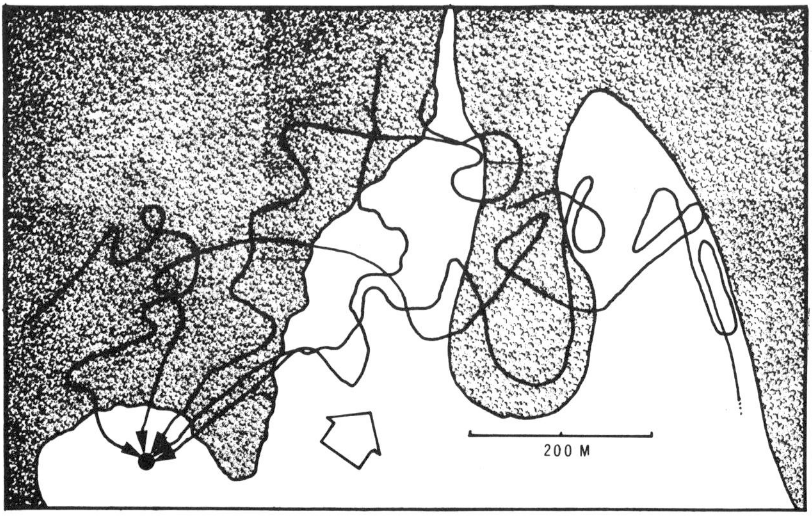

Fig. 9–3. Ora movements (lines with arrows) are less convoluted closer to than farther from carrion (solid circle).

Table 9–3. Number of directional changes with distance from carrion

Distance from carrion (m)	Number of directional changes (>45°) per 100 m	Number of oras
1–100	12	110
101–200	31	45
201–300	46	13
301–400	108	9
400+	201	11

completely of earth. Von Rathgen (1894) believed that small stones may be important in the digestion of prey by *V. griseus*. Small pebbles and coral were only rarely found in ora droppings; larger "digestion stones," as have been reported in other large reptiles, have not been found in either the stomach or the droppings of oras.

Feeding technique.—Darevsky and Kadarsan (1964) stated that varanids are the only living reptiles that *cut* their prey into sections before swallowing it. Their statement is correct, with the exception of turtles. However, turtles snip out pieces of flesh with sharp, enlarged, scissor-like scales, not with their teeth. The peg-like teeth of crocodilians are used to clamp onto the flesh of large, dead prey, which is then torn loose by twisting off sections. Of all the varanids, *V. komodoensis* is most adept at cutting pieces of flesh out of its prey. Several structural features allow oras to feed in this unique but very efficient manner.

Normally oras begin feeding by simply ripping open the body wall and pulling out the intestines; at other times some of the body wall is eaten first. The stomach and intestines are usually violently slung from side to side, scattering their contents (which are never intentionally eaten) over the surrounding area. The diaphragm is then pulled out and eaten. This is followed by the lungs and heart, which are obtained by thrusting the head deep within the chest cavity from the opened ventral end. The viscera are bolted down rapidly, accompanied by a hollow clapping sound produced by the snapping jaws. The ora usually cuts away the remaining portion of the body wall by biting into the flesh and rocking its body forward and backward on its braced legs. All of its weight is thrown into each backward movement, so that the teeth saw and tear through the skin and flesh of the prey. Burden (1928) described the feeding correctly: "It is remarkable to see the ease with which a piece of tough hide can be cut out of the flank of a dead water buffalo by these [lizards]. When a piece of flesh has been detached through the see-sawing motion . . . , the whole slab is gulped down regardless of size. As the morsel descends, the skin of the neck becomes distended. The beast licks his chops with his thin yellow tongue, rubs both sides of his face on the ground, as if to clean it, and lifts his head, the better to observe the landscape." The statements of Piazzini (1960) that the teeth can "crunch" through bone or by Pfeffer (1965) that one can hear the bones "cracking" are without foundation.

The size of the pieces swallowed and the amount eaten have been reported as somewhat astonishing by most observers. Some of these statements are obiously exaggerated. Burden (1927a) mentioned that "the whole half of a buck had been swallowed at one gulp," and in a later paper (1928) he reported seeing an ora swallow the entire hindquarters of a deer "at one gulp—hooves, legs, vertebrae and all." The first is physically impossible, but if in the second Burden meant *a* hindquarter he was probably correct, for I have often seen oras swallow the entire hindquarter of a full-grown goat. We once observed a 250 cm ora (24W) swallow an entire 15 kg hog after it had been eviscerated, and 19W was once seen swallowing an entire month-old fawn without eviscerating it. Hoogerwerf (1954a) reported finding the head of a full-grown wild boar in the stomach of a large ora, and in another statement he described a 3 m ora easily swallowing a 32 cm wild boar's head. In one large mass regurgitated by an ora we found the skull and lower jaws of a young wild boar. Elsewhere in this publication I have mentioned the head of a still hornless water buffalo calf being swallowed. On a number of occasions we watched oras 8W, 19W, and 24W swallow the entire head, neck, and thorax of adult goats. We once saw ora 23W, who weighed about 40 kg at that time, eat most of a 30 kg wild boar, including the entire head, in a very short period (Fig. 9–4). Though food is eaten remarkably quickly, Pfeffer's statement (1965) that an entire deer was eaten by several oras in 10 minutes is probably an exaggeration. Darevsky and Kadarsan (1964) stated that four large oras ate a large deer in 3–4 hours, which is in accord with our observations. Our studies on Komodo and those of Piazzini (1960) and Pfeffer (1965) on Rintja indicate that several adult oras can usually devour a large water buffalo of about 1200 kg in 3–5 days.

Only rarely does a more normal goat- or wild boar-sized carcass last for more than several hours when oras are allowed to feed at a normal pace in the shade. In the direct sun it may last much longer. Group feeding does not reduce consumption time significantly, for much activity is then devoted to various social interactions around the carcass. The average amount of time spent at wild boar and goat carrion by individual oras (N = 103) per day is 80.8 min (range <1–390 min). This is much longer than the 15–20 min reported by Pfeffer (1959) on the basis of his several observations on Rintja. Of this period, the mean time devoted to actual feeding is 6.9 min/day, or 12.4 percent of the period they are within sight of food. The remaining time is expended in social and thermoregulatory activi-

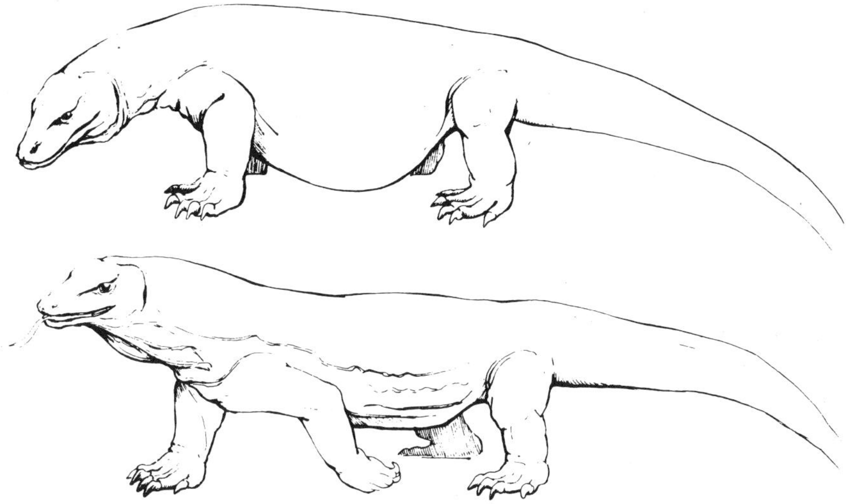

Fig. 9–4. Adult female ora (23W) (42 kg) before (*bottom*) and after (*top*) eating a wild boar carcass of 30 kg in 17 minutes.

Table 9–4. Ora overall range and $\overline{X}$ percent of total time spent eating while oras are at carcass

Ora length (m)	(N = 86)	$\overline{X}$ percent of total time	Overall range of percent
1.0–1.4	13	6.9	0.0–40.0
1.5–1.9	27	7.5	0.0–72.0
2.0–2.4	21	8.7	0.0–69.4
2.5–2.9	11	20.1	0.0–62.5
3.0+	14	18.9	0.0–80.0

ties. Larger animals spend a considerably longer time eating carrion than do smaller ones (Table 9–4).

Pfeffer (1959) reported that a 2.2-m ora on Rintja ate about 3–4 kg in one day, and another about 2.8 m long ate approximately 20 kg in 4 days at a deer bait. Our data show that mean food intake rate for groups of average-sized oras (with 17 groups an average of 7 individuals/day; $\overline{X}$ = 2.7/hr) is 2.4 kg/min (range <1.0–5.1 kg/min). Large, single oras consume flesh at intake rates of from 1 to 2.5 kg/min. Few comparable data are available for other predators, but data provided by Schaller (1967) show that intake of a large tiger is about 3 kg/min and about 0.1 kg/min for a leopard of about the same weight as the ora. A typical feeding sequence is well illustrated for ora 19W in Table 9–5.

Rapid feeding by oras is probably a strategy to forestall appropriation by other individuals, a condition operating interspecifically among the larger African carnivores (Schaller and Lowther, 1969).

As stated above, each sideward or backward jerk of the head usually produces a deep slice in the flesh. The number of slices required to free an edible portion of the carcass obviously varies greatly, depending on many factors, which include the feeding peculiarities of individual lizards. In one feeding sequence, ora 19W had a tearing rate of 9 backward jerking movements per 12.5 sec, or 56 tears per min (Table 9–6). The maxillary is used much more frequently than the dentary, which becomes a more or less fixed point around which the maxillary rotates.

Hunger will drive oras to eat food that is low on the scale of preference or that lies in an unfavorable site. I have seen oras eat the slipped horn sheaths of rotten goats when all the flesh had already been eaten by other oras. I presume this is how kerabau horn sheaths appear in pellets cast by oras. Deer antlers are rarely eaten, though short ones are sometimes swallowed with the head. Hair is usually spit out when it slips from the rotten carcass. However, when attached to large

Table 9–5. Typical feeding sequence of large ora (19W) eating 25-kg goat (observed March 14)

Time	Remarks
940	Ora kills goat; cuts away and feeds on body wall; tears out intestines.
942	All flesh of one hind leg eaten.
943	Exposed hind leg bones torn off and swallowed.
948	Remaining hind leg and pelvis swallowed in one piece.
955	Lumbar area torn loose from carcass and swallowed entire.
1006	Head, neck, front legs, and thorax swallowed in one piece, headfirst, after initially attempting to do so against the lay of the hair.

Table 9–6. Typical flesh-tearing sequence per minute by ora 19W

Number of tears with right maxillary	Number of tears with left maxillary	Amount of flesh removed
7	4	About 5 cm diameter
11	2	Nothing
1	1	Nothing
0	1	Nothing, but tried to swallow entire head of boar
4	4	About 5 cm diameter
0	19	About 7.5 cm diameter

pieces of flesh it is most often eaten. Bones, even when clean of flesh, are usually swallowed if not too dry. Mummified carcasses are rarely eaten. Our observations, as well as those of Burden (1928), DePrato (MS), and Pfeffer (1965), show that food regurgitated during stress is eaten by either the same individuals or by others. The claws are occasionally used to scratch at the carcass hide, usually to pull the slipping hair off a rotting carcass before the ora begins to eat it. Claw use is discussed more fully below.

Disturbance by man, interference by village dogs or other oras, or the unsuitable location of carcasses often prevent oras from feeding, even when it is the kind of food they highly prefer. Deer, boar, and goat carcasses are the foods most eagerly devoured by large oras. As in other scavengers, old, mummified carcasses are of low acceptability. Thus, dried carcasses are not an important reserve food supply, as they are in other scavengers (Koford, 1953).

Wastage.—Other than the trampled grass, the only evidence that oras have eaten a carcass are a few wisps of hair and the spilled stomach and intestinal contents. Approximately one-third of the latter are usually accidentally ingested. Under these circumstances the percent of the total weight of the uningested part of a boar carcass (wastage) varies from approximately 8.75 to 9.25 percent (Table 9–7). If the feeding oras are all rather small, however, the skull and its contents remain uneaten, and the total noningested weight varies from about 12 to 13 percent (Table 9–7). We were unable to obtain comparable data for Komodo deer, *Cervus timorensis*, but in three *Odocoileus virginianus* the stomach and intestinal contents varied from 9.0 to 10.3 percent of the total weight. Uneaten heads of deer would represent much less of the total weight than the proportionately larger heads of wild boar. Thus, the wastage factor in deer is roughly equivalent to that of boar. However, in kerabau the wastage is proportionately much higher because oras do not swallow as much of the skeleton. Estimates place the total uneaten percentage of an adult kerabau carcass (presuming it is not shared with dogs or others) at about 30 percent. The latter is equivalent to wastage percentages in large mammal predators (20–46 percent, Table 9–8). This higher proportion is within the wastage estimates of deer carcasses eaten by feral dogs on Komodo and Padar. In condors the waste factor is higher still. From Koford's (1953) data I calculate the minimum nonutilized amounts are about 24 percent but perhaps as much as 30 percent, since he stated that the viscera were not eaten and then failed to add their weight to the total of the skeleton and hide remaining when the scavenging birds are finished. This same criticism can be leveled at many other prey utilization estimates for large predators, suggesting that many percentages given are too low. Schaller's (1967) data, though few, are of the type and accuracy needed for such analyses. From them I conclude a wastage factor of from 32 to 37 percent when tigers prey upon medium to very large ungulates. In the African wild dog (*Lycaon pictus*) the percentage of carcass remaining after feeding is often as high as 40 percent (Estes and Goddard,

Table 9–7. Wastage factor analysis for Komodo wild boar *(Sus scrofa)*

Total boar weight (kg)	Hair weight (kg)[a]	Head weight (kg)[b]	Rumen and intestinal contents weight (kg)[c]	Tooth weight (kg)	Waste if head eaten (%)	Waste if head not eaten (%)
46.56♀	0.16	3.73	4.43	0.56	9.20	12.13
35.51♂	0.14	2.96	4.00	0.50	9.05	12.44
27.27♀	0.07	2.04	3.56	0.31	8.83	12.58
48.13♂	0.21	4.03	5.01	0.63	9.13	12.31
36.03♀	0.16	3.01	4.17	0.53	9.04	12.49
43.73♂	0.17	3.58	3.92	0.53	9.27	12.07
23.11♀	0.08	2.02	3.31	0.26	8.71	13.00

a. Animal skin macerated and hair later dried and weighed.

b. Skin removed, since it is usually eaten.

c. Two-thirds of actual weight, since about one-third often accidentally ingested.

Table 9–8. Approximate feeding waste (in %) of various large, wild predators

Predator	Usual prey	Approximate $\overline{X}$ carcass not eaten (waste)	Source
Ora	Deer, boar	8.5–13[a]	This study
Puma	Deer, elk	30	Honneker (1970)
African lion	Impala	30	Pienaar (1968)
Leopard	Goat	46	Schaller (1967)
Wolf	Deer, moose, caribou	20	Pimlot (1967)
Tiger	Kerabau	32–37	Schaller (1966, 1967)
African wild dog	Gazelles	40	Estes and Goddard (1967)

a. As high as 30 percent in unusually large prey (kerabau and horses).

1967). Pienaar (1968) reported that lions and cheetahs characteristically leave large amounts of prey when they finish eating.

Thus, except for water buffalo carcasses, the ora has a much lower waste factor than most other large-prey predators studied, mainly because oras tend to swallow all the bones. Even pure scavengers, such as condors and vultures, usually leave most of the skeleton intact (Koford, 1953). Hyenas often smash the long bones to obtain the marrow and frequently eat the rib ends (Pienaar, 1968). I conclude that, with the exception of the large snakes, no other known predator or scavenger exhibits an equivalently high ungulate prey utilization ratio as the ora.

The prodigious appetite, high feeding rate, and low percentage of waste are undoubtedly all important factors in the competitive aspects of scavenging in the islands on which oras occur. Together they could easily account for the lack of other scavengers, such as vultures, in an area that might otherwise be expected to possess them.

Feeding adaptations: teeth.—Of all the feeding adaptations, perhaps the most important are the teeth. Their general shape and disposition have been described by Burden (1928) and Mertens (1942). They bear a greater resemblance to those of flesh-eating carcharid sharks and carnosaurian dinosaurs than to the teeth of living reptiles or mammals.

Perhaps the most obvious adaptation is that ora teeth are more laterally compressed than those of any other lizard known (ratio greatest thickness/greatest length, 1:3.1). The teeth of *V. varius* are similar but are less recurved laterally and flattened externally. The maxillary teeth of oras are more recurved than those on the dentary and particularly the premaxillary. The longest maxillary teeth are directed slightly laterally, whereas those on the dentary are nearly vertical. This may be related to the position in which the head is held while the lizard is slicing off a piece of flesh. In lateral view the outer form of the *tooth row* is specialized in its general shape, which is essentially convex, in the form of a curved scalpel blade (Fig. 2–4). Thus when the maxillary or dentary are drawn backward, each tooth that comes into contact with the flesh cuts deeper than the one before. In crown view the cutting edges of the teeth are aligned at an angle so that the flesh will not bind between the teeth, with each cut section being forced lingually (Fig. 2–5).

All the teeth are further adapted to cutting and tearing flesh by having fine serrations on most of the posterior edge and on a small section of the anterior edge near its tip (Pl. 1). Further details of the morphology of the teeth are provided in Chapter 4.

The gum covering the tooth base is spongy and thick, frequently covering at least two-thirds of each tooth in life. When oras bite into flesh the gum is pushed back from around each tooth, exposing much more of the tooth. The continued movement of the gum back and forth over the sharp anterior and posterior edges of the teeth during feeding usually causes it to bleed, sometimes profusely. When the mouth is opened after considerable feeding in particularly large oras, the saliva and blood in the mouth are mixed so that the saliva appears quite thick and pink or even red in color. The combination *may* have some significance as a bacterial culture medium (Chap. 11). The high salivation rate in oras when they are feeding is undoubtedly of considerable importance in lubricating the astonishingly large sections that are often swallowed entire.

Feeding adaptations: skull.—The most important feeding adaptation in the skull of squamate reptiles is the presence of several movable joints. Of all lizards, the skulls of varanids are perhaps most flexible. It is this extreme flexibility, functionally related to mobility planes as in the skulls of snakes, that allows oras to swallow such ill-shaped and large pieces of flesh and bone. The major movements of the bony elements in the skull of *V. komodoensis* are shown in Figure 9–5. As far as is known, all are related to feeding problems.

In varanids the dorsal surface of the skull has three major hinge planes (Frazetta, 1962): (1) prokinetic—between the nasal and frontal bones—a hinge often well developed in *V. komodoensis*, usually more so in older than in younger specimens; (2) mesokinetic—between the parietals and frontals—a condition often found in specialized burrowing lizards but also characteristic of the platynotids; and (3) metakinetic—under the posterior edge of the skull roof at the juncture of the parietal and occipital bones.

De Jong (1927) reported that a functional metakinetic joint is absent in *V. komodoensis*. However, I found it in all nine *V. komodoensis* skulls I examined for this character (condibasilar length 11.7–20.1 cm). Since both metakinetic and mesokinetic joint planes are found in ora skulls, oras are *amphikinetic*, though the mesokinetic fronto-parietal hinge is apparently the more important

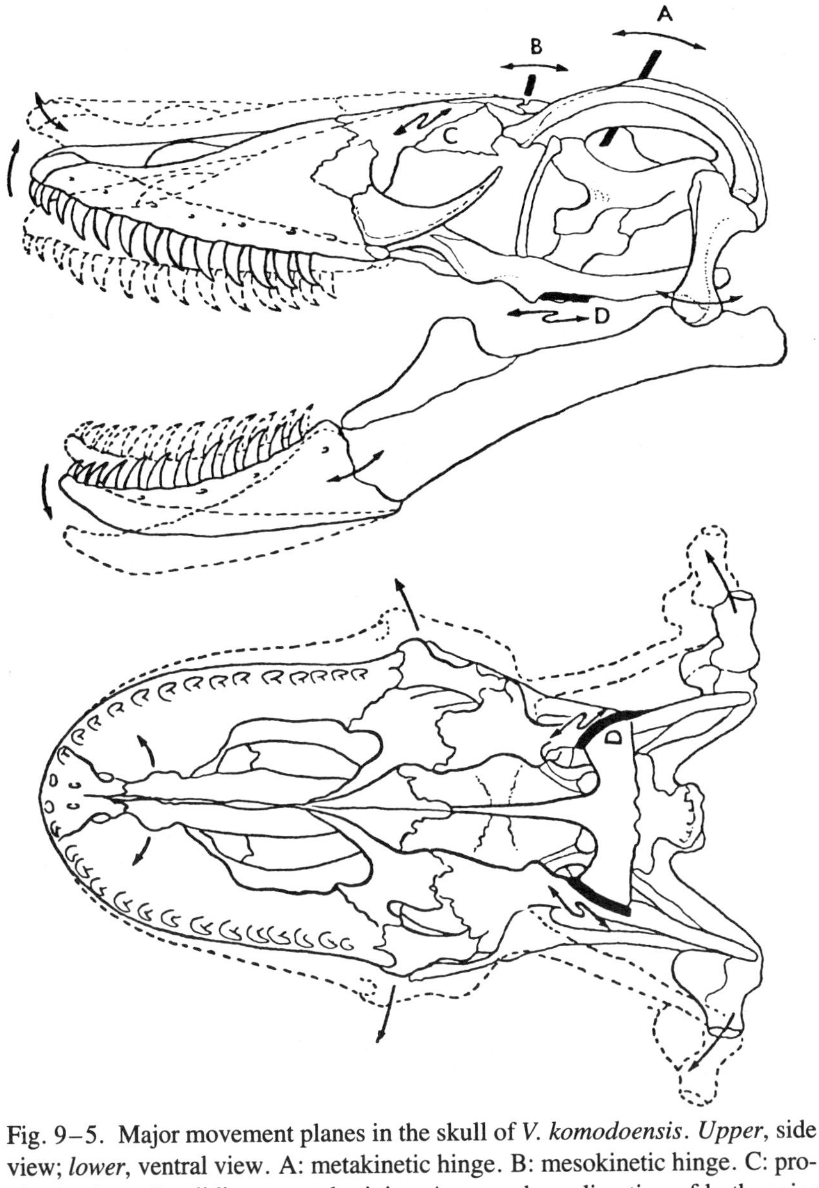

Fig. 9–5. Major movement planes in the skull of *V. komodoensis*. *Upper*, side view; *lower*, ventral view. A: metakinetic hinge. B: mesokinetic hinge. C: prokinetic plane. D: sliding expander joint. Arrows show direction of both major and minor skull movements during feeding.

one. Two other movable joints of the skull are undoubtedly feeding adaptations: that between the pterygoids and the sides of the braincase, allowing the former to slide backward and forward (Fig. 9–5 D) and the extremely movable quadrate. Both have a considerable range of independent movement. Quadrate mobility is apparently particularly important in movement conferred on the lower rather than on the upper jaw and allows an increase in the gape (Robinson, 1967). It also allows the lower jaw to be shifted backward and forward, helping to pull large pieces of flesh into the mouth. Transverse widening of the esophageal passage in *Varanus* (including *V. komodoensis*) is accomplished by a tendency during retraction for lateral movement of the basal palatal units and the quadrates.

In *V. komodoensis* still another device for enlarging the passage into the mouth is the vertically oriented intramandibular hinge in the lower jaw. Frazetta (1962) correctly stated that while externally the hinge-axis corresponds approximately to the dentary-subangular line of contact, lingually the axis passes through the splenial and part of the angular. This type of joint allows outward bending of the middle part of each ramus, with the result that larger pieces of flesh can be manipulated and swallowed.

The details of kinetic movements of all the joint planes in the skull and lower jaw are extremely complicated, and there is disagreement about their precise nature, particularly in the upper jaw. In *V. komodoensis* the major movement witnessed in large, feeding specimens is the slight flattening of the front part of the head and the raising of the maxillary segment as the jaws open.

The functions of most of the planes remain largely speculative. Many workers (e.g., Bellairs, 1970; Frazetta, 1962) believe that of all the movements, those of the palate are most important and that the other planes are simply meant to facilitate them.

One generally attributed planar function is to increase the gape, allowing larger prey or pieces of prey to be eaten. However, Frazetta (1962) maintained that the gape of *Varanus* is decreased rather than increased by kinetic movement, because when the maxillary segment is lifted the lower jaw must be raised as well by the forward movement of the quadrate. On the other hand, Impey (1967) maintained that varanids obtain an additional 10–15 percent gape by virtue of their kinetic movements. Parrington (1955) suggested that the gape of an animal depends on the inclination of the quadrate. For example, if inclined anteriorly from the lower jaw the gape will be increased; if inclined posteriorly the gape will be decreased—both will occur provided the retroarticular process and quadrate are held in the same relation to each other. The former modification to increase the gape has been adopted by many snakes. Frazetta (1962) studied *V. indicus*, a species in which the quadrate slants backward, and this explains his conclusion that kinesis decreases the gape. Small- to medium-sized oras are identical to *V. indicus* in this regard and the effect on the gape should be the same.

However, the quadrate in large oras is often nearly verticial (Fig. 9–5), suggesting the condition in snakes. Thus it is possible that the gape of adult oras is increased by the same kinetic mechanism that restricts the gape in younger specimens. I hope, with Bellairs (1970), that the matter will be resolved by further study.

Frazetta concluded that kinetic movements in the skull produce adjustments allowing the alignment of the upper and lower tooth rows so that prey can be firmly gripped by the maximum number of teeth, and the real significance of kinesis may lie in the assistance this gives in the task of shifting the prey to the back of the mouth where it can be swallowed. This can be accomplished by a process which Gans (1961) called inertial feeding; with the prey remaining relatively static and the jaw shifted over it in a succession of snaps, each accompanied by a quick forward dart of the head. The prey, or part of the prey, may be worked from side to side during the operation. Impey's work on *Varanus* and our observations on *V. komodoensis* make it clear that the inertial movements are very complicated, since the head is first jerked sharply backward, pulling the prey with it, and the jaws are then thrust forward over it in the opposite direction; this cycle of movement is repeated until the food is disposed of. This is the movement that causes the snapping or popping sound made by the closing jaws when *V. komodoensis* is feeding on relatively long, small-diameter food items, such as intestines. As utilized by the ora, inertial feeding is much more rapid than in most lizards, and its efficiency may be enhanced by kinesis, though movement in the skull is not absolutely required by this method of feeding.

In some species of varanids the head is turned from side to side in such a manner as to advance the jaws over the prey first on one side of the head and then the other (Gans, 1961). Oras have not been seen feeding in this manner. However, when large oras tear flesh from a carcass, they usually move the head forward on only one side. At this point they close the jaws slightly and then jerk backward swiftly and forcibly on the same side, causing a great wound in the form of an arc. Because of the slightly canted orientation of the maxillary teeth (Fig. 2–5), as well as the great relative distance between the midline and the tooth row when compared to other, narrower-snouted varanids, the wound produced is one in which each of the more anterior teeth follows closely the incisions made by the more posterior teeth as they move caudad. Trial slashes made in clay clearly indicate that in younger, triangular-headed oras when the head is drawn back in an arc with the largest maxillary teeth on one side as the pivot, almost every tooth cuts just barely external to the previous one. Each tooth slices the flesh somewhat independently of the others. However, in wide-snouted adult oras the radius of the arc produced is greater, so that the teeth follow one another more closely, and the cutting of each tooth is less independent and a deeper wound is produced.

The utility of the uniquely broadened snout of adult Komodo monitors becomes obvious when compared to the narrower-snouted species, such as *V. salvator.* Presuming that the pivot point of the cutting arc is located at the position of the longest teeth on one side of the upper jaw, the arc produced by the cutting teeth of the opposite side as it is pulled posteriorly is much greater in *V. komodoensis* than if the same cutting technique is used by a narrow-snouted species. The teeth tend not only to follow one another more closely in *V. komodoensis*, but when slices are alternately produced by the right and left sides of the jaw, a larger piece of flesh is cut out of the prey by the Komodo monitor than by other monitor species of equal size.

Oras swallow their prey by means other than this combination of slashing and inertial feeding with kinesis. One very simple but quite effective way is by pushing the prey against the ground and so forcing it into the mouth.

Frazetta (1962) suggested that the independent mandibular depression and muzzle elevation in lizards may permit the jaws to operate close to the ground when capturing prey, where otherwise the skull would have to be raised to provide space for mandibular depression. However, in *Varanus* this is not a major problem, since the long neck holds the head well above the ground surface. Small terrestrial animals are attacked essentially from above by at least the larger species of *Varanus*. Ungulate prey are not close to the ground, and so the problem has no relevance in the ora.

Frazetta's point that the functional interdependence of both the upper and lower jaws assures at least some simultaneity of movement is well taken. He reasoned that it permits the two elements to close on the prey at nearly the same time. Furthermore, the rotatory movement of the upper jaw, which he believes changes the orientation of the tooth tips with respect to the prey, helps to impale the prey when the upper jaws first strike the food. As the jaw is rotated downward, the tooth tips are turned caudally to restrict escape movements of the prey. However, this explanation does not suffice in the case of oras, since their teeth are much too recurved to impale the prey. The curvature is probably not to restrict escape movements of the prey but rather to cause a massive wound as the teeth move through the flesh.

Feeding adaptations: feet and claws.—The powerful feet and claws of oras are often said to be important in feeding (e.g., Mertens, 1942; Darevsky, 1963). One Flores villager stated that during an attack oras use their front limbs to grasp and hold the prey while they bite it. Sumbawa villagers, completely unfamiliar with the ora, sometimes maintain that prey is killed with the large claws. There is no substantiating evidence for either tale. Zander (1895) and Thilenius (1897) stated that the front feet of *V. griseus* are used to guide food into the mouth.

The front feet and claws are often used in digging burrows, in climbing, and in courtship. In regard to feeding, the claws of oras are sometimes used in excavating rodent burrows and megapode nests (Pl. 5), digging into rotten stumps, and turning over refuse such as bark, rocks, and leaves in search of small prey. The

Pl. 5. The nests of the megapode bird (*Megapodius freycineti*) are regularly excavated by oras searching for the buried eggs.

front feet and claws are often used to scrape the slipping hair from rotting carcasses before starting to feed. Sometimes one, rarely two, of the front feet are used to hold carrion down while the monitor tears off the flesh. This seems more common in small than in large oras. Attenborough (1957) reported that a large lizard used a foreleg to tear the intestines out of a dead goat. We have occasionally seen both large and small oras do the same during our studies.

Feeding adaptations: legs.—The legs of particularly large specimens are proportionately stocky; those of young specimens are much more slender. The great development of the musculature of the brachium is of considerable importance when large specimens brace their feet far apart and rip at the carcass with powerful backward jerks of the entire body.

Ingestion Behavior

Direction of prey ingestion has been shown to be important in the feeding strategy of *V. bengalensis* by Loop (1974), as it is in *V. komodoensis*. Both shape and hair direction are important factors in efficient feeding of large pieces and/or prey. Proportionately small prey are swallowed either headfirst, tail first, or even sidewards. We witnessed 19W fold a dead young piglet double as it was swallowed, and a month-old fawn was swallowed tail first. However, most entire mammalian or avian prey are ingested head first. Entire limbs are usually swallowed small end first. When doubled they are swallowed only with great difficulty. Comparative studies of *V. komodoensis* in the field and *V. bengalensis* in the laboratory show that the former is capable of swallowing much larger entire prey and/or pieces than the latter.

Drinking.—There is very little useful information available on drinking in varanid lizards, except that they can sometimes go for long periods without water. Tofahr (1908) reported that a specimen of *V. griseus* was without food or water for months with no apparent harmful effects. Father Schmutz (in a letter) reported that an ora was emaciated but alive after having remained in a wire snare without food or water for 1.5 months. Fat storage in the tail of several species (Boulenger, 1898; Mertens, 1942), as well as in the extra-peritoneal area (Mertens, 1942; Darevsky and Kadarsan, 1964), may be important in producing metabolic water during drought. On Komodo the dry period is 8–10 months long, so that water conservation is extremely important for animals living there, particularly those inhabiting low, coastal areas. Here the problem must often be very acute, for standing surface water is sometimes rare even during the wet season.

Oesman (1970) saw these lizards touch their tongues to the surf on the beach and so presumed that they sometimes drink sea water. However, whether they were drinking remains unproven. Most platynotid lizards and snakes suck up water while immersing their heads in it, and usually do not use their tongues to lap it up as do certain other lizards and many mammals. On the other hand, some of my captive *V. bengalensis* did lap up small amounts of water with their tongues. Lederer (1942) stated that his captive oras always sucked up water. During our 11 months camped next to the bay on Komodo we saw many oras on the beach; most of them flicked their tongues at both the sand and the water, but none *drank* sea water. The oras drank only fresh water, usually from the isolated water holes produced by wild boar or kerabau. When drinking, their heads are usually immersed—often up to their eyes. In this position they suck great quantities into the mouth and throat, their masseter muscles noticeably knotting periodically with each sucking movement. After a short while (.25–2 min, according to Lederer, 1942) the head is held up as a chicken does in drinking, letting the water run down the throat.

The first standing water we found at low elevations on Komodo after we came to the island in July was in January. Throughout the rainy season water remains in shallow hog wallows for only several days after each rain and several weeks in the deeper wallows. In our Komodo study area there were only eight hog wallows from which oras might drink (Fig. 1–6). Even after heavy rains, water in the creek beds at low elevations rarely flows more than a few hours—usually only a few minutes. Flash flooding is common. Between flow periods a few shallow puddles are often found near some of the creek mouths. In the foothills these same creeks sometimes contain small, isolated, semipermanent potholes several feet deep. These are used by water buffalo, deer, and boar living nearby, and undoubtedly by the oras as well (though population densities are often low at these elevations). In the middle and particularly near the end of the long dry season, these pools are often dry or unpotable. Water scarcity is sometimes the cause of much anxiety on the part of the human population, and several previous expeditions have faced serious water problems (Hoogerwerf, 1954a). At elevations above 500 m there are a few relatively flat areas containing shallow ponds up to 30 m across. Most disappear during the height of the dry season. They are most common in the valley leading to Sabita.

Padar possesses three small permanent fresh water springs, all obviously important to all the larger resident vertebrates, including the ora. Water resources on Flores are much greater than on any of the other islands on which the ora occurs.

Normally, large oras drink deeply after heavy feeding. While protein digestion produces water (particularly important to oras in the dry season), it also requires water. My notes of February 21, 1970, read, "19W left the bait area after having eaten almost all of an approximately 15 kg wild boar. It moved across the savanna to the same wallow from which it previously drank water after feeding. It entered the water at 1603 hr and remained there, drinking deeply, until 1610 hr, when it moved toward hill 19. . . ." On April 20, a 1.2 m ora, temporarily maintained in captivity for activity studies, stretched itself on the ground during a small rain. It flicked its tongue at drops of water hanging on the metal sides of the pen but drank only from a small pool.

Digestion.—All the prey, except bones, teeth, etc., is usually completely digested. However, when feeding is particularly heavy, small bones and cartilage (both normally digested) are sometimes found in the droppings in various degrees of digestion. Unless crushed, eggs sometimes pass through the stomach and intestines without being digested (Lederer, 1942). The same authority also reported that ticks and mites are undigested. The digestive pause is from 3 to 5 days (Table 9–9), though seasonal variation of the type described by Fukada (1977) for the snake *Rhabdophis tigrina* is expected.

Fecal pellets.—The fecal pellets of captive oras have been described by Lederer (1942). They are distinguished from those of mammals on the basis of many characteristics, including color and consistency. Except for size, they are not easily distinguished from fecal pellets of other "large" lizards, such as the tokay (*Gekko gecko*)—even those of hatchling oras are considerably larger. As in all reptiles, the nitrogenous wastes of oras are excreted in the form of a white (occasionally slightly pinkish) semisolid (uric acid) at the same time as digestive residues. In hatchling and small monitors the uric acid is in the form of a small,

Table 9–9. Digestive pauses of ora 19W after known feedings at Loho Liang, Komodo, bait site

	Intake		Evidence		
Date	Carrion type	Amount (kg)	Date	Pellet type	Digestive pause[a]
Nov. 17	Boar	16.8	Nov. 22	Fecal	5 days
Nov. 20, 21	Dog	17	Nov. 24–25	Fecal	4–5 days
Feb. 1	Boar	12	(No evidence) Feb. 24	Fecal	Less than 23 days
Feb. 21	Boar	14	Feb. 25	Fecal	4 days
April 1	Goat	3	(No evidence) April 6	Fecal	Less than 5 days
April 2 (?)	(Killed live boar)	20	April 6, 7, 8	Fecal	4 days
May 5	Goat	8	May 10	Gastric	5 days

a. Minimum digestive period from feeding date until first evidence in scat.

pasty nitrogenous cap on the digestive residue. In larger oras the nitrogenous materials are more fluid, with the consistency of whitewash or a thin plaster mixture.

Following the release of this material, the digestive residue is expelled as several (1–4) pellets. These fecal pellets constitute the "droppings" of the analyses referred to throughout this publication. They may include many items that are rarely or never digested, such as hair, feathers, partly digested bones, hooves, teeth, claws, and the outer epidermis of reptile, skin and legs of bird, and bird and reptile eggshell fragments. Seeds from the crops of birds, rumen contents, and endo- and exoparasites are also not digested (Lederer, 1942). The droppings may also include earth apparently accidentally ingested. Sometimes the pellets are composed almost entirely of hair. When the stomach is empty, the droppings contain little or no digestive residue and only fine, sand-like granules of uric acid. These droppings were tabulated as "empty," the assumption being that the ora that had cast the pellet had not eaten for a long time.

When completely dry, the powdery uric acid is often quickly blown or washed away. The droppings themselves are fairly resistant to abrasion and may last well into the rainy season. However, by March the droppings in almost all the sites except the most sheltered are washed away.

Individual fecal pellets vary greatly in size. Eight randomly selected fecal pellets from western Flores have the following dimensions: 18.5 × 4.8 cm, 12.4

× 2.8 cm, 14.9 × 6.1 cm, 7.3 × 2.1 cm, 10.0 × 2.1 cm, 4.1 × 0.7 cm, 16.6 × 4.4 cm, and 20.4 × 6.7 cm.

Droppings are normally cast in the morning, after the lizard has awakened, basked (not always), and just begun to move out on the morning hunt. As a result, for at least large oras, the majority of the fecal pellets are found on flat hilltops near the thermoregulatory sites. In fact, such sites are often characterized by the concentrations of droppings nearby (Fig. 9–6). Fresh droppings of an ora that has recently eaten a large meal are often quite odoriferous. Thus, thermoregulatory sites can sometimes be detected simply by the temporary odors associated with certain hilltops. In spite of this, we repeatedly found deer resting on the hillsides remarkably close to habitually used thermoregulatory sites of large oras.

Droppings are also commonly cast in other open places, such as along game trails and dry river beds. These locations, their white color, and their scent all tend to make the fecal pellets conspicuous (Auffenberg, 1978). However, the droppings of smaller oras do not show the localized distribution of those of larger animals. This is largely because smaller individuals often change the location of their nocturnal shelter and tend to be scattered throughout their entire foraging area.

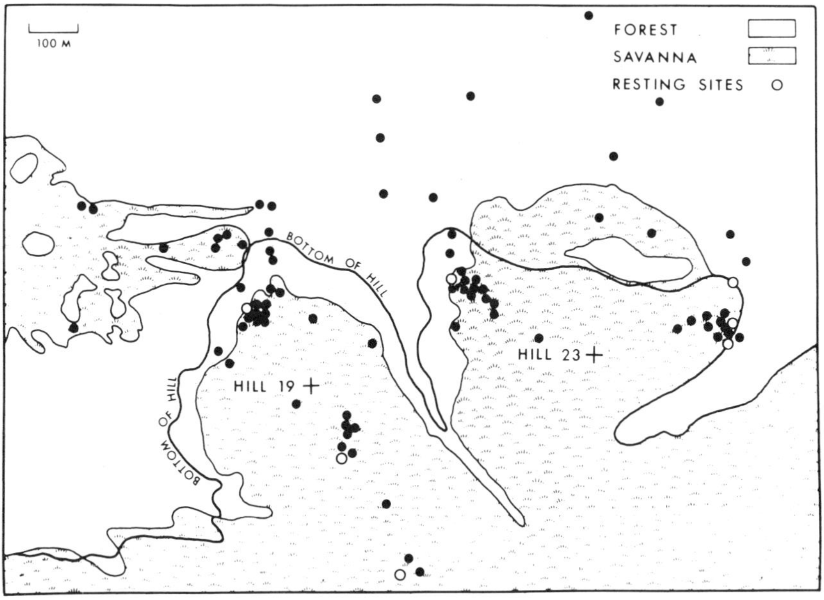

Fig. 9–6. Ora droppings (solid circles) are often concentrated near the ora hillside thermoregulatory sites (hollow circles).

Fecal pellets are apparently used for chemical communication between oras (Chap. 12).

Gastric pellets.—After feeding on large meals snakes often disgorge the undigestible portions of their prey in the from of a gastric pellet containing mainly hair, and occasionally teeth, claws, and feathers. This habit is also known for captive varanids (Hediger, 1934, for *V. indicus*; Lederer, 1942, for *V. komodoensis*). Gastric pellet formation in oras seems particularly important in large individuals. However, this conclusion may be based on the relative difficulty of finding smaller rather than larger gastric pellets in the field.

Ora gastric pellets are much larger, longer, and more uniform in diameter than fecal pellets. They are comprised almost entirely of feathers, hair, hooves, teeth, and some undigested portions of remaining bones and are never associated with the whitish nitrogenous powder covering the fecal pellets. They are rarely accompanied by any other material. The beetle *Trox* commonly feeds on ora gastric pellets after they are cast.

In the wild the major portion of each gastric pellet is hair, normally packed very tightly and aligned along the longitudinal axis of the pellet itself. Bones, teeth, and hooves are often found throughout the pellet. Large pellets occasionally contain items as large as the lower jaw and skull of young wild boar and horn sheaths of kerabau. On November 11, 1969, a single pellet of a large ora from Galo Matavelok, Komodo, comprised partly of hair and claws, contained the horn sheath of a water buffalo that was 42.12 cm in length by 7.42 cm in breadth. A large gastric pellet disgorged by 23W at the base of hill 19 on March 13 contained the remains of at least two meals: hair from a goat fed to this specimen on March 7, six days previously, and the hooves, neck bones, and tusks of an adult boar killed and eaten by this individual on March 2, eleven days previously. This pellet was 45.72 cm long and 10.16 cm in diameter. Smaller gastric pellets were commonly found in our study area on both Komodo and Flores.

All the gastric pellets we found in the Komodo study area were located along trails often used by the large oras, frequently at the base of the low hills with which individual specimens were often associated. Occasionally they were found in dry stream courses or on the valley floor. Most pellets seem to be cast in the early morning, apparently after the lizards descend to the valley floor in preparation for their afternoon hunting sequence. The matted hair, of which most pellets are comprised, is extremely resistant to separation and destruction and is sometimes "glued" together with a mucoid material. As a result, gastric pellets of large oras several years old can be found strewn about the surface, unlike the fecal pellets that are normally broken apart and washed away with the first rainy season.

The pellet is disgorged in a manner similar to that found in snakes. The lizard stands high on all four legs, with body and neck arched up off the ground and in line with the tail. The hyoid apparatus is expanded and the neck and the mouth opened wide. The neck and head are then thrown from side to side in a series of violent undulations as the stomach is pulled in. A small pellet may be thrown a great distance laterally. When a large pellet emerges from the throat, the lizard sometimes backs up and the undulation become less severe. Large gastric pellets are ordinarily cast in one, rarely two pieces. When fresh they are covered with a fetid mucous. After disgorging a large pellet the lizard normally rubs its face on the ground or vegetation and/or cleans the lips with its tongue.

10

The Ora as a Predator

Considerable confusion exists in the literature regarding the extent to which oras attack and kill living prey. That young oras probably capture and eat small living animals has never been contested, though few data were actually available (Table 10–1). It is the feeding of large oras that has generated considerable differences of opinion—often based on intuitive reasoning of what such lizards are probably capable of accomplishing. It is interesting that the earliest literature specifically emphasizes the utilization of large prey, which later remarks tend to refute (Table 10–2). This study makes it very clear that the earlier workers were correct in stating that large, live prey is often attacked, killed, and eaten. When combined, the earliest reports (Burden, 1928; Lallemant, 1929; de Jong, 1931) contain much of what we have since been able to establish firmly.

However, the entire question of the population dynamics of large predators and their prey species has been investigated only recently. Most of this work is based on the availability of prey remains or carcasses. These are used to determine the number and type of prey animals utilized by the predator. Of the recent studies of this type, the most significant in terms of numbers of carcasses examined is that by Pienaar (1968). This excellent summary of prey preferences of large predators in Kruger National Park, South Africa, is based on the study of approximately 43,000 carcasses. In the case of the ora, there is rarely enough of the carcass left to make its discovery possible. When finished with a carcass an ora leaves only a few traces—a smudge of blood, the intestinal contents, and often a few tufts of hair torn from the hide. Thus to study predation dynamics in this predator, an approach other than carcass analysis is necessary.

Fecal Pellet Analyses

The fecal pellets (droppings) of many species of *Varanus* have been reported to contain remains of prey organisms (Mertens, 1942). Those of oras are no differ-

Table 10–1. Records of the food of young *V. komodoensis*

Remarks on feeding	Authority
Captive hatchlings and young were fed small pieces of meat, eggs, and a few dead chicks.	de Jong, 1944
"...undoubtedly feed omnivorously [*sic*] on geckos, snakes, and bird's eggs, as well as on carrion."	Kern, 1968
"It is not impossible that small lizards...may discover a bird's nest and eat the eggs or the helpless fledglings."	Oesman, 1970

ent and often contain hair, teeth, claws, hooves, and occasionally bones (Mertens, 1942; Darevsky and Kadarsan, 1964). The few ora droppings examined by these earlier workers do not constitute an adequate sample of foods eaten. For this reason we collected 4,267 droppings of oras on Komodo and 120 on Padar during October, 1969. During June, 1971, we collected 341 on Flores. Vegetation is never deliberately eaten, though accidental ingestion is common (Chap. 9). Identification of vertebrate remains was made on Komodo after skeletal and hair or skin samples were obtained of all the land vertebrates on the island.[1]

More droppings contain remains of prey animals (N = 3,014) than are empty (N = 1,014), that is, comprised of only uric acid or earth. This suggests that starvation is not an important factor in ora biology. The distribution of prey individuals and/or species per dropping is a typical Poisson type (Table 10–3), with a maximum of five. This indicates that hunting or scavenging success is probably very low because the remains of one prey animal tend to be voided before a second is obtained. As in many of the larger *Varanus* species, the spectrum of prey animals is very broad (Tables 10–4, 10–5, and 10–6), providing a diversity index of approximately 1.51. More important, certain prey animals are common; others are not. The percentage distribution of the more common prey species in each of the samples is shown in Figure 10–1, arranged in three sizes of dropping diameters: small, from 1 to 2 cm; medium, from 2.1 to 3 cm; and large, >3.1 cm. No droppings in the largest category were available from Padar, since the oras there do not attain the large size of those on other islands.

Several conclusions are evident. When the droppings with diameters smaller than 2 cm are compared with those of the largest category, it is obvious that prey utilization is size dependent, with smaller droppings containing fewer remains of deer, pigs, oras, and water buffalo and many more of rats and birds. There are also more empty droppings among the smaller ones.

In addition to the droppings discussed so far, 15 were less than 1 cm and evidently cast by oras less than 1 m in length and less than 1 year. All these

1. The conclusions regarding the reptile and amphibian collection have been published in the *Bulletin of the Florida State Museum, Biological Sciences*, Vol. 25, no. 2.

Table 10–2. Literature records of the live prey attacked by adult *V. komodoensis*[a]

Remarks on feeding	Authority
"Deer and wild pigs are killed and eaten on Komodo...horses on Rintja"	Horst, 1926
"Deer and wild boar form their staple diet, though birds and eggs no doubt often make an addition to the menu....*Varanus komodoensis* deliberately attacks and kills these herbivores....a large lizard ...had, while chained to a tree,...jumped upon an old pony....According to the [Komodo ?] natives, wild horses are plentiful on the island of Rintja where they form the chief food supply of the giant lizard. I can hardly credit the tale, but it is interesting, if true. On Padar [they] live almost entirely on turtle eggs...."	Burden, 1928
"According to Rintja natives, they attack and kill horses on that island, though the prey are said to be primarily the sick or weak individuals. Lame wild pigs are caught along game trails. Most prey are probably attacked near water holes."	Lallemant, 1929
On Komodo a young water buffalo had apparently been attacked by an ora. The tail had been bitten off and there were claw marks on its back.	de Jong, 1929
"They eat kerabau and have been known to attack humans."	Anon., 1934
"...prey includes water buffalo, monkeys, and horses, though mainly deer and wild pigs. They prefer dead food to living prey." He found a dead, still warm fawn that had been killed by an ora.	de Jong, 1937
They describe the way an ora in the Surabaja Zoo attacked and killed several live, 2-month-old pigs.	Tanzer and van Heurn, 1938
Kepala told him healthy as well as wounded deer are attacked from ambush and that a young man had been attacked and killed on Komodo.	de Voogd, 1950
"...attack and kill wild pigs, deer and small water buffalo. They probably obtain them at their water holes." [Probably taken from Lallemant, 1929.]	Delsman, 1951
Deer poachers report living prey attacked along game trails and while sleeping.	Hompes, 1952

(Continued)

droppings were collected in Loho Liang, Komodo. None was empty. The food remains contained in them are listed in Table 10–7. These data and the known activity patterns of these small, tree-dwelling individuals suggest that during the first 8 months of their life (approximately May–December) oras feed primarily on arboreal geckos. At the beginning of the rainy season (January) they apparently spend most of their time hunting grasshoppers, and other insects. Near the

Table 10–2. *Continued*

Hoogerwerf doubts that oras can overpower large pigs, deer, etc., in nature. He describes the feeding of a wild ora on a live monkey. On Padar they feed on turtle eggs.	Hoogerwerf, 1954a
[Three Komodo men] "...had been chased by a big lizard which caught one of them and killed him. The monitors may wait for pigs and deer in scrub or long grass by animal trails, or in water holes with just their eyes and nostrils above the water. ...Other food sources are monkeys (in Rintja only), cats, small species of lizards, turtle and megapode eggs, dead marine fauna...cast up by the sea, and...eggs of tree- and cliff-nesting birds."	Collins, 1956
"Primarily carrion that appears more or less regularly on the island as a result of natural death....Also attack live prey up to deer and pigs." They found a rat and a boar's head in stomach of a large ora, and believe they eat megapode eggs as well as fish and invertebrates thrown upon beach.	Darevsky and Kadarsan, 1964
Abdoessoeki describes attack of wild ora on adult, drugged goat.	Abdoessoeki, n.d.
"A fisherman from Rintja told me the villagers lose some of their goats to dragons...."	Kern, 1968
"So until someone sees one pull down a deer or pig (or macaque or horse on Rintja), we should consider adult oras to be primarily carrion feeders."	King, 1968
"Some authorities maintain that the Komodo lizard will follow and hunt down live prey, but I doubt this....it must nevertheless be very hard for the lizards to stalk such prey. It is my opinion that *Varanus komodoensis* in its wild state is not able to stalk and catch wild game, and that its food is confined to dead, dying, or helplessly wounded animals."	Oesman, 1970

a. Abbreviated statements of only those writers who had actually visited the Lesser Sunda Islands.

Table 10–3. Individuals and/or species per ora fecal pellet

Number of individual prey or species represented in each pellet	Number of pellets
0	1,014
1	2,728
2	264
3	26
4	4
5	2

Table 10–4. *V. komodoensis* prey represented in fecal pellets (N = 4,267) collected on Komodo

Prey eaten	Number of occurrences
Invertebrates	
Land molluscs (*Asperites trochus*)	3
Marine clam (sp. ?)	1
Beetles (several sp.)	15
Grasshoppers (several sp.)	56
Crabs (several sp.)	3
Fishes (several sp.)	5
Reptiles	
Ora (*Varanus komodoensis*)	66
Gecko (*Gekko gecko*)	24
Gecko (*Hemidactylus* sp.)	3
Skink (*Sphenomorphus florense*)	3
Bamboo viper (*Trimeresurus albolabris*)	8
Rat snake (*Elaphe subradiata*)	29
Whip snake (*Dendrelaphis pictus*)	43
Cobra (*Naja naja*)	3
Russell's viper (*Vipera russelli*)	15
Hawksbill sea turtle (*Eretmochelys imbricata* + eggs)	3
Birds	
Megapode (*Megapodius freycineti* + eggs)	48
Jungle fowl (*Gallus varius* + eggs)	4
Miscellaneous small bird species	68
Doves (sp. ?)	3
Mammals	
Shrew (*Crocidura* sp.)	1
Bats (sp. ?)	2

(Continued)

end of the first rainy season and throughout the following dry season they feed primarily on rodents, and birds.

Thus it is clear that ora predation activities can be grouped into three rather broad categories, reflecting the correlation of predator and prey sizes:

(1) Hatchling oras feed exclusively on insects and small lizards. These prey are found under bark, in crevices of stumps, logs, trunks of trees, and in the grass.

(2) Small- to medium-sized oras feed largely on rodents and birds. These prey are normally obtained on or under the ground surface.

(3) Large oras often feed on animals as large as or larger than themselves. As will be shown later, these are obtained through both stealth

Table 10–4. Continued

House mouse (*Mus musculus*)	36
Black rat (*Rattus* sp.)	196
Feral dog (*Canis familiaris*)	8
Feral goat (*Capra hircus*)	26
Palm civet (*Paradoxurus hermaphroditus*)	24
?Palm civet (*Arctogalidia* sp.)	1
Wild boar (*Sus scrofa*)	808
Sunda deer (*Cervus timorensis*)	1979
Feral water buffalo (*Bubalis bubalis*)	15

Table 10–5. *V. komodoensis* prey represented in fecal pellets (N = 120) collected on Padar

Prey eaten	Number of occurrences
Invertebrates	
Grasshoppers (several sp.)	12
Reptiles	
Skinks (*Sphenomorphus florense*)	3
Whip snake (*Dendrelaphis pictus*)	1
Rat snake (*Elaphe subradiata*)	3
Hawksbill sea turtle (*Eretmochelys imbricata* + eggs)	3
Birds	
Unidentified sp.	3
Mammals	
House mouse (*Mus musculus*)	1
Black rat (*Rattus* sp.)	33
Sunda deer (*Cervus timorensis*)	61

and surprise along established game trails and in thickets. In all cases, oras locate live prey by both sight and scent.

A comparison of the food resources utilized on the three islands from which samples are available shows that the feeding habits vary considerably. Thus, though rats and birds are found on all three islands, they constitute a major part of the diet only on Padar—probably because this island lacks the variety of food resources found on the other islands. Both rats and birds are apparently very rarely eaten on Flores, where the food resources are most varied. This correlation is discussed further in Chapter 11. It is also significant that water buffalo are much more frequently eaten on Flores than on Komodo, though this difference is probably entirely density related.

Table 10–6. *V. komodoensis* prey represented in fecal pellets (N = 341) collected on Flores

Prey eaten	Number of occurrences
Invertebrates	
Grasshopper	1
Reptiles	
Bamboo viper (*Trimeresurus albolabris*)	1
Birds	
Unidentified sp.	2
Mammals	
Crab-eating macaque (*Macaca fascicularis*)	4
Palm civet (? *Paradoxurus hermaphroditus*)	2
Porcupine (*Acanthion brachyurus*)	2
Wild boar (*Sus scrofa*)	91
Sunda deer (*Cervus timorensis*)	146
Water buffalo (*Bubalis bubalis*)	28

Table 10–7. Food remains in fecal pellets (N = 15) of oras less than 1 year old

Prey eaten	Months collected	Number of occurrences
Insects		
Grasshoppers	January, March, April	4
Beetles	February	2
Reptiles		
Geckos (*Hemidactylus* sp.)	May, July, Nov.	9

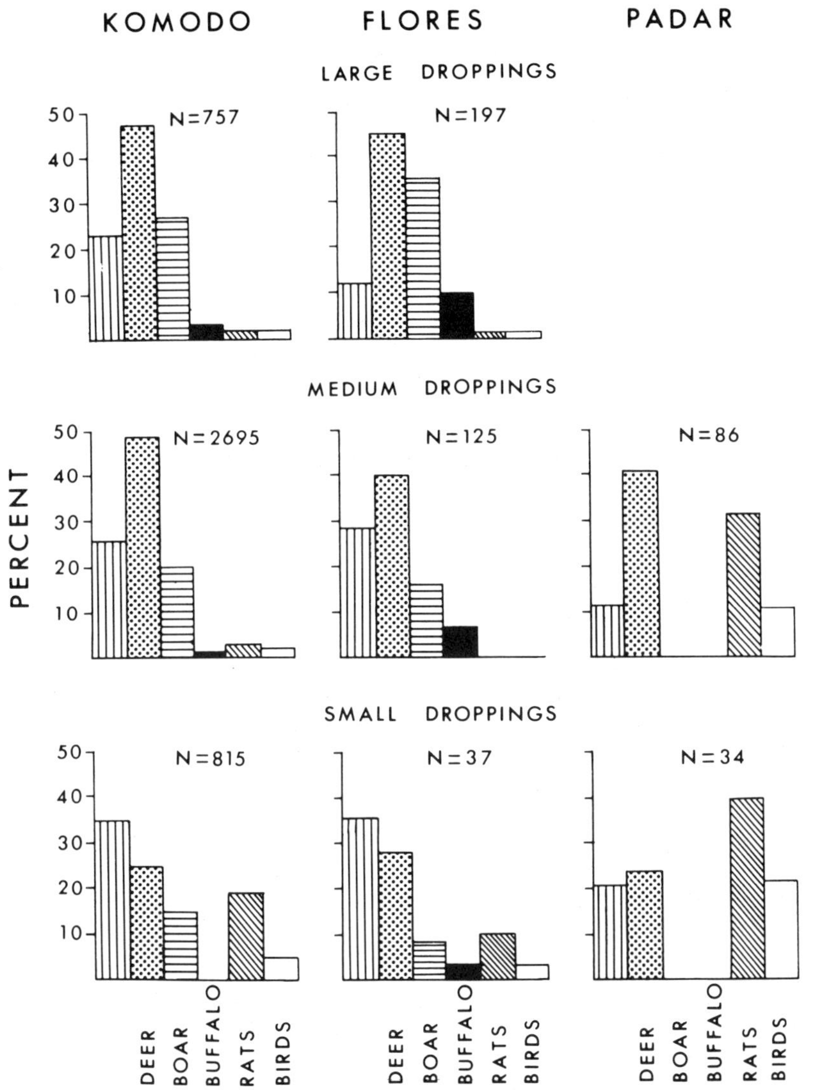

Fig. 10–1. Percent distribution of more common prey species represented in three dropping size classes from Komodo, Flores, and Padar. Vertical shading represents empty droppings.

Droppings are available from 51 localities on Komodo (Fig. 10–1), representing three major habitat types: quasi-rain forest, monsoon forest–*Zizyphus* savanna, and steppe–*Borassus* savanna. Prey frequency in small, medium, and large droppings from each of these habitats is shown in Tables 10–8, 10–9, and 10–10. The relationship of these data to the number of droppings available from

Table 10–8. Prey occurrences in ora fecal pellets from steppe and *Borassus* savanna associations, Komodo (approx. sea level to 250 m elevation)

	Pellet size groups		
Prey remains	Small (N=435)	Medium (N=1675)	Large (N=250)
Empty	120	293	54
Deer	147	671	96
Boar	53	208	72
Rats	69	9	0
Oras	4	19	22
Lizards	2	1	0
Insects	5	3	0
Birds	11	16	1
Snakes	10	10	1
Goats	4	0	0
Civet cats	0	3	1
Fishes	0	7	0
Buffalo	0	0	3

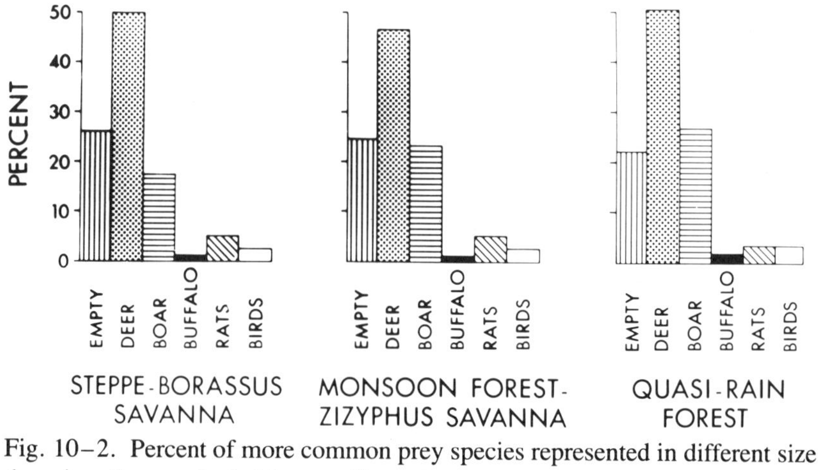

Fig. 10–2. Percent of more common prey species represented in different size droppings from major habitats on Komodo.

Table 10–9. Prey occurrences in ora fecal pellets from monsoon forest and *Zizyphus* savanna associations, Komodo (approx. sea level to 350 m elevation)

Prey remains	Pellet size groups Small (N=348)	Medium (N=990)	Large (N=229)
Empty	115	185	50
Deer	82	487	101
Boar	44	205	63
Rats	34	24	3
Goats	6	4	6
Civet cats	0	12	0
Birds	10	16	2
Snakes	9	15	1
Insects	39	5	0
Oras	2	20	1
Lizards	7	1	0
Fishes	0	1	0
Snails	0	3	0
Turtles	0	2	0
Dogs	0	1	0
Buffalos	0	8	2
Clams	0	1	0

Table 10–10. Prey occurrences in ora fecal pellets from in or near quasi-rain forest, Komodo (approx. 300–600 m elevation)

Prey remains	Pellet size groups Small (N=53)	Medium (N=426)	Large (N=302)
Empty	16	92	62
Deer	10	207	161
Boar	9	102	67
Rats	7	10	0
Birds	6	7	1
Goats	4	2	1
Oras	0	1	2
Insects	1	0	0
Fishes	0	1	0
Snakes	0	1	0
Civet cats	0	2	0
Buffalos	0	1	8

each habitat is shown in Figure 10–2 This comparison shows that there is a positive correlation between the number of prey species represented in droppings from the monsoon forest–*Zizyphus* savanna and number of droppings, so the prey spectrum is sample size-dependent. Continual addition of "new" prey species as dropping sample size increases and the random distribution of the number of individual prey per dropping indicates that oras select not only large available prey but also species that are less abundant or less available. This shows that oras are opportunistic predators that will feed on anything they can catch and kill.

In steppe–*Borassus* savanna the prey spectrum is not significantly increased with larger sample size, undoubtedly a reflection of smaller total number of potential prey species in this habitat. The prey spectrum–sample size correlation is less clear in the remaining habitats.

The percentage distribution of prey types in small, medium, and large droppings in each of the three major habitats on Komodo is compared in Tables 10–11, 10–12, and 10–13. It is highly relevant that within each ora size category there is no significant difference in the number of droppings that are empty, i.e., contain no evidence of differential hunger level in the three major habitats. With deer the most significant factor is that the high utilization is by large and medium oras in quasi-rain forest and monsoon forest–*Zizyphus* savanna associations (the difference between them is statistically insignificant), with low deer utilization for steppe–*Borassus* savanna.

Utilization of wild pigs is similar, except that in quasi-rain forest areas the utilization level is not significantly higher than the others.

Rats are utilized mainly by smaller oras, most commonly fed upon in both this and the medium-size categories in the steppe–*Borassus* savannas, where rodent densities are high. It is not known if the rodent fed upon to a significant extent in quasi-rain forest (13.2 percent) is *Rattus rattus* or another as yet unidentified species restricted to higher elevations.

In general, birds are not a major food source of even small oras, but this is not the case in quasi-rain forest, where birds constitute 11.3 percent of the food of small oras at these higher elevations.

Snakes are frequently eaten by all but the smallest oras, though less often in the quasi-rain forest. There is no indication that snake species diversity or density is inversely correlated with increasing elevation. In general, snake and lizard species diversity tends to increase with elevation in Southeast Asia.

Cannibalism is not practiced by small oras. However, it occurs in medium-sized lizards and becomes very important in the feeding biology of large oras—particularly in the steppe–*Borassus* savannas, where this behavior accounts for 8.8 percent of the total food in terms of individuals of prey taken (see Chap. 11).

Insects (largely grasshoppers) are eaten mainly by small oras (Table 10–11) and are apparently taken in a ratio that reflects their availability. Highest utilization is in savanna habitat of either the *Zizyphus* or *Borassus* type.

Table 10–11. Percentage of common prey species in fecal pellets of *small* oras from major Komodo habitats

Prey remains	Quasi-rain forest	Monsoon forest–*Zizyphus* savanna	Steppe–*Borassus* savanna
Empty	30.2	33.0	27.6
Deer	18.9	23.6	33.8
Boar	17.0	12.6	12.2
Rats	13.2	9.8	15.9
Birds	11.3	2.9	2.5
Snakes	0.0	2.6	2.3
Insects	1.8	11.2	11.5
Oras	0.0	0.6	0.9

Table 10–12. Percentage of common prey species in fecal pellets of *medium-sized* oras from major Komodo habitats

Prey remains	Quasi-rain forest	Monsoon forest–*Zizyphus* savanna	Steppe–*Borassus* savanna
Empty	21.6	18.7	17.5
Deer	48.6	49.1	40.1
Boar	23.9	20.7	12.4
Rats	2.3	2.4	5.3
Birds	1.6	1.6	1.0
Snakes	0.2	1.5	1.7
Insects	0.0	0.5	0.2
Oras	0.2	2.0	1.1

Table 10–13. Percentage of common prey species in fecal pellets of *large* oras from major Komodo habitats

Prey remains	Quasi-rain forest	Monsoon forest–*Zizyphus* savanna	Steppe–*Borassus* savanna
Empty	20.5	21.8	21.2
Deer	53.3	48.3	38.4
Boar	22.2	27.6	28.9
Rats	0.0	1.3	0.0
Birds	0.3	0.9	0.4
Snakes	0.0	0.4	0.4
Insects	0.0	0.0	0.0
Oras	0.7	0.4	8.8

Prey Characteristics

The availability of live prey is largely dependent on the interrelationships of the activity patterns and habitat preferences of the lizard and the prey species. For example, nocturnally active prey can be captured during the daytime only when the oras surprise them in their resting or sleeping places. In this case the prey is generally available during all hours of the day. However, other prey species, such as deer, because of their particular activity patterns, are available only during certain periods of the day. Additional factors important in availability are prey size and density; these and others are discussed below.

Insects.—Insects are commonly eaten by young of several species of *Varanus* (e.g., *indicus,* Dryden, 1965; *gouldi,* Pianka, 1970; *gilleni* and *caudolineatus,* Pianka, 1969; and *bengalensis,* Sharma and Vazirani, 1977). Some of the smaller species are specialized to feed on them. Mertens (1942) suggested that the tongue of *V. rudicollis* may be used to feed on ants or termites, and I have seen *V. bengalensis* use their tongues to lap up ants.

Several different types of insects are eaten by small oras, but the overall variety is much less than in the other lizard species that have been studied. Four species of locusts (Family Acrididae) are commonly eaten: *Gastrimargus lumbokensis* (Oedipodinae), *Stenocatanops exinsula* (Catanopinae), and *Valanga nigricornis lombokensis* and *Austracris g. guttulosa* (both Cyrtacanthacridinae). All are captured by stealth in the grass-covered savannas, where small specimens seem to spend a great deal of time searching for them. We have seen two small oras hunt for locusts; they move quickly through the grass and attempt to catch the insects mainly when they alight on nearby grass blades. The success ratio seems very low. On one occasion I watched one small ora hunting *Stenocatanops exinsula* near the observation blind. He bit at alighting locusts 32 times and was successful only twice. The length of the hunting "run" through the grass was 14 min. The area being hunted had a very high locust density. Most savanna areas have a lower density, so that the usual hunting success ratio is probably even lower than that observed.

Lizards other than Varanus.—Several species of varanids regularly feed on lizards. Examples include *V. indicus* (Dryden, 1965), *V. salvator* (Taylor, 1922), *V. griseus* (Smith, 1935), *V. gilleni* and *V. caudolineatus* (Pianka, 1969), and *V. gouldi* (Pianka, 1970). In the last species, lizards make up 29.6 percent of the total diet.

In captivity, oras are known to have fed on a water lizard (*Physignathus lesueri*) and a lacertid (*Lacerta lepida*) (Lederer, 1942). Several species of ground- and tree-dwelling lizards are regularly eaten in nature by *V. komodoensis* (Table 10–4). Of these the tokay (*Gekko gecko*) is the most frequent prey. This and several other genera (*Hemidactylus frenatus, Platyurus platyurus*, and *Lepidactylus intermedius*) are all common under the loose bark of large dead *Tamarindus*

Pl. 6. After hatching, young oras spend much time foraging for insects and small lizards under the loose bark of dead *Tamarindus* trees.

trees. On several occasions small monitors were observed slowly scouring almost the entire tree to heights of 15 m, evidently searching for lizards. Usually they moved from one large bark slab to another, flicking their tongues here and there, disappearing under the bark for a few moments, and poking their heads in every crevice and hole encountered (Pl. 6). No geckos were ever caught as I watched, but that the monitors eat large numbers of them is evident from their droppings. The density of *Gekko gecko* is about 77/km^2 in Loho Liang on the basis of both calling individuals and the distribution and the number of appropriate trees for them to live in. Densities for the other geckos were not estimated. Cagle's (1946) estimate of 500 adult individuals of *Hemidactylus garnoti* in 2,400 ft^2 is much too high for Loho Liang.

Several skinks occur in Komodo. The one found in the droppings seemed to be referable to *Sphenomorphus florense* only on the basis of dentary and tooth shape. This species is active at night and during the day and often climbs trees to several meters. Most of those eaten by the oras are probably caught in the trees under bark. Though no density estimates were made, this is undoubtedly the most common of all the lizards on Komodo.

Snakes.—Several species of *Varanus* have been reported to feed on snakes, even poisonous species (Spatz, 1931). As far as is known, oras eat any snake they are able to subdue. On the basis of fecal pellets examined during this study, snakes eaten by oras are the palm viper (*Trimeresurus albolabris*), Russell's viper (*Vipera russelli*), whip snake (*Dendrelaphis pictus*), rat snake (*Elaphe subradiata*), and the cobra (*Naja naja*). The absence of several species of pythons (*Python reticulatus*, *P. molurus*, and *P. timorensis*) from the range and habitat of the ora, though they are found in adjacent areas or habitats, may be due to predation by the ora. Lederer (1942) reported eggs of *Python reticulatus* were eaten by a captive ora.

Rodents.—Rodents are one of the most common prey animals of varanids. Almost all of the medium to large species eat rats and mice (Mertens, 1942). Lederer (1942) reported that small mammals (rabbits, guinea pigs, and rats) are often swallowed alive.

Though Burden (1928) found no rats on Komodo, two species of rodents are definitely known from the island, *Mus musculus* and *Rattus rattus*. Wendt (1956) stated that there were no rats on Komodo before European ships stopped there, but he apparently based this on a local legend that the *ancient* people (*orang dulu dulu*) are said to have brought rats to the island. The sometimes repeated story in popular accounts that Komodo means "Island of Rats" in the local dialect is incorrect according to both the local natives and expert linguists. Hoogerwerf (1954a) suspected that a third but unidentified rodent lived at higher elevations. On Flores a large bamboo rat (*Papagomys armandvillei*) lives at higher elevations, and there is probably no good reason why it could not exist in equivalent habitats on Komodo. *Rattus rintjanus*, a probable synonym of *Rattus rattus*, is reported from Padar and Rintja. On Komodo, *Mus musculus* is most common at low elevations, often along the coast. *Rattus rattus* is more widely distributed, has a higher population density, and is larger than the mouse. The average weight of 35 rats (*R. rattus*) taken during this study in Loho Liang, Komodo, was 199 g. Pfeffer (1959) believed that rats were the major prey of young oras on Komodo, though in another part of his report he stated they are absent from Padar and Komodo. However, our studies show they are common throughout the range of the ora (Chap. 11). Darevsky and Kadarsan (1964) reported a single *R. diardi* in the stomach of a young ora killed on Komodo.

Though mice were found in ora fecal pellets, they were rare, reflecting their generally low densities in areas removed from the villages themselves. Small- to medium-sized oras on both these islands eat many rats, some of which they dig out of their burrows under logs and trash. However, some are undoubtedly captured in the early morning and late afternoon when they are quite active.

The Sunda porcupine (*Acanthion brachyurus*) is absent from Komodo, but it is reasonably common at all elevations on both Flores (including western Mangarrai) and Sumbawa. On Flores large oras occasionally feed on them (Table 10–6). Adult Indonesian porcupines are from 78 to 90 cm in total length, with a weight of 11–15 kg.

Since porcupines are largely nocturnal, most likely they are captured while they sleep during the daytime. Near Nggoer the animals apparently favor rocky hillsides and bamboo thickets; thus they are found in both open and forested areas. By day they sleep among rocks or in deep burrows they dig themselves, although sometimes they enlarge ora burrows. Porcupine burrows can be differentiated from those of oras by their great length (to 18 m according to Gee, 1964) in proportion to their diameter, an enlarged chamber at the end, usually with several bolt holes. Chewed bones and antlers are often common near the entrance, and a strong, distinctive musty odor emanates from the burrow. Porcupines are probably occasionally dangerous prey; there are records of tigers dying by accidentally impaling themselves on the spines, or being maimed by the spines, and even of porcupines attacking leopards at a water hole (Gee, 1964). However, there are no records of injuries to oras from the spines of porcupines.

Bats.—A few teeth and bones of an unfortunately unidentified small bat species were found in several ora droppings. We occasionally flushed bats of the same general size from hollow trees, and I presume it is from such places that oras obtain those they sometimes eat.

Shrews.—Two crocidurine shrew genera, *Crocidura* and *Suncus*, are known through various islands in the Lesser Sundas as far eastward as Timor. However, Darlington (1957) stated that both genera include species often associated with man, and their natural range limits are doubtful. *Crocidura* sp. are found on Komodo and were sometimes captured in the drift fence traps we used to survey reptilian and amphibian faunas of several ora habitats. While they seemed to be fairly common, they were found in ora droppings only once. *Suncus* is apparently regularly eaten by *Varanus bengalensis* (Minton, 1966).

Turtles.—There have been many reports of turtles and their eggs being eaten by monitors (e.g., *V. varius*, Berg, 1913; *V. salvator*, Laidlow, 1901; *V. griseus*, Zakhidov, 1938). Deraniyagala (1931) reported that a *V. salvator* 207 cm long swallowed a *Geoemyda* with a carapace length of 16 cm. Captive oras have eaten the eggs of the soft-shelled turtle *Trionyx triunguis* (Lederer, 1942).

As early as 1928, the Douglas Burden Expedition had reported that *V. komodoensis* ate sea turtle eggs on the island of Padar. On that island we also found evidence, on the basis of remains in ora droppings, of egg and even adult sea turtle predation. Along with rats and deer, sea turtles apparently constitute the most common food resources of medium-sized monitors on that island. The same kind of evidence was found in the manure of medium-sized oras from Komodo. Apparently young monitors do not feed on the eggs (perhaps because they are buried too deep). However, 1.5–3 m specimens feed on both eggs and adult turtles regularly. The possible search for turtle nests may explain the rather regular visits to the beach by certain individuals (Chap. 5). Young sea turtles are probably not eaten, perhaps because hatching in the rather scattered nests on the Padar and Komodo beaches is of relatively short duration in terms of the search

Table 10–14. Presence of sea turtles on the beaches of Komodo and surrounding islands

Date	Remarks	Place
Sept. 12	*C. mydas* hatchling dead on beach	Pulau Longo
Oct. 7	*C. mydas* eggs laid on beach	Pulau Padar
Nov. 27	*C. mydas* hatchlings alive on beach	Pulau Longo
Jan. 16	*E. imbricata* remains in fresh ora droppings, presumably from nesting on beach	Pulau Padar
Feb. 12	*E. imbricata* hatchlings alive on beach	Pulau Komodo
March 29	*C. mydas* captured laying eggs on beach	Pulau Padar
April 29	*E. imbricata* hatchling dead on beach	Pulau Komodo

time required of the ora. *Eretmochelys imbricata* nests in small numbers on both islands. Our data confirm native reports that *Chelonia mydas* does not nest on Komodo, but it does nest in fairly large numbers on Pulau Longo, a small islet southeast of Kampung Komodo, and on Padar.

Hatchling sea turtles were found on the beaches, in the surf, or at our base camp on several dates (Table 10-14). Presuming a green sea turtle mean incubation period of about 56 days (Hendrickson, 1958; Carr and Ogren, 1960; for nonmonsoon period) and a hawksbill mean incubation of 58 days (Carr et al., 1966), *C. mydas* eggs would have been available to oras from about mid-July to early December, *E. imbricata* eggs from mid-January to early June. Thus, sea turtle eggs are available on at least certain of the beaches the entire year, making it clear that regular forays onto the beaches by oras undoubtedly represent a reasonably successful hunting technique.

The Asian box turtle (*Cuora amboinensis*) does not occur within the range of the ora, though it broadly overlaps that of *V. salvator*. It is possible that this is due to heavy predation pressure by *V. komodoensis*.

Crocodiles.—The saltwater crocodile (*Crocodylus porosus*) occurs within the range of the ora, and its eggs and young undoubtedly fall prey to the monitors. The animals do not attain a length of 40 ft and are not as rare as Piazzini (1960) reported. Broughton (1936) reported a 2-m crocodile captured in a noose that was intended for capturing oras on Komodo, located about 1 km from the beach. We found bones and teeth of some very large *C. porosus* that had been killed on Komodo by Chinese hide hunters from Flores. Though apparently no longer found on Komodo, individuals are often seen within the range of the ora on Flores, particularly on the north coast.

V. niloticus is especially well known as a crocodile egg predator in at least the central and northern parts of its range (Hesse, 1889; Roosevelt, 1910; Pitman, 1931; and Cott, 1960). Lederer (1942) reported that a captive ora in the Frankfurt Zoo ate eggs of the West African crocodile *Osteolaemus tetraspis*.

Birds and bird eggs.—Many *Varanus* species feed on birds and their eggs (Mertens, 1942), and some, such as *V. praesinus* and *V. salvadori*, apparently feed on them to a great extent. In captivity, hen's eggs are swallowed entire and crushed in the esophagus by bending the neck in a sharp angle (Lederer, 1942; Mertens, 1942). Birds swallowed whole are often manipulated so that they are passed into the gullet headfirst (Thilenius, 1897; Lederer, 1942).

On Komodo, Padar, and Flores several types of birds are caught and eaten, particularly by medium-sized oras. Eggs of all species are eaten wherever they are found. The most common feathers and eggshells found in the droppings collected in Loho Liang belong to Freycinet's megapode (*Megapodius freycineti*). Jungle fowl and their eggs are also common. Next in abundance are ground-nesting doves of several species (Table 10-4).

The green jungle fowl (*Gallus varius*) is widely distributed in Southeast Asia, where it is most common in savanna and open monsoon forest. It roosts in trees, is most active in the early morning and late afternoon, and feeds on shoots, tubers, berries, small snakes, lizards, and many species of insects. Nests are made in thickets, usually under vine clumps, brush piles, or fallen logs, and in treetops. Most nesting seems to take place in the rainy season, though the breeding period is quite long. Some nesting may occur throughout the year, but data are inconclusive. The hen lays 4–15 eggs and seems to incubate them alone. Most birds are seen singly, rarely in small parties of two or three indivduals. The mean weight of 15 birds from Loho Liang is 683 g (range = 400–1,100 g).

Darevsky and Kadarsan (1964) were the first to report that oras eat jungle fowl. This is positively confirmed by our study, which also shows that these birds are most often captured on or near their nests. However, kill sites were found in small openings and along wide trails in the monsoon forest. In one case a kill site was found under an overhanging Nyolo vetok shrub (*Azyma sarmentosa*). The lizards often search both for the eggs and the adults during the late morning and afternoon. They apparently know the exact positions of the most likely nesting and resting locations and move from one area to another with little deviation between.

Jungle fowl are very territorial and exhibit a predictable daily activity pattern. Site attachment and behavioral tenacity of these birds are apparently important in determining the success ratio of oras hunting them. After calling from their tree roosts about 0630 to 0700 hr, the jungle fowl normally fly to the ground, usually scratching under the tamarind trees for most of the morning. During the middle of the day they are rather inactive, remaining in the thicker sections of the forest. Near twilight they again ascend to their regular roosts and call before going to sleep.

Occasionally smaller lizards chase and catch domestic chickens (*Gallus gallus*). Tied chickens are obviously more often eaten than those allowed to run loose. On several occasions during our year on Komodo 1–1.5 m oras chased several loose chickens in our base camp. The oras approached the chickens with great stealth and, with mouth open, attacked from a distance of about 1 m.

Freycinet's megapode (*Megapodius freycineti*) is largely restricted to the Lesser Sunda Islands. The mean weight of 14 individuals from Loho Liang is 696.4 g (range = 514–961 g). Perhaps its most characteristic attribute is the size and nature of its nest. On Komodo these large sand and vegetation mounds are 2 m high with a diameter of 8 m. The shape is usually a truncated cone (Pl. 5). It is generally constructed in sandy, open areas near the forest edge, often along stream courses. The eggs (1–4) are laid singly 0.5–1.0 m deep, and usually scattered in the mound. The adult birds visit the nest regularly throughout the day, removing or adding vegetable trash and sand, presumably as a thermoregulatory control. The adults, usually seen in pairs, are active throughout the day. They often approach the nest very carefully together, sometimes taking as much as ½ hr to move the last few meters. One often acts as a sentinel by perching in a nearby tree or by flying over the nest while the other approaches the nest.

Megapode birds are apparently usually captured on or very near their nests. Oras often visit nests regularly in their hunting circuits, lying on the edge of the nest for several hours at a time. The nests are often dug into and the eggs eaten (Pl. 5) (Darevsky and Kadarsan, 1964; Pfeffer, 1965). Lincoln (1974) provided a good description of megapode egg predation by oras. During a specific study of such predation in a 2-month period (March–April) ora 28W visited one large megapode nest in the overflow area of Kali Besar, Loho Liang, a total of 13 times, at an average rate of once every 3.1 days (range 1–9, mode 1). Visits occurred during the day, from 0800 to 1630 hr. On some of these days 28W dug up the nests. On the remaining days the lizard was either trying to catch the adult birds or tonguing the nest (for the scent of newly laid eggs?). The nests were checked rather regularly, sometimes by several oras in a single day; however, there was no evidence that any of the oras were successful in capturing adult birds during the time we studied the nests.

Masses of jungle fowl, dove, and even brahminy kite feathers were found in fecal pellets, and other signs, such as tracks, prove that these birds are attacked and eaten. All attacks on adult birds seemed to have taken place on the ground, apparently by stealth, and almost always in some type of opening where the predator could be seen but could also maneuver easily. In every case the bird appeared to be feeding on the ground. None except megapode birds are known to be killed at their nests, though this must occur.

Oras do not always attack birds, even if there is a good chance of success. As an example, on November 20, 1969, at Vai Nggulung, Komodo, "the 0.7-m ora walked within 0.5 m of a ground dove, which was walking around in the open, sandy area with scattered grass just a few feet east of our blind. However, neither the ora nor the bird seemed to pay any attention to one another." On July 8, 1971, a 1-m ora was seen chasing a chicken under the stilted houses at Loök, Flores, while we camped nearby. The villagers of Nisar, Flores, complained that many chickens are killed each year in their village.

Oras.—Cannibalism was first reported by Pfeffer (1959) but is generally rare. However, 8.8 percent of all droppings of large oras from steppe and *Borassus* contained remains of small individuals (Chap. 11). All other samples contained ora remains in less than 2 percent of the droppings.

As in other carnivores, cannibalism is most common during the periods of high population density. The usual ecological separation of size groups in *V. komodoensis* tends to keep densities uniform over fairly large areas. However, this density uniformity breaks down during feeding aggregations, when various-sized individuals are crowded together in rather small areas. At these times aggression is common (Chap. 12). Though such aggregations provide opportunities for cannibalism, we have never observed it under those circumstances.

However, 19W killed two smaller oras (18W and 27W) and undoubtedly would have eaten both if he had not been frightened off. In both instances the smaller lizard was in Trap 1 when 19W entered it. Tracks on the ground in both cases suggested that the smaller lizards tried to get away but were apparently cornered, seized, and killed. In both cases a dead bait was available to 19W, but it was ignored in favor of the smaller, live lizard. In one instance 19W was frightened while he was still in the trap, and he left the dead lizard immediately and tried to scramble out of the trap. In the other case, 19W trapped 18W in the trap release box. The struggle there was so fierce that a hinge of the release door was sprung and the 4-mm nylon cords holding the door closed were torn loose. The larger ora then apparently dragged the smaller one through the box and out of the trap. There the widely splattered blood indicated he had thrown the lizard from side to side very violently. (The same technique is also used on smaller prey by other varanid species [Loop, 1974].) It was at this time that the vertebral column was apparently severed and the abdominal cavity opened by a massive wound extending across the belly from side to side. The carcass was being dragged into the brush as we approached the trap. While we spent a few minutes reconstructing the events on the basis of the tracks and signs, the large ora had eaten the viscera. He left the carcass when he saw us as we trailed him into the brush. Lizard 27W also had its vertebral column severed, probably by the same technique. It is clear that under certain circumstances a large ora will attack and eat any smaller ora it can manage to corner. This probably happens most often in ora burrows and along earth and rock cliffs. Overall density of oras and other potential food resources in the general area are obviously important factors. Cannibalism is largely restricted to both the very young and the old. The former are captured alive and eaten, but remains of large monitors are probably the result of scavenging (see Chap. 11). The pattern of cannibalistic activity being largely restricted to larger specimens is identical to that found in *Crocodylus niloticus* by Cott (1960).

Crab-eating macaques (Macaca fascicularis).—This ordinarily common monkey in the Lesser Sundas is not found on either Komodo or Padar, probably because neither island supports the variety of fruits needed throughout the year. Many macaques occur on Rintja and Flores. In western Flores large troops are very common near garden areas and where water is available all year. In the jungle they usually keep to the edges, rarely penetrating into the depths; they have a decided preference for open country. Near the village of Nggoer they occur in a number of different habitats but never at a greater distance from Nanga Loök, the only permanent stream in the area. In addition to the immediate garden areas, near the coast they are quite common in drier parts of the mangrove forests, particularly the *Lumnitzea* swamps. Here they feed on various small ground and arboreal animals, particularly crabs. Like most macaques, this species feeds mainly on the ground. Here, in addition to crabs, they eat fallen fruit, ground plants, insects, and spiders. During our visit there in June the monkeys fed largely on the ripe fruits of *Zizyphus jujube*. Since these were located on the edges of the mangrove, there was movement back and forth through the swamp. Occasionally they came onto the beach and the exposed reefs to look for food. Other group movement of significance was to the few, scattered pools of fresh water in drier parts of the mangrove forest. It is during these movements over the surface that monkeys are apparently most often attacked. In the inland savannas they are caught by oras as they move single file along game trails to water or other forest patches. There were about four groups in this habitat, varying from 8 to 32 individuals ($\overline{X} = 15$).

Farther inland they are often concentrated along watercourses during this time of the year, so densities along Nanga Loök were unusually high when we were in the area. Near Nggoer ora droppings in the mangrove area contained more monkey remains than in any other habitat (monkey remains in droppings from mangrove area 18.5 percent, in nonmangrove area 1.5 percent). Hoogerwerf (1954a) mentioned that a 2-m ora was hunting some monkeys near the "camp" at Loho Gringgo, Rintja, during his visit there and described and presented a few photos of a 2-m ora feeding on a young macaque in Loho Gringgo. He stated that the macaque was slung from side to side, which seemed to break its neck. In view of my experience with the kill technique of oras, the 20 min reported to have taken to kill the monkey seems unduly long.

Not all monkeys eaten are small, however. Large Flores macaques have a head and body length of about 50 cm and weight about 5–7 kg. Some of the ora fecal pellets from Flores contained the canines of large adults. Darevsky and Kadarsan (1964) stated that monkeys allow oras to come close to them and that the ora may strike the monkey with its tail, but they gave no indication of the evidence on which they base these statements. When oras are seen by a troop of macaques from the trees, the macques often cry out and throw down leaves and sticks.

Palm civet (Paradoxurus hermaphroditus).—The palm civet is very widely distributed and has been reported from almost all the islands surrounding Komodo. Hoogerwerf (1954a) reported this species from Komodo only on the basis of footprints and excreta. No specimens were taken during our year of study at Loho Liang, though we saw several civets that seemed identical to *P. hermaphroditus* in color and size. Hair, claws, and teeth of this species are occasionally found in ora feces.[2] The palm civet attains a body length of about 0.7 m, plus an equal-sized tail. The adult weights of 19 museum specimens from the Lesser Sunda Islands vary from 2.7 to 4.5 kg ($\overline{X} = 3.1$ kg). Throughout their range they are most abundant in dry forests. Much of the day is spent in trees, curled up on the branches or in holes in the trunks. Near Ntodo Klea, Komodo, they were very common in *Borassus* palms and rock piles. They are mainly nocturnal and seek their food primarily at night. Because of this pattern, I presume most, if not all, of them are caught by oras during the day when they are asleep in rock piles, hollow logs, etc. They protect themselves by scent glands and often leave a powerful odor over the trail where they have passed (Prater, 1965).

Feral and domestic dogs (Canis familiaris).—Most of the dogs on Komodo Island are pitifully thin village animals. The village dogs near Nggoer, Flores, seemed in much better condition. Feral dogs, often in extremely poor physical condition, inhabit almost all the islands on which deer occur in the entire Sunda area, including Komodo, Padar, Rintja, and Flores. Most of these have been left to run wild at the end of poaching activities.

Several authors have reported that oras attack and sometimes kill and eat dogs (e.g., de Jong, 1929, 1937; Darevsky and Kadarsan, 1964; Pfeffer, 1965). I know of only two cases on Komodo in which an ora was seen to catch a dog, though dog remains are common in ora droppings. In one instance a group of six men was walking from Sabita to Kampung Komodo, accompanied by a village dog that ran back and forth through the high grass bordering the trail. Suddenly the dog howled. The men saw him lifted completely off the ground by a large ora and carried away, presumably to be eaten nearby. The monitor had evidently been lying quietly next to the trail and attacked the dog as it passed. Another time a 1.5-m ora entered the village and approached a sleeping dog lying under one of

2. Dr. S. D. Webb (Florida State Museum) identified some of the juvenile teeth in Komodo droppings as nearest *Arctogalidia* sp., suggesting a second species of civet may occur on Komodo. I have seen what I believe to be *Arctogalidia* near Djarek, Mangarrai, Flores.

the stilted houses. The dog suddenly jumped up but was struck down by the tail of the ora and then quickly bitten and killed. The ora was driven off before it could eat its prey. On Flores, Father Schmutz was told that in 1965 a party of men with dogs had just killed a young deer when a large ora approached the freshly killed animal, attacked and injured three dogs, and dragged the deer off into the forest. The village teacher at Nisar, Flores, says that he saw a dog struck by the tail of a large ora with such force and in such a way that the dog was knocked into the jaws of the ora, whereupon the dog was killed and dragged into the nearby forest. On June 24, 1971, a 1.5-m ora unsuccessfully chased a bitch and her two puppies near our base camp on the beach at Nggoer, Flores.

Feral and domestic goats (Capra hircus).—All the villages in the Lesser Sunda Islands possess at least a few goats. In some areas there are feral populations. On Komodo I know of feral goats only near Kampung Kechil (on Gunung Kampung Kechil, Gunung Galantang, and Gunung Amativa). Ora droppings from this area contain goat hooves and hair, so it is obvious feral animals are occasionally killed. However, most of the live goats attacked and eaten by oras are domestic animals kept in and near villages. To obtain them, monitors must enter the villages in full daylight, often attacking the goats under the stilted houses while surrounded by barking dogs and shouting villagers. Tethered goats are killed more often and more easily than free ones. In some western Flores savannas, goats cannot be tied and left unattended without danger of attack.

On August 26, 1969, a 2.5-m monitor entered our base camp in Loho Liang, Komodo and killed two newly dropped kids while the mother goat strained at the rope to escape. After the kids were eaten, the ora started to attack the female but was driven off by our camp assistant. During the rest of that day and all of the next two, the lizard remained near the camp, repeatedly entering it to kill the goat, though the latter was being moved from one place to another in hopes of discouraging the monitor. This was finally accomplished only by taking the goat to Kampung Kechil, 2 km down the beach.

Most of the smaller islets near Komodo and Flores are regularly stocked with goats. They are allowed to breed and then are periodically harvested by the owner. There are several such islets along the east and south coasts of Komodo. Of these, the situation on Nusa Mbarapu is particularly interesting: when its goat population is fairly high and prey populations low on nearby Komodo, large oras have been known to swim to the islet and live there, killing and eating the goats until the population is reduced to a level where the chance of further success is very low. They then swim back to Komodo. The distance between the islands is 450 m, and the tidal currents are exceptionally swift.

In order to observe the killing technique of oras, Abdoessoeki (n.d.) drugged an adult goat and offered it to oras on Komodo; but his description did not clearly identify the killing technique and its efficiency. For this reason several live, tethered goats were offered as prey in Loho Liang near a hanging rotten bait (to attract oras into the immediate area as an economy of time). The stalking and killing techniques were photographed and carefully observed on these occasions. The following abbreviated notes were taken from my field notebook:

March 14, 0815 hr. One live 25-kg goat tethered to ground anchor with a 10-m lead at the edge of a large thicket. Attractant, a dead 21-kg goat pulled 5 m into a tree.

0910 hr. Ora 19W approached bait area from downwind. It had seen the grazing goat almost immediately after stepping out of the thicket. At this point it was 15 m from the live goat with its head held so it could peer over the grass, watching the goat out of first one eye and then the other.

0927 hr. Moved to 3.0 m from the goat, flicking out tongue occasionally. Head now held low and neck slightly bowed to side.

0927:39 secs. Ora stepped to within 1.2 m of goat, bowed its neck into an S-shape with its head slightly to one side, and continued to stare at the goat.

0928:00 secs. Lunged when goat turned to side, grasping it over the ilial projection of the pelvis with the upper and lower jaws on opposite sides of the body. Held posterior part of goat against ground, while goat struggled to get to its feet.

0928:45 secs. Ora released goat, but immediatley bit it again in lumbar region. The monitor is now better able to hold the struggling animal against the ground with its neck arched vertically.

0929:45 secs. Goat tried to struggle to its feet and was almost successful, but ora released goat, grabbing left front leg and jerking prey back to ground.

0930:00 secs. Ora changed hold to lumbar region again and jerked posteriorly to throw it back to the ground. Goat stopped struggling; apparently in shock.

0930:20 secs. Goat bleating feebly; no movement other than sporadic breathing.

0930:30 secs. Released goat and flicked it with tongue. Bit out section of abdominal wall and then pulled out most of intestines, shook out their contents and swallowed them. Goat died from massive abdominal hemmorhage.

0938:00 secs. Ora bit into goat's back, then easily lifted and carried it to the tether end, where it was placed on the ground, and the ora continued eating the viscera.

March 14, 1430 hrs. Arrived at blind. Live goat of 27 kg tethered to ground anchor by 10 m line (see Fig. 10–3).

Fig. 10–3. Transparency tracings of attack sequence of 42-kg monitor on 27-kg goat, photographed in Loho Liang, Komodo, April 1970. A: instant before attack from a distance of less than 1 m. B: first contact, goat grasped by neck (arrows show direction of lizard's head movements). C: head of goat forced downward. D: goat violently jerked from side to side, then yanked toward monitor. E: goat pushed down onto haunches, then violently jerked back and forth in opposite directions. F: goat pulled laterally toward monitor and starting to go into shock. G: violent lateral jerking with posterior movement by monitor, with apparent intention of throwing goat to ground. H: sudden change of tactic by moving counterclockwise and then pushing goat to ground with downward movement.

1520 hrs. Ora 8W crashed through thicket edge into open savanna, 10 m downwind of goat, assumed alert pose, saw goat and stared at it.

1521:21 secs. Ora moved in semicircle about 2.5 m from goat, flicking its tongue into the air and watching goat closely with head held to one side. Ora faced goat from about 1.2 m, with head tilted up slightly (Fig. 10–3A) and made a few very short feints toward neck.

1522:20 secs. Monitor suddenly lunged at goat and grasped it by the neck with upper jaw above and lower jaw across the bottom of the throat (Fig. 10–3B). Ora lifted goat up off the ground easily and shook it from side to side very quickly like a dog with a slipper. It then slammed the goat to the ground and held it down very firmly (Fig. 10–3C).

1523:20 secs. The ora loosened its hold and suddenly yanked backward with its head (Figs. 10–3D–G). This was done several times in rapid succession, apparently to cut deeper into the neck flesh with its teeth. It then quickly and firmly pushed the goat to the ground again (Fig. 10–3H).

1525:05 secs. Ora accidentally cut through 3/8 in. nylon line around goat's neck, lifted it up easily and carried it about 6.5 m into the thicket edge.

1525:20 secs. Ora shook goat from side to side so hard that the hind hooves of the goat were sometimes as much as 1.2 m above the ground (broke neck ?).

1526:20 secs. Goat obviously dead, with no movement of any kind discernible and ora started to eat intestines.

In both cases it is obvious that the oras (each about 2.5 m long) grasped the goat on the dorsal surface, away from the thrashing hooves. The prey animals were easily held down, dragged, and even lifted up and carried. These and other sequences show considerable modification of attack behavior, apparently largely determined by prey size and its movement. Ganci (MS) has shown that modifications of the same type are not random in *V. bengalensis* when it attacks and kills prey but are related to predator-prey weight ratios, prey species, and prey movement, all interacting with hunger level and predator fright. Death occurred quickly, in about 2.5 min and 4 min, respectively, from the initial attack. Though the goats tended to try to stay on their feet (and were surprisingly quiet), they were thrown and held down so they could not push or pull away with their feet. In the second sequence the goat was violently shaken both laterally and vertically, increasing the severity of the wounds, breaking the neck, and hastening physiological shock. In both sequences death probably occurred as the result of massive visceral bleeding associated with evisceration.

The wide, staring, glazed eyes and protruding tongue of the immobilized prey all suggest that shock is an important factor in a successful attack. In addition to its prey immobilization characteristics, shock is also undoubtedly important in bringing about the rapid death of the prey animal, since the ora's technique of tearing open the viscera seems to be an adaptation to certain physiological factors of shock. Thus lacerations in this region during shock tend to cause more extensive hemorrhaging than would be the case if the final attack were directed elsewhere on the body.

Wild boar.—The subspecies of wild boar found on Komodo (*Sus scrofa vittatus*) ranges from the Malay Peninsula to Timor. Though the Asiatic mainland subspecies *S. s. cristatus* weighs as much as 265 kg (Prater, 1965), the largest Komodo individual we weighed was 48 kg. On this island the height of a large male is about 60 cm at the shoulder; total length 130 cm, tail 25 cm, head 32 cm. Hoogerwerf's (1954a) measurements for specimens from Rintja are comparable. The boars I saw on Flores were of the same approximate size as the Komodo specimens. There are none on Padar.

On both Komodo and Rintja the wild boar is most commonly encountered in savanna or open monsoon forest, rarely in deep parts of the forest unless it is hunted by man. The animal is omnivorous, feeding on roots, tubers, grasses of certain types (see Appendices 4–6), insects, fruits, snakes, and carrion. Most of the feeding takes place in the savanna and in the monsoon forest edge. During low tide they are often seen on the exposed rocky flats searching for marine invertebrates, particularly crabs which they eat commonly. They usually feed early in the morning and late in the afternoon, unless disturbed, when they feed mainly at night. Their sense of smell is acute, but their eyesight and hearing are only moderately developed. Thus, when actively rooting they are often easy to approach very closely. In parts of their range they are known to breed at all seasons of the year (Prater, 1965), though in most areas they are seasonal breeders, with either one or two breeding periods. The sow constructs a heaped-up grass nest, where she keeps and protects the young for several days. Komodo sows have from 2 to 6 young per litter. They are born from December to March and are less vividly striped than the young of *S. s. scrofa*, though some retain faint brownish stripes until the end of their first year of life. Repeated weighing of semiwild individuals at our Komodo base camp shows that by the end of one year they weigh approximately 15 kg.

It has often been reported that oras eat wild boar (Burden, 1928; de Jong, 1929). The analysis of droppings collected in this study indicates they are eaten commonly (Tables 10–2 and 10–3). Many of those devoured were undoubtedly attacked and killed by the monitors. Tanzer and van Heurn (1938) described ora attacks on several live, 2-month-old boars in the Surabaja Zoo. All were grasped by the throat and head and violently jerked from side to side. They were later bitten in the belly so that the intestines fell out, after which they rapidly died. Hoogerwerf (1954a) doubted they could be caught or overpowered in the wild, in spite of the fact that the villagers repeatedly described the boar- and deer-hunting techniques of oras to all the earlier expeditions (Table 10–2).

I have never seen oras kill a wild boar, though I have seen a 1.8-m ora eating at carrion lunge for the legs of a nearby boar of about 45 kg from a distance of 1–2 m.

The following observation of three Komodo villagers certifies that oras do kill boar and describes at least one of the techniques utilized.

> Nuhung, Hussain, and Ibraham were walking in the forest near Gunung Inakar in Loho Liang. They saw a large boar ambling along, with its head down, eating fallen tamarind fruit. An ora about 2 m long slowly approached the boar from behind. When he was approximately 1 m from the boar, the monitor lunged at it, grasping the boar on the right thigh. The boar squealed and tried to turn around and bite the ora, but the latter hung on. The ora then jerked back hard and the boar fell onto its side. Releasing the leg, the ora bit the boar in the soft flesh of the belly. Much blood was lost from both wounds. The ora started to eat the intestines before the boar had stopped moving. The men claim it took one-half hour for the boar to die, but it should be pointed out that the villagers have a poor sense of time periods less than 1–2 hours.

Lallemant (1929) suggested that boar are attacked along game trails and near water holes. Although we have no evidence of this, it is certain that oras lie along game trails often used by boar, presumably to attack passing prey. He stated the ora usually lies "over and obliquely" to the game trails. Our observations show that the position of the ora in respect to the trail is not consistent. The positions assumed along game trails by several large hunting oras on Komodo are shown in Figure 10–4. Lallemant's suggestion that oras hunt at water holes was undoubtedly based on mammalian predator habits more than on any definitive data of ora hunting techniques (however, see section on macaques). Occasionally oras do come to water holes, but they never seem to stay longer than required to drink. There was no evidence that oras hunt around these water holes. At low elevations there is water in the holes only from January to April. Lallemant's statement that oras lie in these water holes during moonlight nights to wait for prey to drink from them is unverified. Pfeffer's (1965) statement that oras sometimes hunt at night is discussed in Chapter 5.

Though we checked the grass nests of female wild boar regularly, we found no evidence that the oras visited them before or after the young were born. Perhaps the sow is too alert and defensive at this time. Most boar are probably captured on the savanna edges and in open monsoon forest, probably early in the morning or late in the evening; the known attacks witnessed by natives and our own crew would indicate the latter. After feeding in the forests during part of the night, the boar are often active on the savannas until 0900 or 1000 hr, especially during the rainy season when certain grasses on which they feed are very common. They are again found on the savannas from 1630 to 0100 hr, with an activity

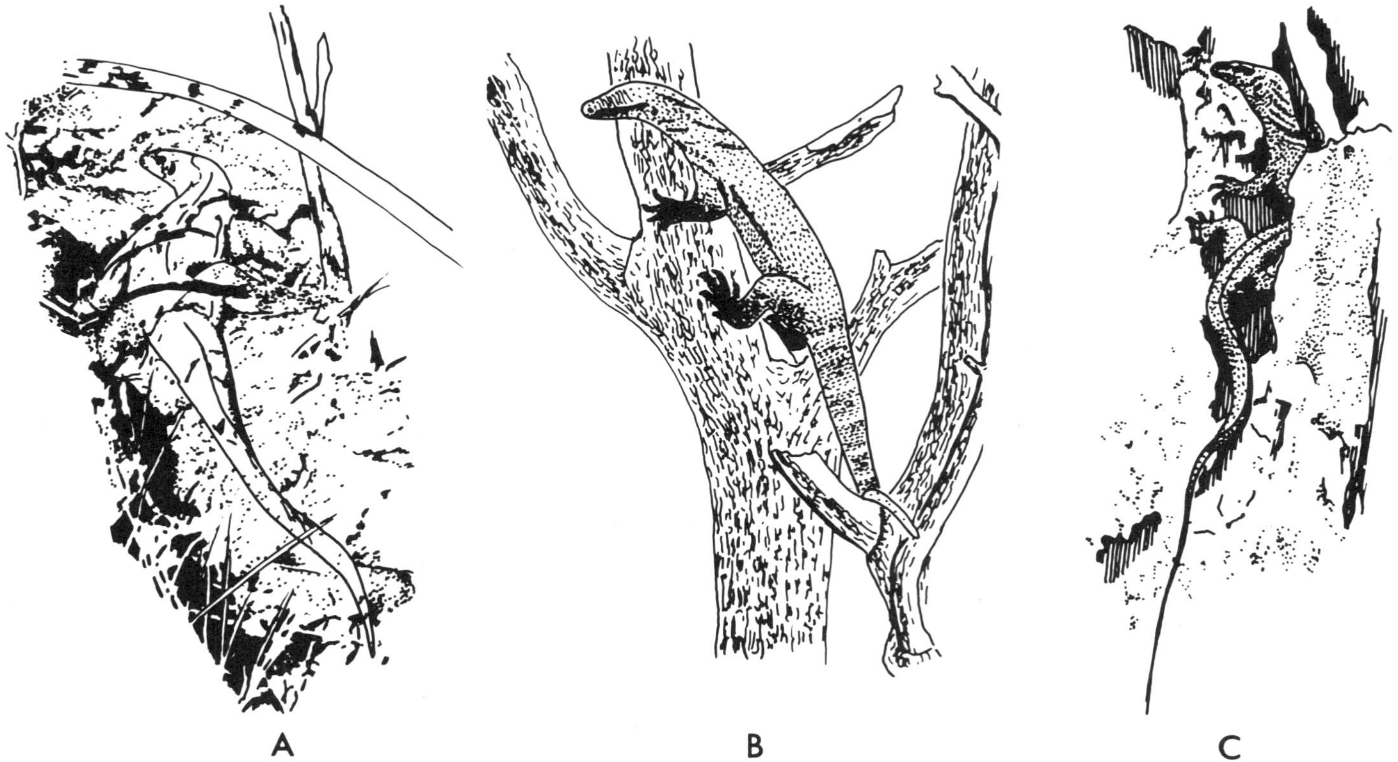

Fig. 10–4. Typical hunting behaviors of *V. komodoensis*. A: 2-m ora lying in game trail ambush site. B: 1.2-m ora actively searching for moving arboreal prey. C: ora hatchling actively hunting for inactive prey under loose bark.

peak from 1800 to 2300 hr. Most of the boar we obtained entered our live traps between 1700 and 2300 hr.

The only actual boar kill site we found was in November, 1972. In the *Hypoestes* brush west of Kali Besar and north of Sugito's Trail a large boar had apparently been attacked as it walked down a game trail. Broken and pushed down brush and grass testified to the extent over which the initial struggle and later feeding took place. All that remained of the boar were a few wisps of hair and the intestinal contents. The contents were composed almost entirely of freshly eaten *Sterculia foetida* fruits, suggesting that this undoubtedly large individual had probably been killed in the morning after his evening feeding activity.

Deer (Cervus [Rusa] timorensis).—The species of deer that inhabits the Lesser Sunda Islands,[3] including Komodo, is closely related to the Asiatic sambar, *Cervus (Rusa) unicolor.* The Sunda sambar deer is not well known. The following remarks are based on data obtained during our study of the Komodo monitors, unless specified otherwise.

Like most deer the sense of sight in this sambar species is moderate, though its senses of smell and hearing are very acute. Individuals communicate by means of several types of scent glands, by stamping their feet, and by an alarm call and the vocal challenge of large stags. They feed on a variety of grasses, leaves, and fruits (see Appendices 4–8). The stags maintain territories and their harems are limited to a few hinds. After the rut the male deserts the females and apparently leads a fairly solitary life until the next mating season. On Komodo pairing seems to take place from June to August, with the fawns dropped in sheltered places during optimum vegetative conditions of the rainy season from January to May.[4] Van Bemmel (1949) reported that on Flores young are born in October and November, which is expected in view of the earlier arrival of the winter monsoon at least at higher elevations. As is typical of the Asiatic sambar (Prater, 1965), the fawn lies alone for 2–5 days and is fed twice a day by the hind. After this time the fawn is strong enough to join the small herd. It apparently takes fawns two years to reach sexual maturity. Though definitely gregarious, the deer are rarely found in large herds, especially at lower elevations; 4 or 5 to a dozen is the usual herd size. Stags and hinds are often seen singly. Parties of hinds and fawns without stags are common.

The subspecies to which the Komodo population is assigned (*C. timorensis floriensis*) is the smallest of the species (van Bemmel, 1949). A large stag on

3. Previously reported from Lomobok, Sumbawa, Flores, Adonaire, Solor, Sumba, Padar, Rintja and Komodo (Hoogerwerf, 1954a), this opportunity is taken to add the island of Gili Mota to the known distribution.

4. The Kepala (= village chief) of Kampung Komodo told me that deer at higher elevations *may* have two litters per year. Van Bemmel (1949) stated that deer inhabiting mesic habitats in Indonesia breed over a longer portion of the year than in temperate regions and that multiple peak breeding periods are possible in a single year. But he also stated that these periods are extremely variable according to region and that in some parts of the archipelago a more or less pronounced breeding season seems to occur.

Komodo weighs about 200 kg, has a total length of about 155 cm, and is approximately 800 cm high at the shoulder. Hinds are considerably smaller. The average weight of Komodo specimens is about 50 kg (Table 11–6). Statistical comparison of two measurements on 48 jaws from Komodo and 6 from Padar suggests very strongly that the two populations are identical in size. Van Bemmel (1949) provided data on Sumba, Sumbawa, and Flores deer, showing that they are also approximately the same size.

Many earlier expeditions have reported that deer constitute an important staple of the ora's diet (Tables 10–2, 10–4, and following). The major questions were whether the deer was actually killed and, if so, under what conditions. It is now clear that most attacks are made under two conditions: stealth of an ora approaching sleeping deer and ambush along game trails. The latter has been reported several times (e.g., Lallemant, 1929; de Jong, 1937; Pfeffer, 1959). Darevsky and Kadarsan (1964) described an ora stealthily approaching a grazing deer in an open savanna while periodically raising its head high above the grass to see its prey. At one point when the deer approached the ora, the latter stopped and lowered its head into the grass. Pfeffer (1959) was correct when he stated that the ora is much too slow to capture large prey by active pursuit. The present information suggests that the ora's usual attack technique for deer is to grasp a leg or the throat, or even the narrow lumbar area if the deer is lying on the ground. The teeth are the major and probably the only offensive weapon. Native tales of bipedal attacks, in which the claws of the front legs are used to lacerate the prey, are undoubtedly based on poor observations of actual attacks in which the bites were high on the body or throat, and the ora stood on its hind legs during the immobilization stage. The tail is completely defensive as far as I can determine and is not used to knock the prey off its feet during the attack, as was stated by Maleyev (1962). Piazzini (1960) disbelieved native reports that deer are sometimes hunted by groups of oras. I never heard of any group hunting by oras, though I talked to dozens of villagers on both Flores and Komodo. In my opinion Piazzini misunderstood their description of ora hunting or the villagers may have been discussing the hunting of feral dogs.

The Sunda sambar, like its Asiatic relative, feeds mainly at night. Many other Asiatic deer are more diurnal. During the day small groups of the Sunda sambar usually are found in thickly vegetated areas within their normal activity range. Here in thickets or under particular species of evergreen bushes (*Azyma, Murraya,* etc.), small herds ruminate or sleep during much of the afternoon.

Normally the deer drop few pellet groups during this part of the day, so fecal scent is not a major factor in the location of sleeping deer herds by oras. There is evidence in some ungulate studies that glandular exudates from the metatarsal glands are important in marking sleeping and resting sites. If this is also true of *Cervus timorensis*, it is highly likely that oras find sleeping individuals by this means. However, data obtained on Komodo suggest that the hunting animal may depend more on past experience and knowledge of the area, checking in a rather uniform manner all the potential resting places on or close to his particular hunting circuit during that afternoon (Fig. 10–5). Komodo deer often sleep in the same thicket for several successive days. It is possible that they move to another only after they have been frightened from the original one by a hunting ora or a pack of feral dogs.

By 1600 to 1700 hr the deer have usually risen from their afternoon resting

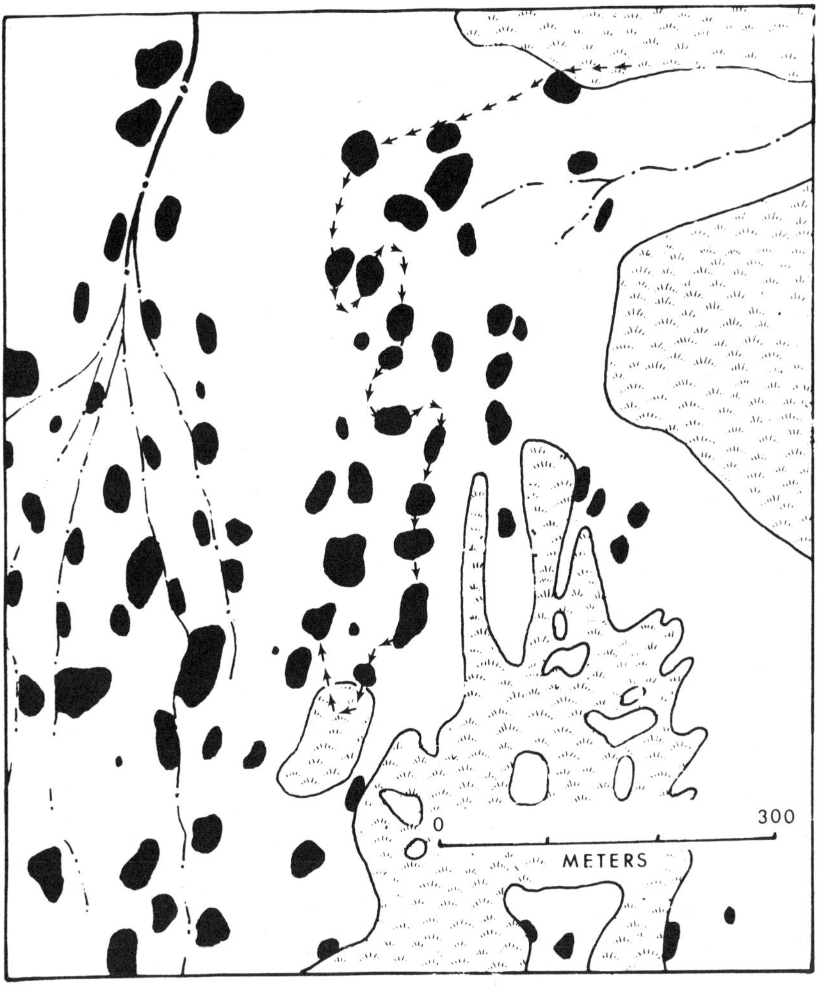

Fig. 10–5. Path of ora 19W (arrows) through brush area in monsoon forest (clear) near Kali Besar. Thickets in which deer frequently sleep during the day are shown in black.

places and slowly move through the forest toward one of the savannas within their activity range. At these times it is conceivable that the ora might hunt for them along game trails, but we have no evidence of very much hunting along these trails at this time of day. The deer slowly move out into the savannas before and during the twilight period and begin to feed. At this time they are not restricted to game trails but are widely scattered as they move toward the savannas. It is nearly dark by the time the deer are on the savannas, and the lizards have already retired for the night. The deer remain on the savanna all night, sometimes in large herds of as many as 54 individuals. At dawn they move toward the bare, grassy hilltops closest to the open savanna in which they have been feeding. This would be a good time for the ora to attack the deer as they move up the slopes on the narrow deer trails, except that the oras are not yet themselves active at this time of the morning. The deer move up the slopes and usually lie down in small groups on grass-covered spurs above the forest valley floor. Normally they lie there, awake and ruminating, until 0900 to 1100 hr, when they move back down the slopes into the thicker valley floor vegetation, presumably because of increasingly hot temperatures on the open slopes. It is at this time that the ora apparently hunts the deer from ambush along the game trails leading to the valley below.

There are usually several major game trails leading from the slopes on which the deer lie to the thickets below. Apparently the lizards know the location of these trails, and large residents sometimes have trails of their own leading to them. Oras lie along deer trails in the morning, often in areas of game trail concentrations (Fig. 10–6), anticipating the small troop of deer to move past. Of course, the possibility of a successful hunt along any one trail in the morning is slight, but the level of success may be sufficiently high to encourage this behavioral pattern, as many of the larger oras hunt this way almost every morning.

After moving down the slopes and into the deeper valley forests, the deer move rapidly to their daytime sleeping places, frequently the larger oras slowly follow them into the thickets to resume their afternoon hunting patterns of attempting to capture deer sleeping under *Azyma* and *Murraya* shrubs.

That deer realize the danger of a large ora nearby is suggested by our observations of February 21. A full-grown doe was seen entering the savanna "cover." The deer saw the advancing ora and immediately stood very still, except for its ears that seemed to "search" the surrounding area for noises of the approaching ora. The predator apparently did not see the doe and walked within 10 m of it. Only after the ora had passed did the doe continue on her way.

On the basis of over 4,000 droppings examined from Komodo and 341 from Flores (Fig. 10–1; Tables 10–4, 10–6), deer constitute the major food source of large oras on those islands. This is the conclusion of almost all the workers on these islands, from Horst in 1926 through Darevsky and Kadarsan in 1964. Villagers from Rintja told me that only larger lizards attack full-grown deer,

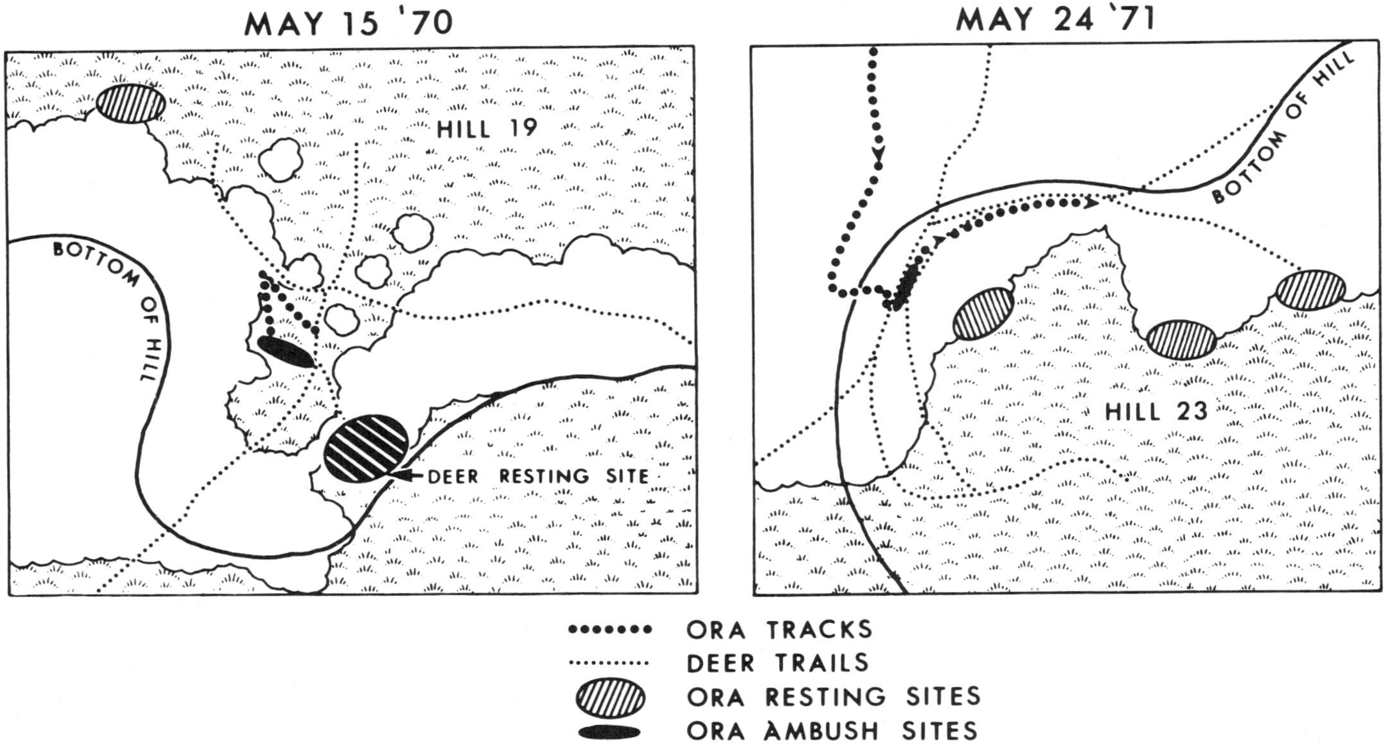

Fig. 10–6. Ambush sites are often located near a concentration of game trails. Both illustrations show the relation of important factors in the ambush hunting of a large monitor (23W) in Loho Liang on two different dates.

rarely by stalking them in tall grass, most commonly by lying in wait next to deer trails. They said that the ora bites the deer wherever it can but normally on the throat or hind legs. Even if bitten well, they stated that it usually takes from 1 to 1.5 hr for the deer to bleed to death. This figure seems much too long on the basis of our own observations.

Sleeping deer are often pulled from beneath cover vegetation: On October 9, I was told that a 1.5-m ora had found a "young deer" under a "nyolo vetok" (*Azyma sarmentosa*) on the far side of Komodo. The ora had grabbed the deer, pulled it from under the shrub, and killed it, but the witnessing villager did not tell me how long it took, or where the ora bit the deer.

Several years ago a Komodo villager named Achmad was working in his garden at Loho Lavi when he heard a deer give an alarm call. Soon thereafter a woman called to him from the forest edge. Running to the place he found a 2-m ora holding onto the rear of a thigh of a full grown stag deer with antlers. The deer was lying on the ground, struggling to get to its feet. The ora held it for a "long time," until Achmad chased the ora away. The deer was bleeding very badly from several large lacerations on the hind leg and could not walk because the Achilles tendon of the left leg had been severed.

De Jong reported (1937) that while he was in western Flores he heard a deer scream. Walking in that direction he soon found a dead, still warm fawn with a large wound on the side of the neck. A large ora stood over it (Pl. 7).

On April 25, along Kali Besar, Komodo, I heard a deer give an alarm call and then scream repeatedly but more weakly at a rate of one bleat per minute for 10 min. I could never determine what had happened but believe that a deer had been brought down by a predator. Since I had heard no hunting dogs, it may have been due to a successful ora attack. The area was within the normal hunting range of 8W.

Though I never witnessed an attack on a deer, I did arrive at the scene of a deer kill just a short time after the death of the animal:

> *October 31, 0930 hr.* While going to the second grass patch several members of our group heard one dog barking and a crow calling from just inside the forest edge. On investigating the disturbance three oras were found to be tearing flesh from the thorax of an adult doe deer of approximately 80 kg, so recently killed that the blood was not yet coagulated. Three oras were feeding from it; one unmarked 1.5-m specimen, Whiteface, and 19W. Before it was completely eaten, the remains of the carcass were dragged away. From signs nearby the entire sequence was easily reconstructed:
>
> During this part of the year five deer slept above the small savanna on either the north or south spurs of hill 23 every morning. Between 0900 and 1100 hr these deer would move down the slope to the savanna below, cross

Pl. 7. Adult monitor feeding from the carcass of a Sunda sambar deer (*Cervus timorensis*).

it on one of several well-established game trails, and enter the forest near the junction of Sugito's Path and the North-South Trail. This same path had apparently been used earlier this morning. The ora must have been lying in grass about 0.7 m high along the game trail [Fig. 10–7] and probably attacked the deer at this point, for here we found much fresh blood and pushed down grass. The deer must have dragged the ora 5 m farther, where we found more blood and another small area of grass pushed down. The deer must have then struggled 3 m more, where it fell or was pulled down for the last time. Here it certainly died. Intestinal contents were scattered over an area of packed down and torn up grass about 1.2 m in diameter and 2 m from the forest edge, and there was evidence of massive bleeding. After the intestines were eaten the carcass was dragged about 16.5 m into the shade of the forest. There much of it was eventually eaten. The two smaller oras, the dog, and the crow had already arrived, probably drawn by the scent of the fresh intestinal contents. The remaining part of the carcass was later dragged for at least 16.5 m, where everything was eaten except the rib cage. Eventually it too was swallowed entire.

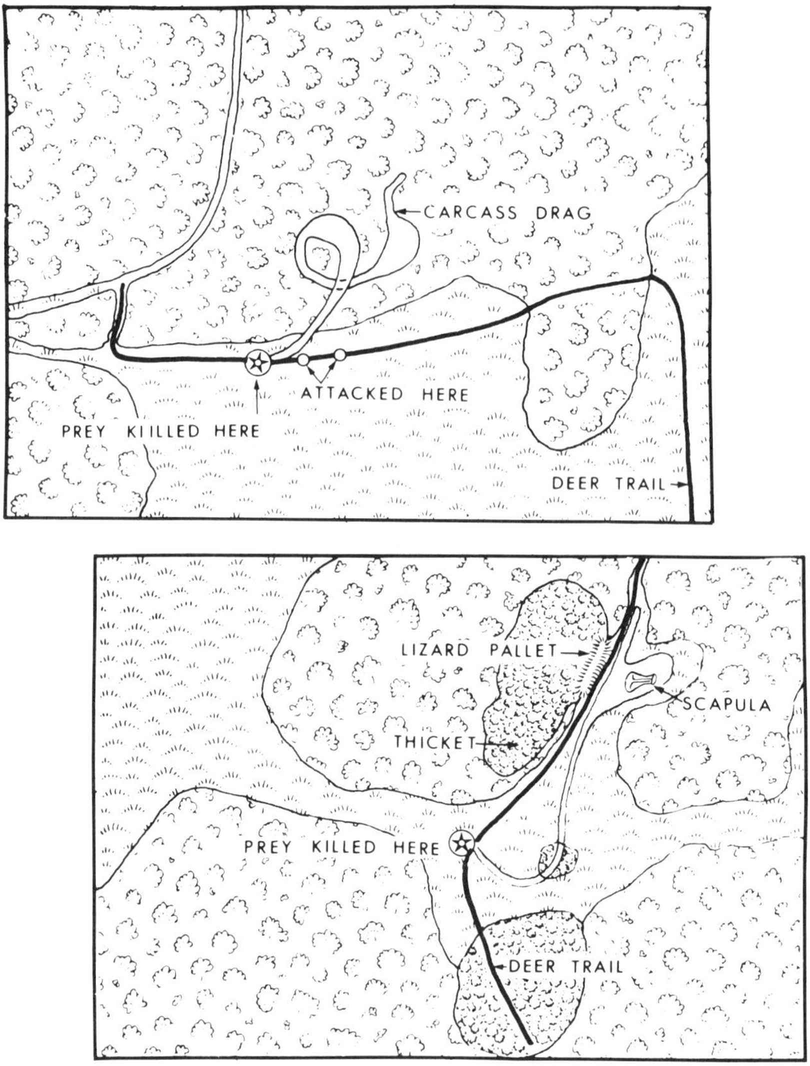

Fig. 10–7. *Top*. Deer attack site. Doe, ca. 45 kg, attacked and killed by a 42-kg monitor as small herd moved along game trail from lower right. Attack occurred as soon as herd stepped through the small finger of forest and back into savanna; location of ambushing ora unknown. *Bottom*. Deer attack site. Ambushing ora lay in pallet along game trail under edge of thicket adjoining trail. Deer herd moved down trail from area at top of figure, and a successful attack on a 27-kg deer occurred along trail between pallet and spot where prey was killed. Clear areas represent places where carcass was dragged and eaten.

A second kill site in the overflow area at the mouth of Kali Besar is also shown in Figure 10–7. Parallel and next to the deer trail was an area about 1.5 m × 0.5 m, scraped free of debris and containing the footprints and belly scale impressions of a large ora. A deer moving down the trail was apparently attacked near this point and eventually killed and eviscerated 3.5 m along the game trail. It was then dragged in a circle to near the attack site, where, with the exception of a single scapula, it was entirely consumed. This element was later compared with scapulae of deer of known weight and proven to be from a subadult deer (probably doe) of about 20 kg.

Horses.—Very little is known of the feral horses of the Lesser Sunda Islands. None occurs on Komodo or Padar, though they are common on Rintja and are found in small isolated populations on Flores (Hoogerwerf, 1954a). Flores is well known for its large number of domestic horses. These are small, standing about 1.2 m high at the shoulder, and have an average adult weight of about 250 kg. Foals are apparently dropped through most of the year and weigh 20–25 kg.

On Rintja, feral horses are most commonly encountered in savanna and open woodland, particularly near perennial water holes in the larger valleys (Hoogerwerf, 1954a). Here during the day they are said to graze in herds of up to 10 individuals, though usually less.

Rintja villagers told me that adult horses on that island are not attacked by oras, but Burden (1928) was told the opposite. He learned that just before his expedition stopped at Sumbawa a captive, tethered ora attacked a domestic horse, which was so badly wounded it had to be destroyed.

During our studies in western Flores we learned that the semidomestic horses of the Loök-Nggoer area are often attacked and killed in *Zizyphus* savannas near the villages (Table 12–1), usually when asleep but also along game trails above the villages. Near Nggoer, an adult horse was unsuccessfully attacked on July 4, 1971, from a distance of 1 m in an open *Zizyphus* savanna.

Mares and foals often live separately from the stallions, and villagers believe this lack of protection may account for the number of foals said to be eaten. However, on the afternoon of July 13, 1971, a large ora ate a stillborn foal I had noticed lying dead on the savanna floor that same morning. I suspect many of the foals eaten are obtained as carrion. However, I have repeatedly been told that oras sometimes remain near a mare nearly ready to give birth, in order to obtain the newborn animal. Flores villagers said they have seen oras beneath a hunched and straining mare waiting for the foal to be dropped, sometimes so close that the mare tried to kick the dodging lizard.

Water buffalo (Bubalis bubalis).—The water buffalo has a long history of domestication by man. Today feral and domesticated individuals are found in many parts of the world where before they were unknown. Indigenous wild herds are known definitely only from the Asian mainland. According to the islanders, the water buffalo on Komodo resulted from the introduction of a single pair of animals from Sapi, Sumbawa, in the late nineteenth century. The villagers estimate that there are less than 100 individuals on the island at the present time.

Though Burden (1928) reported water buffalo at elevations no lower than 650 m on Komodo, we often saw them as low as 300 m in the hills overlooking Loho Liang, and on the northwest coast they are found near sea level, as they are on Flores. Apparently the major factor in their distribution is the presence of permanent water pools. On Komodo they have not yet crossed the wide, dry savanna at the isthmus near Loho Lavi, so they are absent from the mesic forests with permanent water holes in the Gunung Vau area south of the savanna. None are found on Padar, though feral herds are known from both Rintja and Flores. Both domestic and feral water buffalo are found over most of Flores.

On mainland Asia large wild bull buffalo average 170 cm at the shoulder and weigh up to 909 kg. Domestic or feral individuals are generally smaller, weighing about 680 kg. The same relationship is found in the Lesser Sunda Islands, though the adult sizes are generally smaller. On Flores large bulls weigh 590 kg, and full-grown females are about 400 kg. The first calf is usually born to females in their fifth year, when they weigh about 350 kg ($\bar{X}$ 16 specimens = 275 kg). The calves weigh 30–40 kg. Komodo animals are about the same size and weight as those from Flores; Rintja specimens are undoubtedly similar.

The prime habitat of feral herds on both Komodo and Flores is tall grass savanna with intermittent marshy swales and scattered patches of open forest. On Komodo they are most common in savanna and open forest areas containing pools of water maintained by both themselves and wild hogs. They are often found in small herds of up to a dozen individuals (Hoogerwerf, 1954a), grazing on the savanna grasses in the early morning and late evening—rarely at night. Normally they sleep during the afternoon in *Themeda, Imperata,* or other tall grass thickets or in pools. They have moderately good sight and their sense of smell is said to be particularly good. Komodo villagers told me that normally one or rarely two calves are dropped in the early part of the January monsoon. The breeding season is apparently March to May.

Though most water buffalo eaten by oras are probably dead ones, live individuals are often attacked and killed (Pl. 8). The best documentation of such instances is in Mangarrai District, western Flores. Here many domestic buffalo are attacked and killed in "pastures" near the villages, when either tethered or free. Data on numbers killed are provided in Chapter 11. In this region attacks have often been witnessed by the villagers. During June–July, 1971, there were two attacks on buffalo near the village of Djarek, though we failed to see them. Father Schmutz, of Missi Nunang, Mangarrai, Flores, provided the following records from within his parish:

During the month of September 1969, one male and three female water buffalos are known to have been attacked by a single large monitor near the village of Nggoer. All the buffalo were bitten on the lower limbs [see Pl. 3 and Auffenberg, 1978], and in one case the buffalo could not stand and had to be killed because the abductor tendon was bitten through. All four legs of one buffalo had been severely bitten. Two of the buffalo were tethered. Early in the morning of October 6, 1969, a pregnant female buffalo, which had already had four calves (about 400 kg and 9 years old), was tethered in the savanna near the village at the forest edge. At 0900 hr some boys were to bring it to water, but found it dead, with a large ora next to it. The Achilles tendon was bitten through and the belly had been ripped open. Recruiting villagers with spears and hounds, they returned, surrounded the ora, and killed it.

The total length of the ora (measured by Father Schmutz) was between 275 and 280 cm. It was distinctively marked with a peculiar group of white flecks

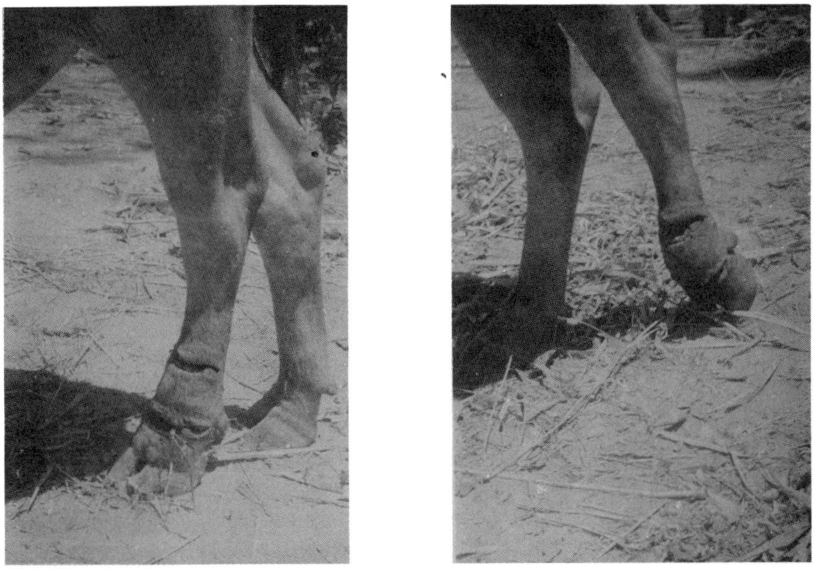

Pl. 8. Photo of 320-kg female water buffalo recently attacked by an adult ora approximately 2.8 m in length, Nggoer, Flores. Note the nature and severity of wounds, plus edema caused by infection. Wounds of the upper jaw appear above the base of the hooves on the anterior surface (*left*) and those of the lower jaw on the posterior surface (*right*). The adductor tendons had been severed, and the buffalo could no longer walk. From photo by Erwin Schmutz, S. V. D., Flores.

around the narial openings, and had been given the name "Seloba." This particular individual was well known to the villagers and was said to have inhabited the same general area for at least 15 years, during which time it was known to have attacked and killed a number of water buffalos.

Domestic buffalos as heavy as 590 kg have been killed by oras between 2.5 and 3.0 m long. On the basis of both the buffalo remains found in the Komodo ora droppings and the statements of Flores villagers, buffalo calves are more commonly attacked and eaten than adult animals. In some small savannas near the villages of Nggoer, Loök, and Djarek an average of as many as 5 buffalos are killed each year (Table 12–1).

The Komodo villagers have apparently never seen oras attack a live water buffalo, but it is well accepted that they are eaten. The manure of large oras living at high elevations on Komodo often contains hair, hooves, phalanges, carpals, or tarsals of these animals. One large mass of water buffalo hair disgorged by an ora at Galo Matavelok contained a horn sheath 24.12 cm long and 7.42 cm in diameter at the base. This animal must have been very nearly adult (four adult horn sheaths from Komodo measure 33.8 × 8.6, 36.2 × 9.2, 35.5 × 12.7, and 32.0 × 8.3 cm), with a probable weight of about 318 kg. Several of the hooves were from half-grown animals, about 226 kg. A group of buffalo calf teeth in one dropping from Banu Pei suggests the entire head of an animal about 54 kg must have been ingested.

Buffalo calves accompany the herd (Hoogerwerf, 1954a) and so cannot be attacked in isolation, as is possible with fawn deer. Both feral adults and calves are probably most often attacked along the meter-wide buffalo trails connecting commonly grazed savannas at higher elevations.

Cows (Bos taurus).—Wendt (1956) incorrectly stated that "Bali-cows" occur on Komodo. Records of cows having been attacked, killed, and eaten by oras are only available from Flores (cows occur on no other island on which the ora is found). Such attacks are apparently rare, perhaps because the cow is a fairly uncommon domestic animal on Flores, particularly in that section of the island inhabited by the monitors.

Man.—Human beings are occasionally attacked and on rare occasions killed and even eaten. Details are given in Chapter 12.

Foraging activity.—Sample foraging patterns are shown and discussed in Chapter 5 (Figs. 10-8, 10-9).

Possible Selection of Pregnant Prey Animals

Perhaps one of the most interesting facets of predation by this animal is the often-repeated statement of villagers throughout its range that large oras monitor the movement of female ungulates nearly ready to give birth. They state that full-term individuals are often attacked and killed and that the fetus is often snatched from between the legs of a female giving birth. Most knowledge of this type of predation is obviously associated with domestic animals, particularly goats. However, it has been mentioned in regard to horses and water buffalo as well. During our year on Komodo we witnessed two instances in which a large ora (19W) repeatedly attempted to attack pregnant goats tethered in our camp. In one case the ora visited the camp for four successive days. Each time it sought out the pregnant animal (the only pregnant one in camp at that time), though the goat was often moved to foil the attempts of the monitor to attack it. The goat gave birth to twin kids, which were eaten by the monitor. The monitor then tried to kill the female but was driven off by people in the camp. The goat was finally moved 2 km down the beach. On Flores, villagers at Loök-Nggoer state that mares carrying near-term fetuses are more often attacked than stallions (though less than colts?). These and similar statements need verification.

A possible solution to the stated high frequency of attacks on pregnant ungulates is provided by one of our observations at the base camp on Komodo in April, 1970. Ora 19W had come into camp and evidently "worried" several tethered goats. It was driven off just as it was about to attack a tethered and very

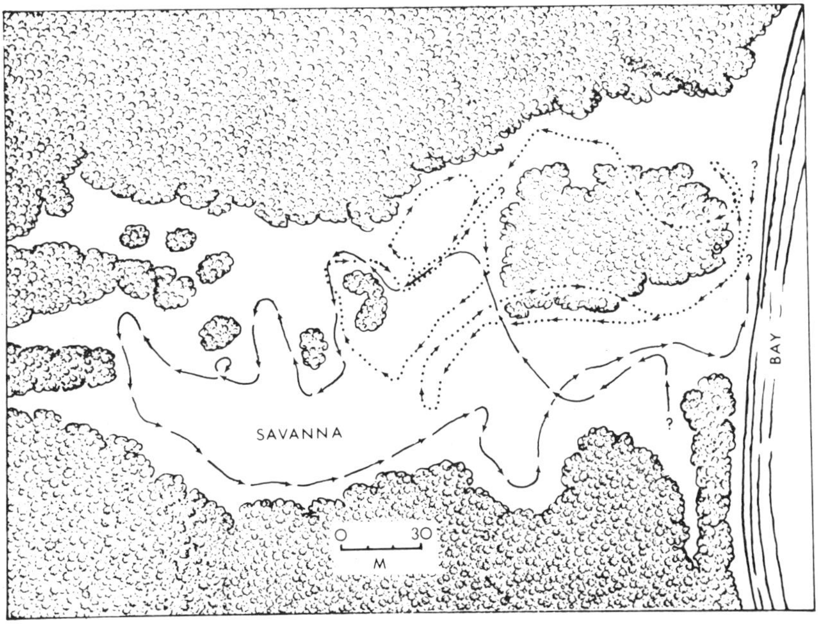

Fig. 10–8. Approximate diel foraging pattern (immature ora 50W, 52.7 cm snout-vent length). Two consecutive days are shown. Note concentration of activity in savanna (clear) area, where largely locusts were being hunted. Compare with adult ora foraging pattern in fig. 10–9.

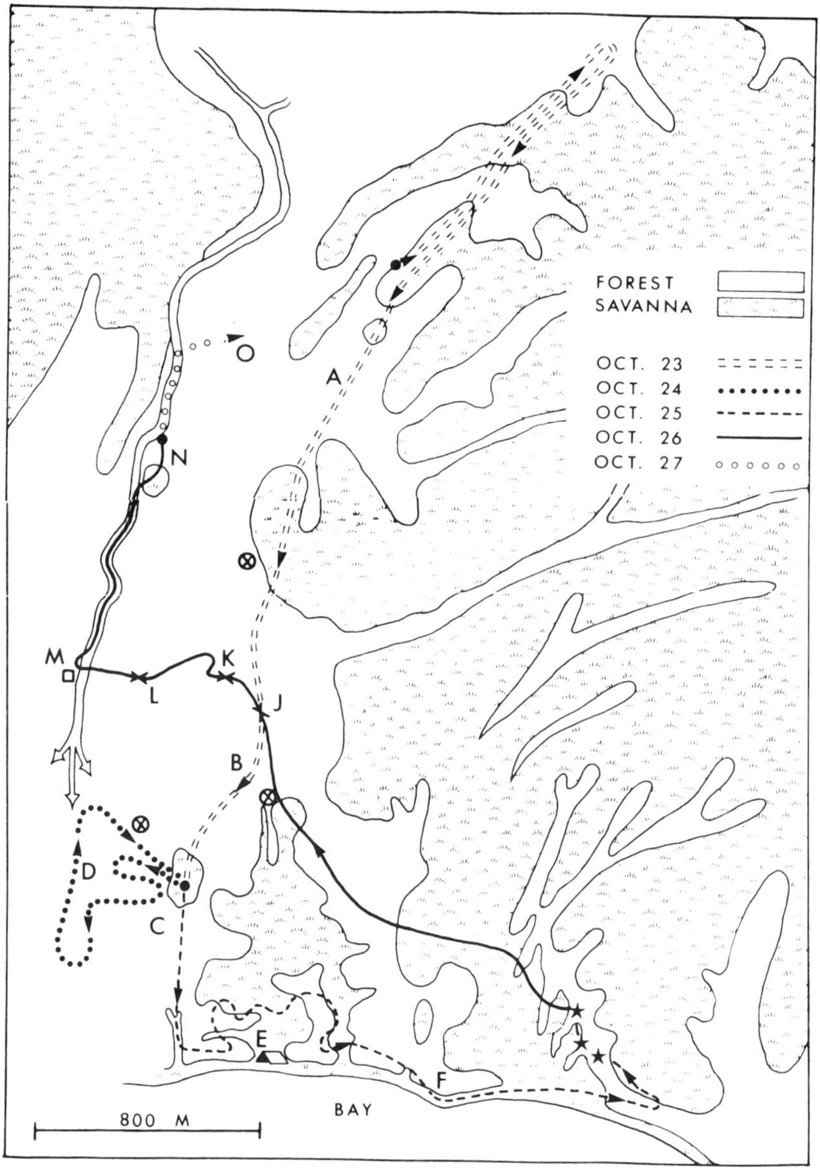

Fig. 10–9. Five-day foraging pattern of ora 19W (3-m snout-vent length). Oct. 23: Spent almost entire morning and early afternoon (A) moving northeast through the monsoon forest to Kali Inaloah, then doubled back along a major game trail paralleling the base of hill 23. Did not seem to be searching for food. In late afternoon it left game trail (B) and hunted the overgrown garden area, paying close attention to thickets, apparently searching for sleeping ungulates.

(Continued)

frightened young female. Later in the day the same animal showed signs of severe stress at the approach of the returning ora. Several hours later we noticed she had miscarried an approximately half-developed fetus. The miscarriage was probably caused by the stressful situation of the near attacks and presence of the ora.

Perhaps even minor repeated attacks on pregnant animals produce a sufficient number of miscarriages to encourage identification of pregnant prey and persistent subsequent annoyance of it by a large ora.

The fetus and afterbirth may not be the only reward, since the miscarriage itself tends to incapacitate the female for several hours following the accident. If actually accomplished, the identification of pregnant prey by oras must be based on their ability to distinguish between the scent of pregnant and nonpregnant prey. This is not impossible since the prey animals themselves make such distinctions by scent, and the oras are obviously extremely efficient in their own ability to distinguish important scents from one another.

Evening spent in large *Azyma* thicket (C) near Trap 1, having covered a distance of approximately 1500 m during the day. Oct. 24: Seemed to be hunting for deer and wild boar while moving about in the tangled overflow area west of Trap 1, slowly covering a distance of about 980 m (D). Evening again spent at (C). Oct. 25: Early in morning it checked Trap 1 for carrion, then went to beach by walking down a dry creek bed. It then paralleled the bay, circling the base camp. By midday it was searching the beach wrack for marine carrion (F). At 1700 hrs it lay on a small promontory overlooking the bay and just before nightfall dropped into the forest edge, spending the night under the vine-covered tangle of a fallen tree (G). Total distance moved during the day was 1160 m. Oct. 26: By 0730 hrs it had already searched for jungle fowl nests under two tangles similar to that in which it had spent the night (H and I). At 0830 hrs it crossed a small savanna and a spur of hill 19, past burrow 17, where it was known to have spent the night on other occasions, moved through the monsoon forest west of the hill, then lay down next to a major game trail (J), where we had found it in ambush several other times. Here it was accidentally frightened by some villagers and moved almost west through an overgrown old garden area, where it frequently hunted for sleeping deer. This day it checked two *Azyma* thickets (K and L) and then moved to the megapode nest (M), remaining there for almost three hours. Late afternoon it moved up Kali Besar, spending the night in a large burrow it had used several times previously (N). Total distance moved during the day 1700 m. Oct. 27: By 0920 hrs it had already left the burrow area, walked up the creek bed, and entered the forest, moving to a zone of high deer density (O). Trail lost at this point. During entire period it used game trails sparingly when moving from one area to another. Very little movement took place in the savanna area (monsoon forest clear).

Possible Trailing of Prey Animals

The scenting efficiency of oras is of possible significance in predation in another very important way. Lederer (1942) showed that a 2-m captive ora could follow without any difficulty the spoor of rabbits allowed to run free in large rooms, unseen by the ora. This suggests that oras may be able to trail deer in the wild as accurately as hunting dogs do. Though there is considerable confusion and much lack of data regarding the scents utilized by predators in trailing ungulate prey animals, most authorities are of the opinion that scent organs in the skin produce the olfacients used as orientation cues by a pursuing predator.

However, available data on skin glands of artiodactyls is fragmentary and often contradictory. Several conflicting statements by early workers regarding the glands of the feet make it obvious that the entire problem of the distribution and type of pedal glands in Asiatic deer needs further study. It is now clear, though incompletely documented, that there is marked interspecific variation. There is some evidence of sexual differentiation (Cowan, 1947), and changes with age and season are highly likely in many species (Quay, 1959, 1966). Four body areas are of primary interest: tarsal, metatarsal, interdigital, and preorbital. In most species all these areas contain both sebaceous and apocrine sudoriferous glands (Quay, 1959, 1966). Although no conclusive data are available for *Cervus timorensis*, my cursory examination suggests that all four glandular areas are present and that those between the hooves of both the front and hind feet are well developed, as they are in the closely related *C. unicolor* (Pocock, 1910; Ogilby, 1936; Schaffer, 1940).

The metatarsal gland contains an area of enlarged scent-producing glands (Quay, 1959) that is usually brought in contact with the ground when the deer are lying down. Thus, it is suspected that secretions from this area may be important scent markers of individual or group resting and sleeping areas and are undoubtedly important cues by which oras find sleeping deer. Sometimes the scent produced around some of the deer sleeping areas is extremely strong, and if the wind was right, even I could orient myself in respect to them from distances up to about 150 m. Figure 10–10 shows the location of a series of deer sleeping forms on a Komodo savanna and a large ora's exploratory movements around them (see also Fig. 10–5).

The most important glands are probably the interdigitals. These very complex structures (see Quay, 1959, for a description of their histology) are located in the dorsal interdigital skin, and are better developed in the hind feet than in the front feet. While they may not have the same function in all artiodactyls, it is highly likely that these glands are used as scent organs by both sexes and that their function is to leave an olfactory signal identifying particular groups or individuals or to leave a warning scent when frightened.

Deer hounds are believed to follow three "kinds" of scents: foot scent, bush

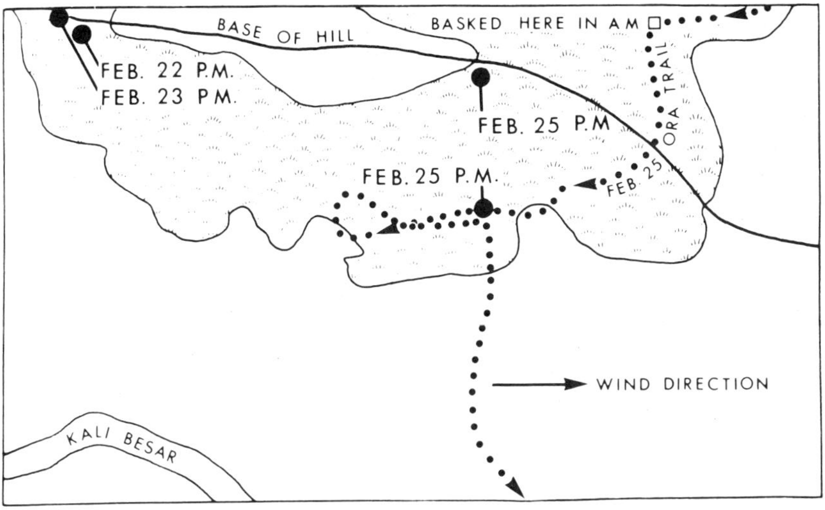

Fig. 10–10. Sleeping sites of deer (solid circles) on different evenings and track of hunting 2-m ora (dotted line).

scent, and air scent. A predator following close behind jumped quarry has no time for the part on the ground: "He runs openmouthed, sucking the hot scent in out of the air with the breaths between his cries" (Carr, 1962). But this trail is short lived and of use to only a rapidly moving predator. The more durable spoor are the bush and ground scents, particularly the latter. As every hunter knows, a good hound presses his nose into the individual prints regularly. He loses interest if they are old. More important, he knows which way the deer was going. This is evidently accomplished by recognizing a scent gradient along the trail and selecting the direction of the greater concentration and strength of the downtrail end of the segment. I see no good reason why a highly specialized, scent-oriented organism, such as the ora, could not do exactly the same. There is at least a little field evidence on the function of interdigital glands of cervids bearing on trailing of predators. Dugmore (1913) believed frightened barren-ground caribou left a trail which wolves followed. Pruitt (1960) described an "excitation jump," resulting in the deposit of an alarm scent from the interdigital glands. Though much work remains to be done to obtain proof, a similar function can be imagined for the digital glands of the sambar deer group. Such a signal would serve primarily to alert other deer of a potentially dangerous area. This scent would be expected to be deposited mainly at attack sites, though some would undoubtedly be carried along the escape trail itself. A predator able to detect this scent would be able to track the frightened animal by moving down the scent gradient as deposited in the spoor of the escape trail. Furthermore, the ora's afternoon foraging pattern for sleeping deer may not be as random as initially suggested by the data. Thus, the directions taken by hunting oras may reflect the movements of deer in seeking a thicket for the afternoon nap.

Sweeney, Marchinton, and Sweeney (1971), who studied the responses of deer being chased by hunting dogs, showed that the evasive behavior of the deer fell into several categories. Of these, one was particularly interesting from the standpoint of ora movements. In this escape pattern the deer follow a circuitous zigzag course in which they often pass over their own trails or the spoor of other deer. The authors found it to be behavior that managed to lose the pursuing dogs in over one-half of all the trials without necessitating a great expenditure of energy or movement out of the deer's activity range. The resultant tangled trail is remarkably similar to the movement patterns of some large oras and could be responsible for producing them. The entire matter of ora hunting techniques deserves careful study. Information is needed on (1) the presence and development of interdigital glands in *C. timorensis,* (2) how the pedal gland scent is used, (3) the degree to which a large ora can separate it from other scents, (4) whether a hungry ora would follow such a trail, and (5) the simultaneous, detailed movements of a deer and an ora hunting it after an unsuccessful attack, *or* the daily movements of a deer herd and a hunting ora.

Transmission of Incapacitating Infections

Many medium to large species of *Varanus* have the reputation of possessing poisonous bites. Smith (1935) reported that in India it is the young that are regarded as venomous. Normally, even uneducated hunter-gatherers, such as some of the Australian aborigines, realize that it is an infection that is transmitted by the bite of some *Varanus* individuals. An Australian folk saying reflects the severity of some of these infections by stating that a monitor bite will re-erupt every year for seven years (Cogger, 1959). However, several writers doubt this infectious quality of the bite and present as evidence *Varanus* bites they themselves have experienced with no ill effects.

Varanus komodoensis has the same reputation, but on the basis of two aseptic bites known to me (by 1-m and 1.2-m oras), I might be led to the latter conclusion—that the infectious feature of an ora bite is a folk myth. However, at least one person is known to have died from massive sepsis (bacteremia?) four days following a severe ora bite in the biceps. Another human death two years after an ora bite on Flores sounds suspiciously like a septicemic condition accompanied by metastic abscesses. Other less severe cases of definite infections following bites are known. In domestic animals septic bites are very common and are considered a serious postattack threat to horses and water buffalo in western Flores. The bites are characterized by severe cellulitis, often breaking down into foul-smelling, purulent lesions.

The same result is expected following an attack on wild animals. Though apparently very few infected deer or pigs are seen by Komodo villagers, it is significant that they *do* see them. During our study on Komodo we saw only one badly infected deer. It was having great difficulty walking due to a large liquefied and infected wound on the thigh.

A recent study of Newfoundland caribou (*Rangifer tarandus*) by Bergerud (1971) concluded that an abscess syndrome, often fatal, was caused by bites of the lynx (*Lynx canadensis*) in which the calf caribou escaped the predator. Dead and moribund neonatal caribou calves were often found with cervical abscesses resulting from the pathogen *Pasturella multicila,* mucoid type A. Death usually occurred from 4.5 to 15 days after the attack.

In November, 1972, we had an opportunity to study the buccal bacterial flora of several wild oras on Komodo. Mucoid samples taken from the external gum surface of the upper jaw of two freshly captured oras (approximately 1.90 m and 2.25 m in total length) were transferred to a liquid growth-retarding media and recultured approximately every 48 hr. Other mucoid samples from the same oras were prepared in the field as acetone-fixed thin smears. Both sample types were studied and identified by the Clinical Pathology Laboratory of the University of Florida Medical School.

Four bacterial species were identified, all capable of producing severe infections: *Staphylococcus* sp., *Providencia* sp., *Proteus morgani,* and *Proteus mirabilis.*[5] With the exception of the first, all are paracolonic bacteria belonging to the family Enterobacteriaceae. They are nonmotile, nonspore-forming types easily killed by sunlight and are usually transmitted by contact. All are common, and the last three are often responsible for the severe infections caused by the bites of many types of animals. While *Staphylococcus* and *Providencia* are severely infective types, the extremely rapid growth rate of *Proteus*, its often high resistance to treatment, and certain other characteristics sometimes produce in humans what have been termed *superinfections* (Wheeler and Volk, 1969). The ability of *Proteus* to digest proteins rapidly and, particularly, its facultative anaerobic nature are important factors responsible for the severity of many infections caused by animal bites. *Providencia* and *Proteus* are often discharged in massive numbers into the bloodstream, producing clinically dangerous cases of sustained bacteremia requiring vigorous therapy.

The ability to produce a severe and septic wound is undoubtedly of more importance to the ora than to the lynx or to most other mammalian predators. In most parts of the world wounds produced in an ungulate by the unsuccessful attack of a large predator may become septic, as in the case of lynx predation. The prey then becomes more susceptible to a second attack—perhaps by the

5. After 7 years in captivity, a specimen in the San Diego Zoo possessed none of these bacteria but *Pseudomonas* sp. (J. P. Bason, in a letter). Thus, the bacterial population in wild individuals may depend on frequent reinfestation from carrion.

same or a second species of predator. If the ungulate dies of the infection, it usually serves as food for specialized scavengers, rather than for the original predator species. The situation is quite different in the Lesser Sundas. The ora is the only predator or scavenger (with the exception of the recently arrived feral hunting hounds). The deer—withered, weakened, or killed as the result of the septicemia produced by one ora—becomes either live or dead prey to it and to other oras. Provided a *Proteus*-dominated infection is transmitted by an attack wound and the infection becomes lethal, or at least very serious, the initially unsuccessful attack eventually becomes successful from the standpoint of the ora population.

Perhaps the greatest adverse effect of the feral dog populations on these islands is that many of these maimed and infected prey animals that would normally fall to the predator oras now fall prey to the dogs. This form of competition perhaps is more important than the number of healthy deer hunted down by the dog packs.

11

Feeding Dynamics

Prey Populations

Only a few of the present day predatory reptiles, such as the boiform snakes, monitor lizards, and crocodilians, attain large size. Of these none has been so extensively studied that we understand the interrelationship of predation and population dynamics in both predator and prey species. The data that do exist are from only small insectivorous lizard species and crocodiles (Cott, 1960). All the reptile studies stress only the feeding habits of the predator and fail to consider the effects of predation on the prey. In fact, except for a few studies on large mammalian predators (e.g., Mech, 1966; Pimlot, 1967; Pienaar, 1968; Kruuk, 1972; Schaller, 1972), such information is almost completely lacking for any of the larger vertebrates. For smaller predators more information is available. The need for additional data on predation dynamics of all predator sizes becomes particularly important in view of the recent development of several optimization models for the analysis of feeding behavior (e.g., Covich, 1971; Rapport, 1971). In this chapter I identify those factors believed to be important in the coactions between individuals comprising the ora population and those of its prey.

Prey Density

Predation generally increases as a function of prey density in a characteristic way (Holling, 1969), that is, as the density of favored prey rises, the amount of less favored, alternate prey drops (see Curio, 1976, for discussion).

Errington (1946) concluded that densities of bobwhite quail (*Colinus virginianus*) and their vulnerability to predators depend more on the comparative density of the prey than on their absolute density. Thus the population density in relation to carrying capacity of the environment, high or low, was the fundamental factor in determining predation level. Murdoch (1971) has further theorized that simple density does not alone determine mortality rate, since the latter remains a fairly constant percentage of prey density over a rather large range of prey densities.

Locusts.—Four species of locusts are commonly eaten, especially by small oras: *Gastrimargus lumbokensis, Stenocatantops exinsula, Valanga nigricornis,* and *Austracris guttulosa*. Locust density was determined by a transect flushing technique, with no attempt to separate the four prey species, since there was no evidence for species selection by the oras. In February (the middle of the rainy season) all four locust species were found on the savanna. We made 11 transects, each 150 m long and 2 m wide on either side of the path of travel, constituting a total flushing width of 4 m (total transect area $= 600\ m^2$). The results are: $\overline{X}$ number locusts/transect = 53, range = 17–78 individuals, density = 1 locust/9.5m^2, or 170,832 locusts/km^2. At the end of the rainy season (March) the same area was used for 18 transects: $\overline{X} = 81$ locusts/transect, range = 27–102, density = 1 locust/4.3 m^2, or 261,621 locusts/km^2 in this habitat (Table 11–1). The average individual weight of a mixed sample of 63 locusts was 1.1 g, and the total biomass of locusts/km^2 of *Zizyphus* savanna near the end of the rainy season was 4,604 kg. There are no locusts on Komodo and Padar during the dry season.

Rodents.—Data on rat densities from Komodo are rather poor, because rela-

Table 11–1. Ora and ora prey densities on Komodo

Species	Average number per km^2	Average prime number per km^2	Courting techniques
Sunda deer	25	60	Plot-prevalence, flushing, pellet count, head count
Wild boar	41	100	Flushing transects
Green jungle fowl	53	72	Calling males × 1.5 females
Freycinet's megapode	0.5	10	Active nests × 2
Oras	8.8	16	Head count
Tokay geckos	?50	125	Calling males × 6.3
Rodents	1,875	4,538	Trap line
Locusts	110,996	170,823[a]	Flushing transects
	183,594	261,621[b]	Flushing transects

a. Mid-rainy season.
b. End rainy season.

Table 11–2. *Rattus rattus* densities on Komodo

Habitat	Individuals (no./ha)	Standard deviation	Biomass (kg/ha)	Biomass (kg/km^2)
Beach-side monsoon forest	2.8	± 0.5	0.55	949
Savanna–monsoon forest ecotone	1.5	± 0.4	0.30	486
Savanna	0.8	± 0.2	0.15	240
Hilltop steppe	0.7	± 0.2	0.05	80
Hillside monsoon forest	0.2	± 0.01	0.03	40
Creekside monsoon forest	0.1	± 0.1	0.05	80

tively few habitats were well studied from this standpoint. A removal trap line of 3 traps per station, 15 stations per line, spaced 15 m apart, was set for two consecutive nights. Lines were run as follows: 50 m back from and parallel to the beach just inside the monsoon forest ecotone; in monsoon forest on the side of hill 19; through the steppe on the top of hill 19 (a parallel line); on the savanna at the base camp; on the brush-savanna edge at the base of hill 19; and in the monsoon forest along Kali Besar. Densities were estimated using the simplified trap line analysis method suggested by Zippin (1958).

In general, rodent densities are higher in steppe and other open habitats than they are in the more dense *Tamarindus* forest. This is clearly shown in Table 11–2, where the beach area has the highest density, and the hillside and creekside monsoon forests have the lowest density. This general tendency is also seen in Loho Kalo, where rodents seem most abundant in the steppe and open *Borassus* savanna near Ntodo Klea, particularly where rocks are abundant. Hoogerwerf (1954a) reported they are common close to Kampung Komodo, but rare elsewhere.

Native reports suggest considerable fluctuations in Komodo rodent populations. Apparently rats were so abundant (cyclic peak?) in 1967–68 on Komodo that they are said to have eaten most of the garden corn and even dug into the ground to eat much of the *ubi kayu* (manioc) in the gardens. Though no rat density studies were made on Padar, villagers say that rats are much more common there than on Komodo. This may explain the predominance of rat remains in fecal pellets from Padar (Fig. 10–1).[1] Table 11–3 shows that Komodo has a large rodent density that is intermediate between more xeric and more mesic habitats. The data would not be significantly changed if mice were added, since the combined mouse-rat density on Komodo, or even Padar, is normally not much more than 44.9/ha. The biomass of rats in all habitats varies from 40 to 949 kg/km^2. The average weight of 35 rats from Loho Liang was 199 g. Table 11–3 also

1. The statement by Piazzini (1960) that "...Padar is...devoid of any living creatures [besides oras]" is completely untrue.

Table 11–3. Biomass of rodent species of similar size in Komodo and other areas

Common name	Scientific name	Biomass (kg/km^2)	Location	Reference
Ground squirrel	*Citellus tridecemlineatus*	328	Kansas	Seton (*in* Mohr, 1943, 1947)
Woodrat	*Neotoma albigula*	398[a]	Arizona	Mohr, 1943, 1947
Brown rat	*Rattus rattus*	2,125[b]	Komodo	This study
Cotton rat	*Sigmodon hispidus*	6,014	Florida	Stoddard (*in* Mohr, 1943, 1947)
Brown lemming	*Lemmus trimucronatus*	4,379[a]	Alaska	Pitelka (*in* Dasman, 1964)
Muskrat	*Ondatra zibethica*	14,637[a]	Iowa	Errington, 1963

a. Cyclic peak.

b. In highest density habitat.

includes some data on biomass calculations for rodents from more mesic areas, supporting the statement by Chew and Chew (1970) that there is a positive correlation between rodent density and annual rainfall and plant cover. However, the comparative *Rattus rattus* data from several habitats in Komodo with different rainfall patterns (Table 11–2) suggests that this generalization is not applicable in the essentially altitudinally zoned habitats of the Lesser Sunda Islands.

Green jungle fowl.—The habitat density of jungle fowl in Loho Liang's prime habitat, based on the number of calling males times an estimated ratio of 1.5 females:1.0 males, is calculated at 71.8 individuals/km^2; in marginal habitat of Loho Kalo the density is estimated as 22.5 individuals/km^2; in all monsoon forest combined the habitat density is calculated as 53 (Table 11–1). The mean weight of 18 individuals from Loho Liang is 758 g (range = 580–986 g), and biomass in Loho Liang is estimated as 40.3 kg/km^2. No evidence of jungle fowl was found on Padar, and near Nggoer, Flores, they are much less common than in Loho Liang.

Freycinet's megapode.—The mean weight of 14 individuals from Loho Liang is 696.4 g (range = 514–961 g), and the density in prime habitat is 10/km^2, based on a pair of birds for each active nest in the study area. Overall density for Komodo, considering all habitats, is 0.5/km^2 (Table 11–1) and biomass is 3.5 kg/km^2.

Wild boar.—Table 11–4 shows density estimates for wild boar, obtained by a flushing technique, in Komodo habitats. The highest density/km^2 is found in monsoon forest and the lowest in steppe. The overall density on Komodo is estimated as 40 individuals/km^2 and the approximate biomass as 1,236 kg/km^2. Densities near Nggoer and Loök, Flores, seem much lower. On Komodo we found that the activity ranges of adults tended to be rather small, but that there was considerable overlap in the feeding areas. When a large female or male was killed, the original inhabitant's area was not invaded by a new resident adult for a week or two.

Deer.—Four techniques were used to estimate deer densities in Loho Liang, Komodo: plot-prevalence, flushing, pellet count, and actual census count (Table 11–5). Descriptions of the techniques are provided in Appendix 9. Of these techniques the most accurate is obviously the actual census or head count. This indicates a population density of 1 deer/2.3 ha, or 60 deer/km^2 in prime mixed monsoon forest and savanna habitat. Because much of Komodo is considerably less than prime habitat, the actual densities for Loho Liang definitely cannot be extended to the entire island. A pellet-count study and one plot-prevalence analysis in the less optimum habitat of Loho Kalo provide indices of 4 and 6 deer/unit area respectively. Using five deer as a base, I calculate approximately 16 deer/km^2 in steppe and *Borassus* savanna. On the basis of distribution of these two habitat types I would estimate the total standing deer crop on Komodo as 7,110 animals (Table 11–6). Hoogerwerf's (1954a) report of approximately 500 ani-

Table 11–4. Population parameters of Komodo wild boar

Range type	Approx. total area (km^2)	Approx. total number of boars	Approx. number per km^2
Quasi-rain forest	14	720	50
Monsoon forest	57	5,700	100
Steppe-savanna	173	3,348	19
Approximate total for Komodo island	244	9,768	40

Average weight, Lesser Sunda populations of *Sus scrofa* (N = 43) = 31 kg

Approximate biomass of boars = 1,236 kg/km^2

Table 11–5. Technique comparison of Loho Liang deer population studies in monsoon forest and *Zizyphus* savanna

Technique	Number of deer per study area
Plot-prevalence	10
Flushing	5–10
Pellet count	9–12
Actual count	20

Table 11–6. Population parameters of Sunda deer, Komodo

Range type	Approx. total area (km^2)	Approx. total number of deer	Approx. number per km^2
Quasi-rain forest	14	990	69
Monsoon forest	57	3,420	60
Steppe-savanna	173	2,700	16
Approximate total for Komodo Island	244	7,110	25

Average weight of Lesser Sunda populations of *Cervus (Rusa) timorensis* (N = 32) = 52.1 kg

Approximate biomass of deer = 1,250 kg/km^2

mals for the entire island is the only other estimate of the deer population on Komodo. He apparently based his estimate on casual observations made in the Gunung Ara–Gunung Satilibo complex. Since no density studies were conducted there, my general impression is that the density of deer on Flores is much lower than it is on Komodo, and lower still on Padar. Apparently Hoogerwerf (1954a) had the same impression for Rintja. He estimated only 60–75 for the entire island of Padar. I suspect his figures are reasonably correct (see footnote 1).

Predator Densities and Competition

In addition to the relative densities and availability of prey populations, the ora's food supply is influenced by the abundance of other predators and the degree to which they compete for the same prey. The major predators within the range of the ora are dogs, palm civets, and the oras themselves.

Dogs.—Estimates of canid densities must include all healthy village dogs and feral individuals, because feral dogs (particularly the sought-after "good hunters") frequently become domesticated, and village dogs regularly hunt in the wild areas near the villages. The "poor hunters" (often diseased or lame) are the only truly feral dogs and are normally found in areas more removed from the villages. As a result there are two significant canid densities—the high, near-village densities of effective deer hounds that hunt for "pleasure" as much as for food, and the low densities of the generally poorer deer hounds in the bush. The distribution is determined largely by human encouragement of some dogs and ostracism of others. Recruitment for both groups is through birth and interisland poaching activities (Chap. 8).

There were approximately 20 resident dogs in Kampung Komodo, though the number was variable because feral dogs were often adopted by villagers. During the first few months we trapped 8 feral dogs within our study area, and I believe we obtained almost all that lived in our entire end of the island. Thus, I estimate the total number of feral dogs found on Komodo is 40, with an average density of about $0.16/km^2$. The densities on Flores and Padar seemed roughly equivalent.

Palm civets.—No conclusive data are available, but spoor distribution in our study area suggests a density of greater than $0.1/km^2$ but less than $3/km^2$.

Oras.—Based on data presented earlier, I estimate that in prime habitat the average number of scavenging oras is $16/km^2$; in steppe it is about $6/km^2$; and for the entire island it is $8.8/km^2$ ($11/km^2$ when ecotonal areas are not considered, see Table 11–7). The density on Padar is certainly less than this and is here subjectively estimated as $4.2/km^2$. Ora density on Rintja and on some portions of Flores (Chap. 2) is essentially identical to that of Komodo, though on other parts of Flores the density is very low and should not even be considered since the oras in those areas are largely solitary transients. Thus the standing crop of scavenging-age oras on Komodo is estimated at 2,245 individuals, Padar 60, Oewada Sami 25, Rintja 792, Gili Mota 40, and Flores 2,448; total standing crop 5,610 scavenging-age individuals, plus estimated 25 percent of total potential annual egg crop (Chap. 8) of 1,500 equals 7,110 oras.

If one considers only large prey, the predator-prey ratio of Komodo is 1 adult (2.5 m or more): 1 kerabau: 40 deer: 64 boar, or approximately 1 ora: 105 ungulate prey animals. In Idaho, Hornocker (1970) found 1 adult puma to 114 deer and 87 elk, or 1 puma:201 prey animals. Leopold (1933) estimated mountain lion to prey ratios in California as 1:200–360. Bourliere (1965) stated that lion-ungulate ratios in African parks are 1:260–360. From the data of Schaller (1967) and Locke (1954) I conclude tiger density in good habitat in India and Malaya is about 1/16 km^2. Ungulate density in Schaller's study was about 22 /km^2. Thus the tiger to ungulate ratio is approximately 1:350, remarkably consistent with ratios of other large mammalian predators from Africa and North America. The high ratio of oras to ungulate prey on Komodo is undoubtedly due to the greater prey spectrum (including many nonungulate prey) utilized by oras when compared with that of the large mammalian pursuit predators. In addition, the lower ungulate variety on Komodo (4 species, including feral goats) compared to Africa (27 ungulate species in Kruger Park) and India (11 ungulate species) suggests considerably less niche segregation on Komodo than on the other savanna areas listed. This type of nonspecialization in predation is particularly important during times of low density for certain prey species, since "switching" to several other prey types is easily accomplished. Thus, as in large crocodiles (Cott, 1960), food shortage is not an important limiting factor in ora numbers.

Table 11–7. Population parameters of scavenging-sized oras on Komodo

Range type	Approx. total area (km^2)	Approx. total number of oras	Approx. number per km^2
Monsoon forest	57	798	14
Forest–savanna ecotone	10	160	16
Savanna	103	927	9
Steppe	70	360	6
Approximate total for Komodo Island	240	2,245	8.8

Average weight of Komodo population of *V. komodoensis* (N = 50) = 7.5 kg

Approximate biomass of oras = 66 kg/km^2

Prey Utilization

Changes in density and survival rate of prey often cause related changes in ecological, biological, and behavioral factors that are important in almost all predator-prey interactions (Leopold, 1933; Holling, 1959, 1969). For example, increased density of ungulates may lead to a decrease of desirable forage plants. Such decreases in both forage quality and quantity lead to a weakened physical condition in the population as a whole. Theoretically, animals in poor physical condition are more vulnerable to predation. There is no evidence that oras regularly prey on animals in poor nutritional condition, though some may have previously been infected by bacteria transferred by other oras during an unsuccessful attack (see Chap. 10). Oras often eat juveniles, most of which are in significantly better health than the adults (Hornocker, 1970). Thus, malnutrition is probably not an important factor in the vulnerability of buffalo calves, fawn deer, or piglets. Hornocker (1969) showed that as elk densities increased, calves became not only more available, but also more vulnerable to mountain lion predation. A similar situation is expected on Komodo.

The ora's hunting technique is an important factor in determining the kind of prey killed. Unlike most large canids and felids which rely on a high pursuit speed to overtake their prey during a chase of at least several tens of meters, the ora must place itself, through stealth, within at least 1 m of its potential prey before it launches an attack. This technique results in a prey composition different in several respects (Chap. 10) from that produced by predators employing an open pursuit attack. Some mammalian predators often identify weak, vulnerable animals by deliberately chasing a herd for short distances (Mech, 1966; Kruuk, 1972; Schaller, 1972). The result of this technique is that young, very old, or diseased animals often make up the bulk of the kill (Murie, 1944; Cowan, 1947; Crisler, 1956; Mech, 1966; Cole, 1972; see Curio, 1976, for discussion). However, Hornocker's (1970) conclusions on mountain lion predation do not show such a clear distinction in prey selection. With few exceptions, any animal placing itself in a vulnerable position is apparently subjected to mountain lion predation. In this case vulnerability refers to more than general health of the potential prey individual. It includes various physical, biological, and ecological considerations that act together or separately to increase the chance of certain classes of individuals in the prey population, or of certain species, to place themselves inadvertently in a position to be attacked. That the types of animals killed by large oras with respect to taxa and age are close to those killed by pursuit predators preying on similar overstocked populations is not coincidental, though the hunting techniques of ambush and pursuit predators are quite different.

Taxa Utilized

The present evidence clearly indicates that the ora is a true opportunist, feeding on whatever species are available. For at least large oras, when the potential prey spectrum is expanded (such as on Flores compared to Komodo; Tables 10–4, 10–6), actual prey diversity is increased. Furthermore, density is not the only factor important in the type of prey utilized. For example, on Komodo the density of wild boars to deer in prime habitat is approximately 3:2, yet remains of these two ungulate prey species in fecal pellets occur in a ratio of 1:2 (Fig. 11–1). Using Pienaar's (1968) prey preference formula of kill frequency of prey (preference rating = kill frequency of prey/relative abundance of prey), one obtains a value of 0.44 for boar and 1.16 for deer, theoretically indicating a preference for deer 2.7 times as strong as that for boar. However, Pienaar's approach assumes a direct relationship to predation pressure only. There is no indication oras *prefer* deer even though they are proportionately more commonly preyed upon. In fact, it is clear that both ecological and behavioral mechanisms make the less abundant deer more vulnerable to ora predation by the predictability of the diel activity patterns of deer; and the ferocity and strength of wild boars, as well as the danger of attacking them (both further discussed below). The latter may cause oras to hesitate to attack a wild boar, even though they may often have the opportunity to do so.

As in most larger reptiles where there is great disparity between the size of the hatchlings and the adults, feeding behavior of oras changes markedly through their life (Figs. 10–4 to 10–6). Predator body length is therefore an important factor in ora feeding ecology, as it is in the Nile crocodile (Cott, 1960), the only other large predatory reptile studied to date. In certain smaller reptilian predators, such as rat snakes (*Elaphe obsoleta*), prey size has been shown to be an important determinant of the feeding response in young (Campbell, 1970). Oras utilize three prey sizes (coincidentally taxonomically distinct) throughout their life: (1) Insects and geckos—the former on the ground and the latter in the trees. Neither are eaten by large oras, but only by the very young. (2) Rats and birds—ingested mainly by medium-sized oras, and usually captured on the ground. (3) Animals as large as or larger than oras—boar, deer, horses, and even water buffalos. These are taken by stealth and surprise, near trails and under bushes (Chap. 10).

Size and Age of Prey

The data available suggest that large oras, and most of the medium-sized ones, do not select for size among prey species that are considerably smaller than themselves. However, large oras select strongly for smaller size classes among their ungulate prey.

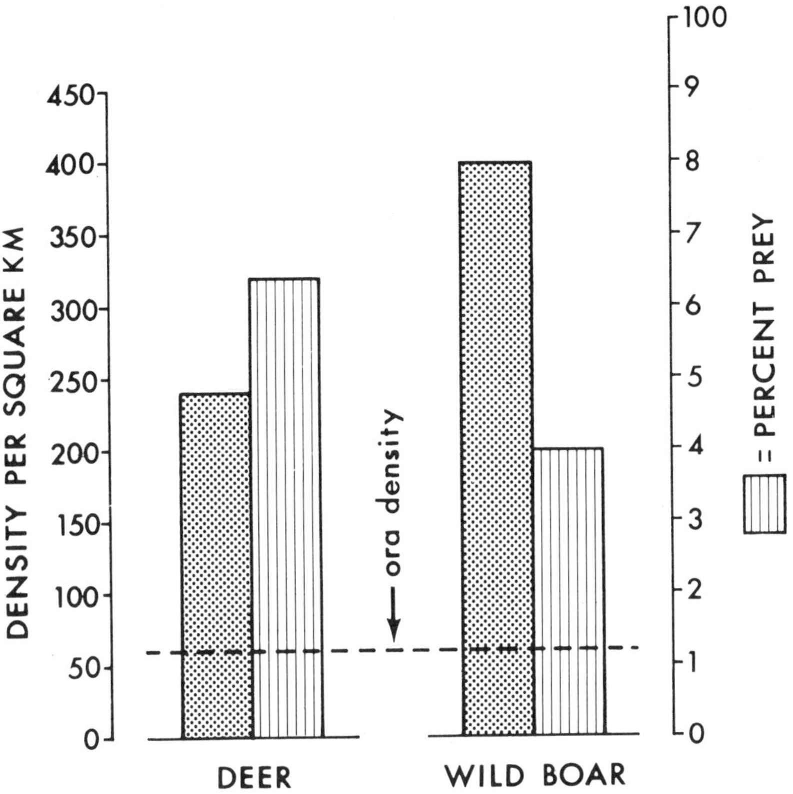

Fig. 11–1. On Komodo the ratio of the density of wild boar to deer in prime habitat is approximately 3:2; yet the remains of these two prey species in droppings occur in a ratio of 1:2.

Deer.—Figure 11–2 shows the size distribution of deer eaten by oras on the basis of hooves found in the Komodo fecal pellets.[2] The mean weight of deer eaten ($\overline{X}$ = 13.6 kg) is much less than the Komodo wild deer average ($\overline{X}$ = 52.1 kg) on the basis of hoof measurements made from prints in the moist mud and clay of several unvegetated regions in our study area. However, the more important statistic is that the modal size of the deer eaten is the smallest size class of all, smaller than the smallest represented in our sample of the living population. A sample of deer prints from Padar (N = 22) suggests the mean size of deer on that island is larger than on Komodo (Padar $\overline{X}$ weight estimate, 63.1 kg ± 6.3; Komodo $\overline{X}$ weight estimate, 52.1 kg ± 2.2). This is perhaps due to a greater

2. The relationship between hoof length and weight was established on the basis of 17 deer obtained on Komodo and from data accompanying 7 specimens in the Bogor Zoological Museum.

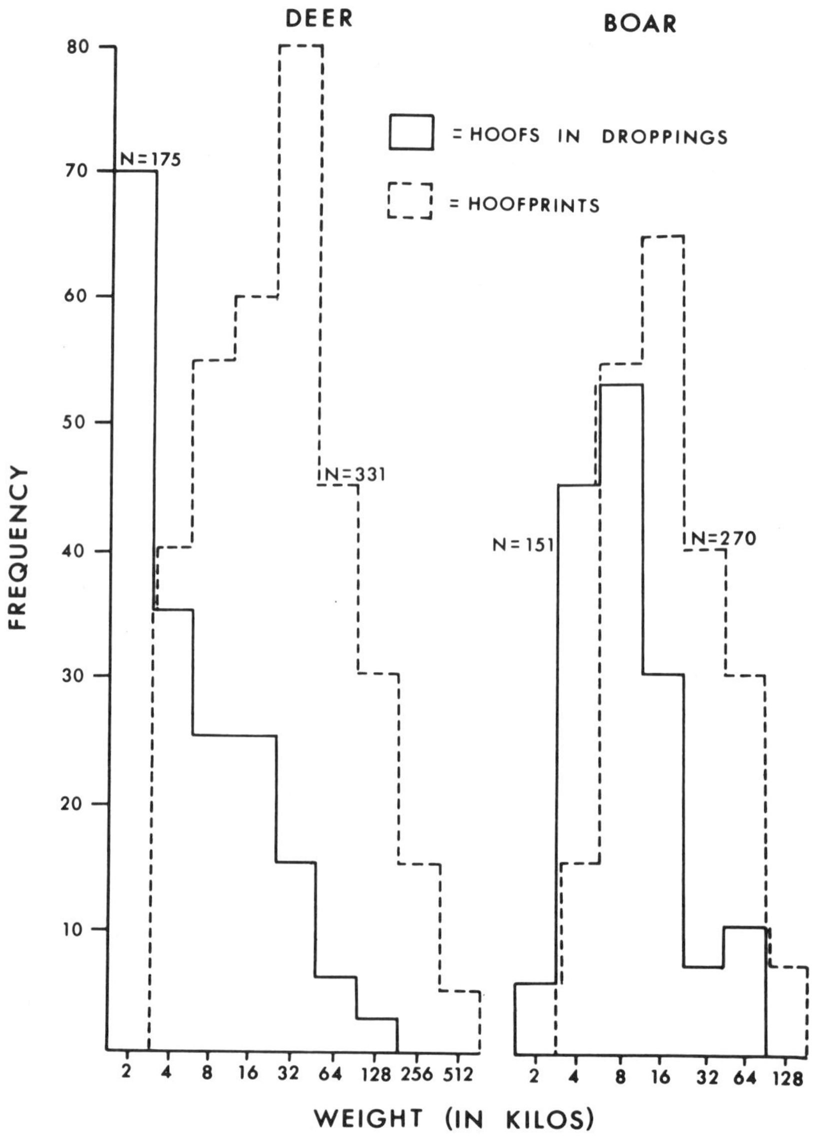

Fig. 11–2. Size distribution of deer and wild boar eaten by Komodo oras on the basis of hooves found in droppings.

predation pressure on the deer of Padar than of Komodo, with a higher proportion of the younger deer falling prey to the dogs and oras on the smaller island.

The only evidence on which to base an estimate of the age of deer eaten is that obtained from the deer remains found in the ora fecal pellets. These remains include hair, certain bones, hooves, and teeth, and of these, only the teeth have been found useful in age estimates. Apparently the most accurate method yet devised for estimating age is based on cementum annuli counts of the incisors (Erikson et al., 1970) and the premolars and molars (Mitchell, 1967). To date these age estimate studies have been conducted on only European and North American deer where strong seasonal temperature changes lead to pronounced annulation. The periodicity of the long dry season on Komodo should produce similar annulations. Unfortunately, no age studies have yet been conducted of Sunda sambar deer. In any event, this technique could not be used in this study, because the cementum was usually digested away in teeth obtained from ora fecal pellets.

Erikson et al. (1970) evaluated several techniques for estimating age. Particular emphasis was placed on tooth wear and eruption criteria. Though there was a steady increase in mean wear ratios for all deer age classes, there is great overlap. Wear patterns of permanent molariform teeth provided an age estimate accuracy of 95 percent for 2-year-old deer and 47 percent for all older age classes. Most errors were underestimates of age. Tooth eruption criteria were particularly useful for separating age classes younger than 2 years. In mule deer (*Odocoileus hemionus*) the tooth eruption pattern is usually complete at 28–29 months. The Erickson et al. study showed that, though none of the techniques was completely reliable, for deer over 2.5 years of age the cheapest and simplest technique retaining reasonable accuracy was visual examination of molariform tooth wear.

The only study of aging as evidenced in the teeth of Indonesian deer is that by van Bemmel (1949), who compared a large museum series of wild deer with a small sample of captive deer of known age. On this basis he concluded that some individuals attained ages of at least 32 years. More importantly, he showed that the first upper molar is the most reliable of all the teeth, and that premolar wear patterns are less reliable. He pointed out that the third molar makes an early appearance in the Lesser Sunda sambar deer. Though valuable, the study leaves much to be desired from the practical standpoint of placing jaws or teeth into size classes, particularly the younger ones. However, the combination of van Bemmel's conclusions, our large sample of deer skulls and jaws from Komodo, and the work of Robinette et al. (1957) forms a useful comparative scale with which to estimate age in Komodo deer. This classification is then applied to eruption pattern and molariform wear in jaws and teeth found in ora droppings (Table 11–8). My age classes differ from those of Robinette et al. (1957), or the

Table 11–8. Age distribution of deer eaten by oras, March–October 1969, Loho Liang, Komodo

Field groups	Characteristics	Number of individuals	Age classes
1	DP_1, DP_2, DP_3 erupted, but M_1 not erupted	31	< 6 months
2	M_1 erupted	36	6 months
3	M_2 erupted	26	6 months
4	M_3 erupted	5	12 months
5	Permanent premolars have replaced deciduous premolars	4	19 months
6	Infundibulum present on M_1 and PM_3; remaining teeth worn	14	ca. 6 years

age determinations of van Bemmel (1949) in that I could not separate my oldest grouping (Class 6) into as many units as they were able to do. Thus, there is a decided age gap between my Classes 5 and 6, representing that age period from 1.5 to about 6 years. These missing age classes are probably not eaten. Moreover, the distributions (Table 11–8; Figs. 11–3, 11–4) are slightly incorrect since some individuals older than those in Class 6 are undoubtedly missing. This is because the teeth of the oldest deer possess only isolated rings and bands of enamel (van Bemmel, 1949) and fall apart when the dentine is digested in the ora's stomach and intestines. In Class 6 the enamel folds are still broadly connected, and the teeth do not fall apart when digestion removes the dentine. Pieces of very worn molar teeth were found, so deer older than Class 6 are undoubtedly eaten. Their number and age remain unknown.

In the sample of 116 complete teeth from the Komodo fecal pellets the majority of deer eaten by oras are less than 1 year old (Fig. 11–3). Deer 6 years and older are also commonly eaten; those from 1.5 to 6 years are rarely eaten.

Schaller (1967, 1972) and Kruuk (1972) analyzed mammalian predation of ungulates in a similar fashion. Though the field age classes are slightly different, the age predation curves show considerable correspondence (Fig. 11–4).

It is clear that many of the prey of oras are very young fawns. Perhaps these are found when the hinds leave the fawns lying alone for the first two to five days of their lives, only coming back twice a day to feed them, and to ingest the urine and feces (after Prater, 1965). By contrast, human predation on Komodo deer (based on 197 deer jaws taken from poachers' camps) is largely in the 19 month to 10 year or more age classes (Table 11–9; Fig. 11–5).

Boar.—The size distribution of wild boar eaten by oras, based on hooves found in the droppings studied on Komodo (see Appendix 12 for boar hoof length–body weight data) is shown in Figure 11–2. The mean weight of boar eaten ($\overline{X}$ = 13.4 kg) is considerably less than the mean weight of the population (22.0 kg) and very close to the mean weight of deer eaten (13.6 kg), although sizes of the two prey species at the same weight are quite different.

Oras.—Cannibalism is apparently rare, with ora remains generally found in less than 2 percent of the droppings (Chap. 10). Whereas skin and teeth in the droppings may originate from the same individual that cast the dropping, claws and bones cannot, and these were used to determine age and size of oras eaten. While the average estimated total length of the oras eaten is 176 cm, it is clear that oras of two sizes are most commonly eaten—the smallest and the largest (Table 11–10). The former are known to be commonly attacked, killed, and eaten. Although large oras attack one another, we do not know to what extent the injuries result in death of one of the combatants. Oras will eat one another in a carrion state, but it is not clear how many of the adults represented in Table 11–10 were actually predated rather than scavenged.

Ora and prey sizes.—As in other predators studied, first glance suggests oras are extremely size-selective, based on the high incidence of fawns and piglets eaten. In the case of the larger Komodo ungulates, the high incidence of young animals, much higher than their relative abundance, suggests that they are sought out. However, as Hornocker (1970) concluded, while selectivity for young may be obvious, vulnerability of fawns and piglets, caused by several interrelated factors (disease, behavior, etc.), must be very important. The matter deserves study by future Komodo game wardens and wildlife biologists.

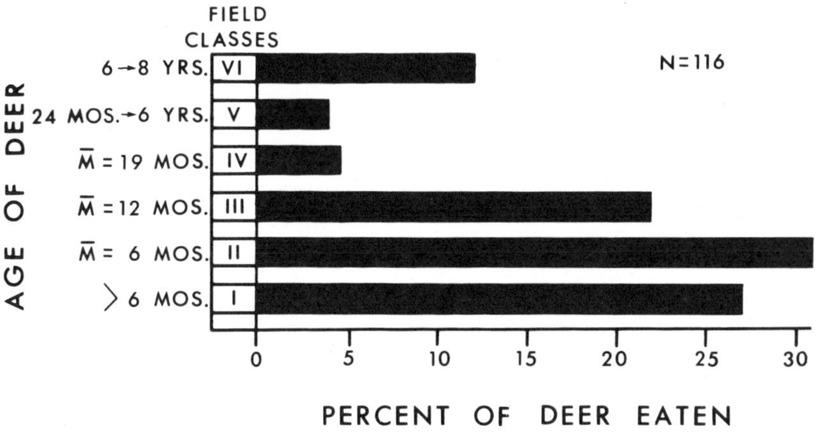

Fig. 11–3. Judged by the deer hooves and jaws found in droppings of oras, the majority of deer eaten are less than one year old.

Although mature bull buffalo and stag sambar deer are sometimes killed, it appears that these larger animals are relatively immune to attack.

Many workers have noticed the reluctance of predators to attack larger, more dangerous animals (Mech, 1966; Schaller, 1967, 1972; Estes and Goddard, 1967; Hornocker, 1970; Kruuk, 1972). Bourliere (1963) noted that most prey on herbivores about their same size and weight. Though this may be an overgeneralization, it does emphasize the fact that relative prey size is a factor in most

Table 11–9. Age distribution of deer killed by humans on Komodo

Field classes	Number of individuals (N = 197)	Age classes
1	0	< 6 months
2	9	6 months
3	9	12 months
4	5	17 months
5	93	> 19 months
6	81	ca. 6 years
7	5	ca. 6 years

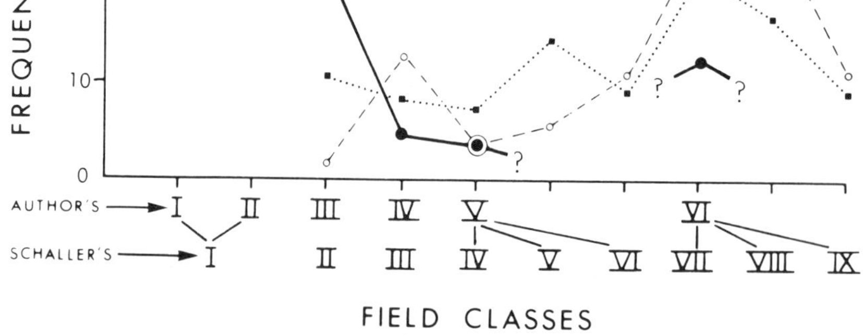

Fig. 11–4. Comparison of predation on deer by oras (this study) and tigers (Schaller, 1967).

predator-prey relationships studied so far (Rosenzweig, 1966). While the average weights of all deer and boar (32.0 and 22.0 kg) on Komodo are significantly different, the average weights of deer and boar (14.8 and 13.4 kg) actually eaten by oras are very close to those of the smallest oras (15.3 kg) that are known to feed on these prey. Furthermore, it is probably not a coincidence that the average weight of the largest ora size class (35 kg) approximates the average overall prey weight. However, the rather uniform relationship between size of mammalian predator species and their prey size (Rosenzweig, 1966) is not found in oras (or in reptiles generally, see Cott, 1960). Adult reptiles often select prey different

Table 11–10. Estimates (based on claws found in droppings) of total lengths of cannabalized oras

Size classes (cm)	Frequency
40–79	5
80–119	3
120–159	1
160–199	0
200–239	5
240–259	2

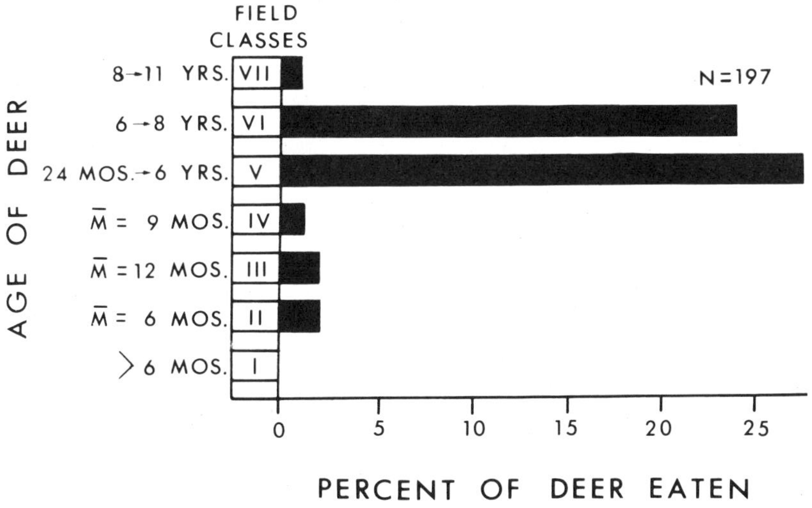

Fig. 11–5. Deer killed by poachers on Komodo tend to be in the age class of 2 to 8 years.

from that of juveniles, with entirely different predator-prey size relationships. Large oras regularly attack and kill prey up to twice their own weight, and in the case of water buffalo as much as 10–15 times their weight. On the other hand, young oras regularly feed on food particles (locusts) almost 2,000 times lighter than they are.

That size is related to predation is obvious. In almost all organisms, optimal predator size is largely determined by the interaction of both the abundance of different prey size classes and the relative energy extractable from them by predators of a given size (Schoener, 1969).

The ora is clearly one of the most important components of the insular fauna with which it is associated. As a highly specialized, almost solitary large carnivore, it tends to monopolize prey energy sources within its range. The virtual absence of important competing scavengers and predators tends to dampen major irregularities in food abundance; thus establishing reasonably consistent conditions for producing an "optimal" predator size (= that size requiring the least time to satisfy energy requirements based on the amount of energy expended to obtain different sizes or kinds of prey [Schoener, 1969]). Optimal ora size is particularly interesting in view of the fact that the animal is the largest living species of lizard in the world today.

Schoener designed and analyzed several models intended to predict optimal size for a number of predator types, defined mainly on the basis of both how they locate and overtake their prey and how their feeding time is allocated throughout the daily activity cycle. For the sake of simplicity and maximum contrast, two extreme types of predators were first considered, neither of which was completely realistic. Predator I was expected to expend time and energy in pursuit, in handling, and in swallowing of items, but not in search; whereas predator II expended time in search, handling, and swallowing, but not in pursuit. Small- and medium-sized oras would largely fall into the category of predator I; large oras in predator II.

Schoener's model predicts an optimal size for different-sized predators, suggesting that large predators expend more energy in pursuing prey over a given distance than do smaller ones and tend to expend less time and energy handling and swallowing prey. The latter is particularly applicable to the ora and may be the most important factor in determining maximum size attained by oras from the standpoint of prey utilization and predation dynamics.

As has been shown above, large oras feed primarily on large prey, most often obtained from ambush, and unsuccessful attacks often lead to a bacteremia in the prey that eventually results in their becoming food for the ora population. In a study of the influence of prey hunting behavior, Enders (1975) showed that in spiders ambush and surprise behavior tended to increase the size of the prey taken. The ora's ambush technique is apparently most commonly used in obtaining deer. These are most often found in small herds, and Taylor (1976) has shown that the evolution of body size in ambush predators is strongly influenced by this pattern of prey distribution. While it may be reasonable to ascribe the body size of the ora to this or related prey behavioral factors, it is important to note that although several Pleistocene fossil deposits are now known from the Lesser Sunda Islands, no deer have yet been found in them (Hooijer, 1967, 1972b). The only larger prey animals known from these deposits are miniature stegodont elephants and giant tortoises. Deer, water buffalo, and perhaps even wild boar may have been brought to these islands by man (see Auffenberg, 1980). Thus the evolution of large size in the Komodo monitor—due perhaps completely to the energetic benefits derived from feeding on large prey—was dependent on the available large prey found in this area at that time, i.e., two species of miniature elephants. These are thought to have been approximately 1.5 m high at the shoulder (Hooijer, 1967, 1972a) and probably weighed about the same as an adult buffalo.

Oras of different lengths should select ranges of food resources enabling each size group to satisfy its own rigid nutritional requirements in the least amount of time, assuming that over a long period the items are encountered in the proportion in which they occur (MacArthur and Wilson, 1967). However, the situation is often much more complex. For example, behavioral differences in deer and wild boar lead to oras encountering these prey species in numbers *not* representative of their actual proportions.

Schoener's model further suggests that variance in prey size selection should be greater in nonpursuit predators, such as the ora, than in pursuit predators—partially explaining the highly opportunistic feeding behavior of the generally inefficient pursuit patterns of monitors in general. This is particularly evident in the fact that the largest oras hunt mainly from ambush. Thus large oras expend the least possible energy in hunting—their size conferring energy consumption advantages in both thermoregulation and feeding, and their large prey representing more efficient packets than the smaller species available in the same habitat.[3] However, large oras can, and do occasionally, feed on small prey species. Thus they utilize a greater prey size spectrum than smaller oras. This may be of importance during food shortages.

The distribution of ora prey sizes versus ora size fits Schoener's model very well; his analysis suggests that efficient pursuers should take fewer prey sizes over a greater hunting range than less efficient pursuers of similar energy requirements, who should take a greater range of food sizes over smaller ranges. The ora is not an efficient pursuit predator and has a smaller activity range than other ungulate predators (Table 11–9). The radius of the ora activity range, while not

3. In view of the possibility that all the extant large mammals may have been recently introduced (deer, boar, and water buffalo, *fide* Darlington, 1957; as well as data provided in Chapter 10) and that large Pleistocene mammals of this region included only two species of small proboscideans (Hooijer, 1967, 1972a), the ora may have been a highly specialized pygmy stegodont predator.

necessarily equal to the maximum of the possible predator-to-prey distance, should vary directly with that distance, especially if activity ranges are composed of a constant number of reaction fields (McNab, 1963; Schoener, 1968; Curio, 1976). Fluctuations in the number of individual prey available per unit area may be partially responsible for variance in activity range and the range of ora prey sizes. According to the model, less efficient pursuit predators should show a greater variance in range of prey sizes accepted, which is true in particularly the medium to large size oras.

Selander (1966), Schoener (1969), and others have shown that in the absence of competitors, and if food is sufficiently abundant (both true on Komodo), most predators will exhibit sexual monomorphism, accompanied by spatial modifications, such as territoriality. These apparently related phenomena are both found in the ora, where sexual dimorphism is not marked, and large oras often show foraging allopatry.

Ora food requirements.—It is difficult to establish the number of consecutive kills made by an individual ora in a given time. The few scanty data obtained during this study, based on the number of boar or deer hooves per month in fresh droppings of well known adult oras while we were trailing them, suggest that large oras make a kill and/or feed on a large carcass about once every month. Hornocker (1970) and Robinette et al. (1959) presented evidence that puma kill big game animals at 6–10-day intervals, and that for some individuals the interval is probably much longer. It is also known to vary seasonally in both pumas and African lions (Pienaar, 1968). The interval is also dependent on prey size, and thus varies with the prey species available.

The ora's food ingestion rate is astonishingly high when compared with that of large mammalian carnivores (Chap. 9). However the amount ingested in a short time gives little indication of the animals' needs on a long-term basis. Both oras and felids appear able to adapt to a "feast or famine" regime. Golley et al. (1965) have found that bobcats maintain their physical condition on widely varying quantities of food and that the percentage utilization of the consumed energy is relatively stable. Hornocker (1970) believed that this is also true for the mountain lion.

On the basis of prey remains found in the fecal pellets of oras continuously tracked for varying lengths of time (3–180 days), large oras kill about one large prey ungulate per month, and would be expected to eat 12 per year. The ratio of deer to boar prey can be expected to be 3:2, or about 7 deer ($\overline{X}$ weight deer eaten = 15 kg):15 boar ($\overline{X}$ weight boar eaten = 13 kg), or a sustenance ungulate rate of about 165 kg/yr. To this must be added a small number of lesser prey organisms, such as birds and rats (18 kg ?). These figures for ungulate prey are compared with those of several large mammalian predators in Table 11–11. Though there is great disparity in comparing the values for oras and the large mammalian predators, when the latter are considered in terms of the proportion-

Table 11–11. Predation dynamics of various large predators

Predator	Adult weight (kg)	Ungulate prey (kg/year)	Average hunting range (km^2)	Predator-to-ungulate prey ratio	Meat required (kg/day)	Source
Puma		864–1,273		1:200	2.2	Hornocker, 1970
African lion	136		36.8[a]	1:165	10.0	Pienaar, 1968
Leopard	57			1:280		Pienaar, 1968
Wolf	27	1,227	16.0	1:300	3.4	Pimlot, 1967
Tiger	136	2,844–3,555	40.0	1:570	5.4	Schaller, 1966, 1967
African wild dog	18	1,394		1:550	2.7	Estes and Goddard, 1967
Oras	45	165	6.4	1:100	0.5	This study

a. Average pride size about 4 in South Africa.

ate difference in predator size and waste precentage of prey, the disparity is not nearly as great. Thus, though the size ratio of an adult ora to an adult tiger is about 1:3, the ratio of the pounds of ungulate prey required per year for each is 1:19. When percentage waste is considered it is 1:15, and when proportionate predator size is considered it is 1:5.

In captivity the required weekly intake of a large (45 kg) ora is about 3 kg (Lederer, 1942). More than this causes obesity in captive animals. However, a greater amount may not be excessive in wild oras in view of the high monthly energy expenditure between feeding periods. During the study year on Komodo, ora 19W (48 kg) ate an average of 6 kg/week, though food at the baiting site was obviously in very good supply. An estimated average of 3 kg/week in the wild under normal conditions provides a total of 156 kg/year, remarkably close to the estimated total of 171 kg eaten in the field (170 kg/year on the basis of 12 estimated kills per year, minus $\overline{X}$ 10 percent waste in large prey = 153, plus about 18 kg of small prey).

Competition

Few scholarly discussions concerning competition among vertebrates are adequately supported by quantitative data. Fewer are without challengeable assumptions. Of the quantitative data bearing upon competition pressures in vertebrate populations, only those small proportions resulting from investigations continued year after year are likely to repay close study. The same criticism can be leveled at this study. It was not possible to continue the present work in sufficient detail for a long enough time to satisfy these requirements. The relatively small proportion of the total range studied and the mobility of the medium- and smaller-sized ora specimens are bound to distort analyses. However, the following is presented as an expression of at least some of the important competitive factors operative in ora populations.

Intraspecific competition.—Intraspecific competition among oras is obviously an important factor in regulating population size and density. The density of oras is partly related to the problem of maintaining individual spacing through territorial behavior. This is undoubtedly less important in oras with undefined home ranges (such as small individuals) than in larger oras that seem to maintain hunting territories (Chap. 5). Competition for the most favorable portions of the habitat in terms of thermoregulatory and sentinel areas is also important and apparently intense in the case of large oras. Its outcome is determined mainly on the basis of the social hierarchy established within the local population. Competition for the available food supply is the most important determinant of ora carrying capacity. As in some fishes, it is possible that as competition becomes more severe, ora density may remain the same but size or weight decreases. The data available from Padar suggest that decrease in total individual size is a possible factor on that island. At present there is no evidence from any ora population that as densities decrease length of individual organisms increase.

Direct competition may, of course, increase mortality through fighting, accidental nest destruction, and loss of food supply. Such competition may serve to lower reproductive potential, increase cannibalism and dispersal, and cause a deterioration of health and vigor. With the exception of the effect on cannibalism, most of these factors remain unknown in the oras.

It is obvious that if no food is available there can be no ora. However, optimum ora density may be determined more by aggressiveness and cannibalism of large individuals than by prey density alone. Whether all ora intraspecific killing is motivated by hunger or by other influences (see Curio, 1976) is unknown, though at last some of the killed oras are eaten. Cannibalism may increase as mammalian prey species decrease, or in direct relation to density. There is no evidence of selection for smaller size prey individuals in intraspecific cannibalism in oras (Table 11–10).

Interspecific competition.—Both direct and indirect types of interspecific competition are included at several life stages. Direct predation on young is discussed in Chapter 12. Apparently the most important predators on young oras are adult oras (Table 10–4). However, adults must compete for this resource with sea eagles, brahminy kites, osprey, civet cats, wild boar, and dogs. Interspecific competition for space is undoubtedly minimal, and, if it occurs at all, must be restricted to the few small animals, such as rats, mice, and bats, found in hollow trees or under loose slabs of bark.

Indirect competition for identical food resources is most important. Geckos (*Gekko gecko*), several species of rodent-feeding snakes (*Vipera russelli, Naja naja, Elaphe subradiata,* and *Boiga cynodon*), and several carnivorous birds are competitors of small oras; man of mainly large oras; and dogs, civets (*Paradoxurus hermaphroditus*), and boar of both large and small oras (Table 11–12). In terms of the number of competitive food items, birds, dogs, and civets are most important, showing a 50 percent overlap with oras. The birds represent a number of species, so that each is proportionately less important than feral dogs and civets. Of the last two, the civet is an important competitor of small oras; feral and village dogs are more important as competitors of medium to large oras (Fig. 11–6). The prey of both civet cats and dogs were determined by analysis of droppings; boar, snakes, and geckos on the basis of stomach analysis of samples from Komodo; birds on the basis of the literature.

Broughton (1936) mentioned that a 2.2 m long crocodile (*Crocodylus porosus*) was accidentally caught at a carrion bait snare set for oras. When present in the Komodo mangrove swamps, crocodiles undoubtedly fed on almost any prey they could manage to capture. This probably included a fair proportion of terrestrial animals, as is true in other large crocodilian species (Cott, 1960). On Sumbawa, *C. porosus* is said to kill many deer as they swim between the main island

Table 11–12. Interspecific competition for ora food resources on Komodo

Major food resources	Major competitors						
	Geckos	Snakes	Birds	Dogs	Civets	Boars	Man
Beetles	+				+	+	
Locusts	+				+		
Geckos	+	+	+	+	+		
Rat snakes			+	+	+	+	
Megapode birds			+	?	?		+
Misc. small birds		+	+	+	+		
Mice		+	+	+	+	?	
Rats			+	+	+	?	
Feral dogs							
Feral goats				+			+
Palm civet				+			
Wild boar				+			+[a]
Sunda deer				+			+
Feral water buffalo							+
Carrion			+	+		+	

a. Only in the few non-Muslim communities.

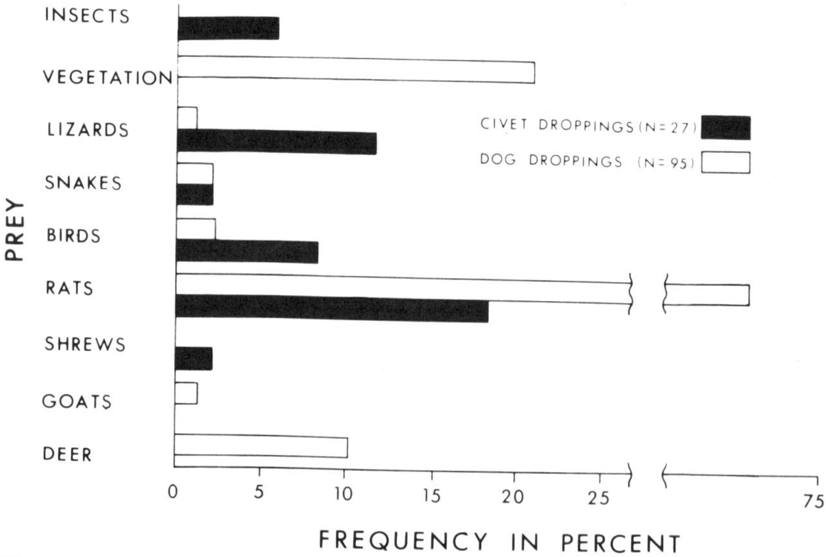

Fig. 11–6. Civet cats (*Paradoxurus hermaphroditus*) are important predators of small oras; feral and village dogs are more important competitors of medium to large oras.

and surrounding islets, particularly during the seasonal migratory movements of the herds (Oesman, in a letter). On Flores it is common knowledge that *C. porosus* feeds on many of that island's terrestrial vertebrates, including man, as they cross or enter river mouths.

However, at the present time, the major competitors of medium to large size oras seem to be dogs and man. Dogs compete mainly for rats, but approximately 10 percent of their food is deer. Dogs sometimes dig rats from their burrows during the day, but most are obtained from late afternoon to about 2300 hr and early in the morning when the rats are active. Small packs of up to four dogs often hunt deer by pursuing them until the deer are exhausted and finally pulling them down. The age of deer killed by feral dogs is unknown, but if the pattern is comparable to that of the African hunting dog (*Lycaon pictus*), they prey heavily on the female segment of the population, killing large numbers of parturient females and newly born young (Estes and Goddard, 1967; Pienaar, 1968). I once watched three Komodo dogs as they hunted for piglets by randomly running through the savannas until the prey bolted from cover. The dogs then gave chase and killed one that weighed about 3 kg. Dogs often compete with oras for carrion, including dead oras (Chap. 9).

Feral dogs are much more numerous in the open savannas than in the monsoon forest, or particularly the mountaintop quasi-rain forests. The correlation here probably stems from the fact that cover denudation through fire or overgrazing makes rodents, ground-nesting birds and their eggs, and snakes more vulnerable and easier to hunt. In addition to large prey, feral dogs in this area feed on locusts (though none was found in our scat sample), almost any small vertebrates they can obtain, including sea turtle young and their eggs, and certain grasses and fruits (particularly *Zizyphus jujube*).

Feral dogs are not, by themselves, important competitors of large oras, particularly since the oras feed on dogs whenever they can and so exert at least some influence on dog densities. However, the combination of trained packs of dogs, man, and fire becomes extremely important to ora densities. The details of the varied interactions between oras and man are provided in Chapter 12. However, in the present context it is important to point out the level and type of competition between large oras and man for prey animals.

The present coastal Malays of Komodo and the surrounding islands on which the ora lives are largely Muslim. Therefore in most areas man does not kill wild boar for food. In pre-Muslim cultures this was not always the case. At present a few boar are killed by villagers when they enter gardens and are generally destructive to crops. Years ago wild boar of Komodo were occasionally hunted commercially, as many as several hundred were taken at one time from near Loho Lavi. However, the overall effect on boar density is extremely low, negligible as far as the oras are concerned.

However, deer poaching is sometimes very significant, particularly when hunts are well organized commercial operations.[4] On islands other than Komodo, such as Padar and parts of Flores and Rintja, the constant hunting pressure and repeated burning of savannas have an important impact on ora densities and feeding habits. At present the impact on Komodo is not great—perhaps even less than it has been in the past. The poaching is largely rather restricted geographically and numerically (about 130 head per year). Furthermore, it is directed particularly to the older deer (Table 11–9, Fig. 11–5), in contrast to the younger deer (Table 11–8) taken by oras. However, it is obvious that if deer herds are allowed to continue to deteriorate, the effect on oras through reduced fawn production would certainly affect monitor populations.

On the basis of the preceding discussion, it is clear that feral dogs, man, and medium to large oras on Komodo compete for the same major resources (deer and wild boars), though dogs are much less efficient in obtaining large ungulate prey than are oras or man. The relationship between the densities of deer and wild boar and the effectiveness of oras, man, and dogs in obtaining these two major prey organisms are extremely important factors in determining not only the densities of both these prey species, but of the competing wild predators as well. For example, it is clear that if oras are more effective than feral dogs in reducing both the boar and deer resources, the oras would keep the resources too sparse for dogs to persist. However, in actual practice, competition in the wild without the interference of man is not that severe, because many deer and boar continue to live and produce young. Feral dogs and oras are not determining prey density through competition between them alone. Other more important factors are operative in actually establishing and maintaining deer and wild boar densities.

Predation Level

Of all the factors to be considered in any predation study, those that affect the number of prey animals taken are probably the most important, because they so clearly influence prey density. These factors and the conclusions regarding them (detailed elsewhere in this publication) are:

(1) *The density of oras*: Data show clearly that ora densities differ from one habitat to another. Average ora density is 8.8/km^2. In prime habitat on Komodo (mixed monsoon forest–*Zizyphus* savanna) the density is approximately 16/km^2, in marginal steppe habitat 16/km^2. The standing population total for Komodo is estimated as 2,824 individuals.

(2) *The density of prey*: The data bearing on this factor have been discussed above and are summarized in Table 11–1.

(3) *Search time*: The only definitive data available on oras show that from

4. All wildlife on Komodo, Padar, and designated parts of Rintja and Flores within the ora's range is protected.

several hours to at least 8 continuous days may be spent in hunting. Additional subjective evidence suggests that large individuals probably make a kill approximately once every month. Thus, about 26 days would be the average time spent by a large ora in prey searching behavior. Small oras obtain prey much more frequently than do larger specimens.

(4) *Searching efficiency*: It is particularly obvious that oras have a high searching ability, with both object and area-concentrated search patterns as part of the behavioral repertory. The use of their highly specialized olfactory organs in their orientation to carrion and the possible tracking of living prey has already been discussed. The searching patterns of both small and large oras are not random. They are affected by prey densities, topography, local cover conditions, ora and/or potential prey sizes, the behavior and psychology of particular individuals (Chap. 12), and the attack pattern utilized (either based on experience or determined by the unique conditions under which the attack is made). Many of these factors facilitate the finding of certain prey, such as the behavioral adaptations of lying in wait along game trails or creeping through the brush in search of sleeping prey animals. Searching efficiency is also obviously increased by avoiding unsuitable habitats. A good example is that, as far as is known, none of the larger oras hunted for large prey animals on steppe-covered slopes or hillsides, for this habitat affords relatively little predator cover and prey have a tendency to wander widely, rather than remaining restricted to a few trails.

Because oras hunt independently of one another, two or more may cover the same areas or visit the same megapode nest. Along commonly used routes of travel, such as Kali Besar, as many as 5 or 6 oras would have traveled exactly the same path in the course of a single day. Each had hunted independently of the other, the same logs and bird dusting sites checked several times. At the same time large tracts of high prey–density territories on either side of the river bed might remain relatively unhunted for days, perhaps because of the ease of moving up and down the stream valleys as opposed to the difficulty of moving through the rather thick brush on either side.

(5) *The hunger level triggering an attack*: The few studies of how hunger affects the feeding responses in reptiles do not make it clear that feeding patterns are totally related to satiation (Curio, 1976). Why oras sometimes feed voraciously and apparently fast at other times is unknown. However, the hunger threshold is often low: oras 8W and 19W attacked live goats after having fed on a great deal of flesh 12 days previously. On October 1, 19W ate 18 kg of carrion and was hunting for more the next day. One lizard ate 14 kg of carrion on February 21 and started hunting again on February 25 (96 hr later). However, in each of these instances, diel rhythmicity dominated over feeding behavior, probably for thermoregulatory reasons.

(6) *Pursuit time*: The adult ora spends very little time in actual pursuit of prey, since most larger individuals apparently make most of their kills lying in

wait along game trails, or by surprising animals sleeping under cover. Juveniles and medium-sized individuals often pursue smaller prey, such as locusts, lizards, rodents, and birds. In these instances pursuit rarely lasted for more than 15 seconds, usually closer to 5.

(7) *Length of time prey is exposed to potential attack*: This is an important factor, largely determining the vulnerability of particular prey species. Sparrowe (1972) found that duration of exposure is more important in determining the number of attacks per unit time than is prey visibility or cover density. In the case of larger, hunting oras, deer are mainly exposed when moving along particular game trails. Thus total exposure time to ambushing oras is probably no more than several hours for the entire deer population, but for any one individual the period is only a very few moments if it happens to pass down a game trail in sight of a waiting ora. Wild boars and deer are also hunted while they sleep. This exposure is certainly about 3 hr/day and perhaps longer. Smaller prey are exposed for shorter periods of time, due to the greater availability of shelter.

(8) *The number of attacks per unit area per unit time*: This figure is probably extremely low for large oras, since, unlike most large mammalian carnivores, large oras do very little hunting by actively seeking out grazing or drinking prey animals. Most encounters are primarily the result of chance. However, the hunting pattern is certainly far from uniform. For young oras feeding on more common small prey, attacks should be more plentiful per unit time. One ora hunting in a swarm of grasshoppers made about two attacks every minute. Prey density was very high and this must represent a very high attack-to-time ratio.

(9) *Per capita effectiveness of the attack*: Most predators have a low rate of hunting success, but quantitative evidence for this belief is not easily obtained. The general activity pattern and movement of 19W is known for 81 days on Komodo. We are certain of only two successful attacks during this period (Chap. 12) and have no evidence of the number that were unsuccessful. Mech (1966) found that the success ratio of *Canis lupus* in attacks on Isle Royale moose was 7.8 percent. Saunders (1963), tracking *Lynx canadensis* in snow, determined that 42 percent of the attempts to catch snowshoe hares were successful. Schaller (1966) speculated that the tiger's success ratio was about 3–5 percent. Hornocker (1970) and Stevenson-Hamilton (1949) stated that the failure of African lions to capture prey exceeded their successes. Rudebeck (1950, 1951) found through direct observation of European raptors that only 7.6 percent of 688 attempts to procure prey were successful. In contrast are the high rates of success reported by Pearson (1964) for feral house cats (*Felis catus*) and by Estes and Goddard (1967) for African wild dogs (*Lycaon pictus*).

It is extremely difficult to observe directly attempts of predatory animals to capture wild prey in dense vegetation. Some authentic records of actual attacks of oras on prey animals have been observed, but these are few. Tracking oras or observing them from afar cannot provide much quantitative data on unsuccessful vs. successful attempts. For one reason, rather than attacking prey wherever seen, oras usually seek particular secluded situations in which they can place themselves within 1–1.5 m of their intended prey. Once they are within striking distance, an attack is undoubtedly imminent, but its duration is short and close to the ground. Success is another matter. It is doubtful that even from short distances the ora attacks are as successful as those of mountain lions (Hornocker, 1970), African lions (Schaller, 1972), or even crocodiles (Cott, 1960).

Unfortunately, attack effectiveness of large oras remains unknown. Only when an ora has actually launched an attack and has failed to succeed can a rigid determination of success be made. In any event, success depends on a number of factors, including hunger level, past experience, size, health, and perhaps even position in the social hierarchy. The disparate effectiveness of attacks on tethered goats by oras 8W and 19W was very noticeable. The attack of 8W was very clumsy and loud; his poor health at that time made it almost impossible to move effectively or rapidly. It is extremely doubtful that he would have been able to make a successful kill of his own on an untethered large prey, even if he had the opportunity to do so. Though 8W regained his vigor later in the year, 19W remained slower in his attack, moved more quietly, and was apparently more often successful. 19W's weight varied less than that of 8W throughout the year, and he grew at a faster rate.

Attacks by large oras on untethered livestock near Nggoer, Flores, have a high success ratio. This is probably because almost all successful ora attacks on large prey (horses and water buffalo, whether tethered or free) are accomplished only by certain individuals that have developed particularly effective attack patterns (Auffenberg, 1972).

For smaller oras preying on insects, at least some data are available. In one case a 1 m ora successfully obtained grasshopper prey in 6.3 percent of 32 attacks, during a total search time of 14 minutes.

(10) *Number of prey animals actually eaten*: On Komodo the only data available to estimate minimum number of large prey animals eaten by oras are the remains found in droppings. For reasons given previously, the Komodo scat sample collected in October represents only that period between the end of the rainy season and the time at which it was collected. As a result it is possible to state that between approximately March and October (roughly 200 days), within the area encompassed by a line drawn from the main village to Gunung Ara, to Satilibo, to Pahme, to Sabita, and then along the east coast to Gunung Klinta (about 31 km^2), the *minimum* number of ungulate prey taken by all oras (minimum number of large oras about 40) was 1 water buffalo, 1 goat, 54 deer, and 26 boar (the actual numbers of hooves for each prey category are: water buffalo 2, goat 3, deer 218, boar 105). The estimated minimum total for the entire year (assuming the same predation rate for the remainder of the year) is 2:2:80:40. Within this area the estimated standing crop of deer is about 2,000 and of wild

boar about 3,000. Thus a minimum of about 4 percent of the deer and 2.5 percent of the wild boar standing crops are eaten by oras each year. Unfortunately, these percentages are extremely crude estimates, and until more data are obtained, they should be considered as minima.

(11) *Time spent eating prey*: Data presented above make it abundantly clear that oras are remarkably adept at consuming most or all of a large carcass in a surprisingly short period of time. The mean intake rate of large individuals may be as much as 2.5 kg/min—higher than any other predator known (with the exception of large snakes). The importance of this rate is perhaps better understood when it is realized that I once saw 23W, a 2.5 m female ora weighing 50 kg at the time, swallow a 31 kg wild boar in 17 minutes. Considering the extremely low prey wastage factor of oras compared with those of other predators or scavengers (Table 9–5), the advantage of these prodigious feeding accomplishments is that the greatest proportion of the prey is consumed in the shortest possible time. Furthermore, if they kill or find carrion relatively infrequently, it is of great advantage to make maximum utilization of the resource.

(12) *Time spent in the digestive pause*: Mertens (1942) stated that a large male ora still contained undigested bones of deer after 4 weeks in captivity. However, this is obviously longer than normal, and the observation was made on a freshly caught individual that might have been suffering some type of stress. Lederer (1942) reported that in a large, healthy, captive female ora the digestive rate varied from 2 to 4 days. This digestive period is equivalent to field data obtained during this study ($N = 16$, $\overline{X} = 4.7$ days, range 3–6 days). These data were obtained by feeding oras 19W and 23W carrion that included dye capsules (methyl green, methyl blue) and then checking for appropriately dyed droppings near the resting sites during the succeeding days.

Biomass

To calculate biomass I determined the number of animals and the weights of sample groups. For the most part weights were obtained from captured or killed animals, using a standard single beam balance (Tables 11–3, 11–4, 11–6, 11–7). Where necessary, sample size was increased with weight records of specimens maintained in the Zoological Museum at Bogor, Java.

The size of Komodo is about 245 km^2. Of this total, approximately 173 km^2 are steppe and savanna, 58 km^2 are deciduous monsoon forest, and 14 km^2 are quasi-rain forest. With the exception of rodents, there is a relatively low resident population of prey animals in the open steppe and savanna. As far as is known, migration is not a major factor, and seasonal variation in biomass on Komodo seems negligible. Highest prey biomass in kg/km^2 is found in monsoon forest and along its edge. In general, quasi-rain forest supports few organisms preyed upon by the ora. A summary of the density and biomass of different ora prey is

Table 11–13. Estimates of densities of ora prey biomass on Komodo

Prey species	Average number per km^2	Average biomass (kg/km^2)
Sunda deer	25	1,250
Wild boar	40	1,236
Green jungle fowl	53	40
Freycinet's megapode	5	4
Rodents (scv. sp.)	4,538	2,125
Locusts (sev. sp.)	261,621	4,604
Oras	8.8	66

given in Table 11–13. The total ungulate biomass of 12,806 kg/km^2 on Komodo is not high when compared to estimates from other parts of the world (Table 11–14).

Effect of Ora Predation

The work of Mech (1966) on wolves and Hornocker (1970) on mountain lions shows that the mere presence of a predator in a locality does not alarm game animals. However, when a kill is made, the reaction is striking. Large prey animals immediately leave the area. Hornocker pointed out that this behavior tends to disperse game animals. Its importance on restricted overused range is obvious. Leopold (1947) recognized that predators might benefit ungulates in this manner, and stated "there is a presumption, but no proof, that natural predators, by breaking up congestions, tend to distribute the deer load more evenly." However, he had earlier stated that animals living within the carrying capacity of their range suffered little harassment (Leopold, 1933). This is apparently also true on Komodo, since there was no evidence that the deer killed at the base of hill 23 on October 31 affected the distribution and movement of any of the deer herds in the immediate area.

In practice, stability between predator and prey populations seems to be the rule. The present situation is no exception. Stability depends on population size adjustments of the prey to the predatory demands of the oras. Murdoch (1971) and others have shown that mortality is a fairly constant percentage of prey density over even a large range of prey densities. The developmental response of a predator to prey density proposed by Murdoch (1971), whereby a growing predator, because of its increasing size, often kills a higher proportion of prey, is of little consequence in the ontogeny of the ora, since the type of prey selected is mainly size-dependent.

Table 11–14. Ungulate biomass data from Komodo and elsewhere

Approximate year-long biomass (kg/km^2)	Approximate area (km^2)	Animals	Range	Location	Source
98,000–140,000	1,600–3,200	Wild ungulates	Savanna	East Africa	Talbot and Talbot, 1963
92,000	3,200–4,800	Wild ungulates	Bush	East Africa	Talbot and Talbot, 1963
29,820–44,000	—	Cattle	Managed savanna (European ranches)	East Africa	Henderson, 1950; Ledger et al., 1961
37,380[a]	—	Domestic livestock	Average of all virgin long and short grass	Western United States	Watts et al., 1936
19,600–28,000	—	Bison and associated wild ungulates	Prairie	United States	Bourliere, 1961; Petrides, 1956
27,580[a]	1,802,400	Domestic livestock	Average of all virgin ranges	Western United States	Watts et al., 1936
15,480–22,400	—	Domestic livestock	Savanna (tribal grazing land)	East Africa	Talbot and Talbot, 1963

Table 11–14. *Continued*

14,840	—	Black-tailed deer	Managed chaparral	California	Taber and Dasman, 1958
12,806	153	Wild ungulates	Average all ranges	Komodo	This study
9,660	—	Black-tailed deer	Average of oak woodland plus chaparral	Pacific coast North America	Taber, 1961
8,120	3.2	White-tailed deer	Woodland, proper density	Michigan	O'Roke and Hamerstrom, 1948
6,160	245	Sunda deer	Average all ranges	Komodo	This study
4,970	6,997	Red deer	Deer forest	Scotland	Lowe, 1961
3,785	104	White-tailed deer	Average all ranges	Florida	Harlow and Jones, 1965
1,904	140,928	Mule deer (5 races)	Average all ranges	California	Longhurst et al., 1952
1,820	—	Mule deer	Average all ranges	Arizona	Swank, 1958
504	—	Caribou	Average tundra	Canada	Banfield (*in* Dasman, 1964)

a. Based on animal units equivalent to one 600-kg steer.

In this study available data offer evidence that predation by oras is ineffective in limiting ultimate numbers of either deer or wild boar, though they may regulate them. Holling's (1959) emphasis of the difference between *control* (exercised by a density independent factor such as weather) and *regulation* (resulting from interaction with a density dependent factor) imposes unnecessary constraints on a discussion of predation. The important thing is that predation by oras on Komodo deer and wild boar does not reach a level equaling the effective birth rate of either prey species.

Though ora predation is not limiting in this sense, the removal of prey animals by a population of oras acts to lower the rate of prey increase before more drastic forces, such as disease or starvation, become dominant. Thus, ora predation dampens and protracts severe oscillations. Removal of two or three hundred or more deer per year on Komodo by oras is obviously significant in reducing fluctuation of numbers. Predation dampening effects have been recognized by numerous other workers (e.g., Pitelka et al. 1955; Craighead and Craighead, 1956; Holling, 1959; Cole, 1972). Indeed, Klomp (1956) considered the dampening of oscillations of animal numbers to be as important as regulation. Dampening can act to reduce range damage, and serious overstocking may be averted until such time as other limiting forces may come into operation (Hornocker, 1970).

The effect of ora predation is not, however, limited to dampening prey oscillations. Predation pressure by the ora undoubtedly shapes the evolution of the Sunda deer in this part of its range. Field and experimental data support this assumption. Popham (1942) has shown experimentally that the selective action of predators tends to remove individuals possessing less successful adaptive characters. They may possess nonadaptive physical or behavioral characteristics or a lower intelligence lever corresponding to Leopold's (1933) "sucker list." Theoretically, individuals possessing such characteristics are more vulnerable to ora attack and are culled from the herd by ora predation. However, since most ora attacks result from chance encounters, selection of this sort is probably weak.

Thus, although predation may have a profound effect on some prey populations and some of these effects are far-reaching, there is no clear evidence that ora predation *limits* ungulate populations. Nor is there any evidence that ora numbers are limited entirely by food supply. For example, intraspecific relationships, manifested in some form of territoriality, may act to limit numbers of oras. From the standpoint of determining ultimate numbers of boar and deer, free from the limitation of time, it may be concluded that ora predation is inconsequential.

The *kind* of animal actually preyed upon has been shown to be the function of different factors acting separately or collectively to increase vulnerability. These include prey density, behavior, age, health, inter- and perhaps intraspecific competition, and the ora's predatory characteristics.

It appears oras select young deer or buffalo, avoiding the prime bulls or stags, and kill wild boar slightly more indiscriminately—even though there is a tendency to kill smaller individuals. In the Komodo study area, where prey populations are dense, oras kill a large proportion of very young and very old animals. This pattern of predation appears universal where prey populations exceed the carrying capacity regardless of the type of predator operating (Hornocker, 1970).

Where prey species remain within the carrying capacity of an environment, a different kind of predation with regard to the kind of prey animal killed might be expected, i.e., more prime animals are killed. Under these conditions, however, the predator population may be limited by the food supply. Wright (1960) found this to be the case in Africa.

In an ecological sense, the influence of ora predation on surviving prey animals may be more important than the actual killing of animals. Many predators have been shown to force the redistribution of their prey animals—particularly on limited ranges. This is doubly important to ungulate species exhibiting weak or nonterritorial behavior that allows them to overpopulate an area and seriously damage the habitat. The importance of predation on such nonterritorial species in Africa has been recognized, and Errington (1946) stated "the less that strong territoriality or other self-limitation enters the population equation, the more that something else must do the limiting."

In its simplest form, predation is similar to grazing, with the prey having no more ability to escape exploitation when detected than have pasture plants in the presence of livestock (Errington, 1946). However, prey movement greatly increases the complexity of predator-prey coactions and is of extreme significance in almost every predator reported so far. Behavioral characteristics of both predator and prey determine the types of interactions possible and the conditions under which they occur (Kruuk, 1972; Schaller, 1972; Curio, 1976). This is also true of the ora. However, in this predator the *frequency* of encounter remains the major determinant in the amount of predation suffered, particularly by the larger prey such as deer and boar. Though this is a function of prey density, other extrinsic factors, such as topography, are also of prime importance (in spite of the general conclusions of Errington, 1946, and others).

Though there is a certain degree of inflexibility in the relationship between adult oras and particularly deer, the population levels of each species in different habitats cannot always be linked to gross differences in food and cover, once minimal requirements have been met. Oras may occasionally starve under even the most ideal situations due to an unfortunate series of encounters with prey. Greater prey density may confer little or no real advantage under these conditions. The biotic characteristics of Komodo make it clear that this island is replete with superfluities of food and shelter for an adequate ora density, and that any excess will be lost somehow. This is extremely important from the standpoint of possible initiation of future management programs. Whether the ora utilizes a temporary surplus of either carrion or live prey, or whether it utilizes a larger or smaller proportion of the food surplus that might be available, most students of predator dynamics believe that both predator and prey populations of vertebrates are ultimately self-limited— the bounds determined by habitat characteristics and the influence of cyclic factors, including diseases.

As most ora prey species have a broader ecological and geographic distribution than *V. komodoensis*, it is important to consider the degree to which oras limit prey in the habitats in which they are, or may be expected to be, sympatric. The only related data are the exclusion of certain larger reptile species (*Python molurus, Cuora amboinensis*, etc.) from areas inhabited by the oras, whereas they occur in the same habitats outside the ora's range.

One of the most important factors in considering the density of either the ora or its prey species is that population levels are often unstable when, as on Komodo, the food web is comprised of only a few species. Under these conditions, a slight change in the population of one species affects population levels of all other species in the community. This is probably even more important on smaller islands, such as Pulau Padar. On Nusa Mbarapu the phenomenon is particularly evident. Any small change in the population number of goats on that islet results in a remarkable change in ora population density (Chap. 4).

Perhaps one of the most important factors in ora predation is that, when compared with other lizards, the predatory pressure exerted by this large carnivore is often extremely flexible, shifting its major impact from one species to another and from one locality to the next as the situation demands. This represents a life style more similar to that of unspecialized predaceous mammals than to most reptiles (Curio, 1976).

The probability of a given food item being devoured by an individual ora is the product of the probability that the ora will find that item during its search times and the probability that it can kill the prey; while the probability of a successful hunt is determined by the grain of the environment relative to the species. If the environment is fine-grained, the predator contacts items of different resources in the proportion in which those resources occur; if coarse-grained, the contact is in a proportion unlike the actual percentage of the resources' occurrence. In the ora both situations apply, depending on the prey species being considered. The random hunting for locusts represents a fine-grained environmental situation, and the proportion of deer and boar actually eaten represents a coarse-grained environment. The fact that both situations apply to the ora is again a reflection of the almost unbounded opportunistic predatory life style of *V. komodoensis*. Thus, the ora's opportunistic hunting strategy differs from the preferential hunting pattern of most large African mammalian carnivores (Bourliere, 1963). However, although oras appear incapable of *limiting* water buffalo, boar, and deer populations on Komodo through predation, the effect and influence of such predation are of great significance in tending to maintain ecological stability in the wilderness environment in which this predator exists.

There is no evidence that oras have a strong vertebrate prey preference, with the exception of boar and deer. In most cases the proportionate differences in prey species taken by the different-sized oras probably reflect differential prey availability or prey density in different areas (Tables 10–8, 10–9). I agree with Emlen (1966) that prey proportions by themselves do not give any indication of predator preferences. In the case of deer and wild boar, the number of each eaten by oras is clearly not density dependent (Fig. 11–1). Approximately twice as many deer are eaten as boar, though the density of deer is about half that of boar. Considering only these data one might presume that the proportionate number of deer and boar eaten reflects the proportionate number of attacks on the two prey species, but this may not be the case. If deer and boar were attacked in direct proportion to their density in Loho Liang, twice as many boar should be attacked as deer. However, attacks on wild boar may be proportionately more unsuccessful than those on deer, since boar, though often smaller than deer, are certainly more dangerous from the standpoint of injuring or even killing the predator. Another probably important factor is that deer are less randomly distributed than boar and tend to have a clustered distribution. They often form small herds. This is usually not the case in wild boar, with the exception of a sow with its suckling young. Perhaps even more important, the behavioral patterns of deer are much more habitual. Thus they have a very strong tendency to follow closely certain game trails, and they tend to sleep and eat in the same places for successive time units. This means that the behavior of deer is more predictable than that of the wild boar. That this predictability factor is used by large oras is demonstrated by the different manner in which deer are hunted during the early morning and in the afternoon. I believe it is the higher predictability of deer behavioral patterns that constitutes the basis of an otherwise anomalous density-predation ratio. This suggests that some type of "training" is taking place in oras.

Training

Though we have no information to what extent training is important in determining the proportion of certain prey eaten by the ora, it is quite likely that inexperience or former successes and failures of an individual affect not only the area in which it hunts but the prey hunted and even the technique utilized. The proportion of different types of livestock taken in savannas on Flores is due partly to behavioral peculiarities of resident oras and not to the proportion of prey species available. While the physical and psychological characteristics of individual large mammalian predators, such as man-eating lions, are recognized as being of extreme importance, the importance of these same factors in reptilian predators, with the exception of the crocodilians, remains largely unappreciated. I suspect this is because most reptile predators studied to date have been small, usually quite common, and relatively short lived. Small size makes it more difficult to recognize slight behavioral differences, abundance tends to mask individual differences, and the short life span tends to keep certain behavioral traits that are ordinarily associated with a long resident life from developing. However, in the case of the ora these factors are not as important. Oras are large, and individual differences in behavioral traits, such as attack patterns, are easily recognized, as discussed above. In all cases the habits of the more successful animals are undoubtedly based largely on past experience. Thus training becomes a major factor in prey selection, attack patterns, etc. According to Murdock (1969) such training can probably occur only when the prey is abundant and exhibits a contagious distribution, so that the individual experiences have some referable meaning. The contagious distribution of deer and their greater behavioral predictability suggest that ora training with respect to deer may be much greater than it is with wild boar.

In addition, "retraining" must occur whenever the often transient smaller *Varanus komodoensis* move to a new area, at least in respect to the kind of country that is hunted. In fact, nomadism itself is probably a very important "schooling" for the young, providing them with a broad range of experiences which can be called upon at critical times later in life.

As pointed out by Rapport (1971), the dilemma of any predator with alternate food resources is "to select the optimal prey combination among the alternative predation possibilities." The behavior of at least large oras shows that they possess the ability to sense the predation possibilities in terms of types of prey to be sought as well as in the techniques most likely leading to success. Thus they possess "rational" preferences, have knowledge of the consumption possibilities, and exhibit what Rapport and others call optimizing behavior. However, the ability to predict an ora's response to changes in density of any prey organism on which it might feed is very poor. There are two reasons for this: data have not yet been gathered on the actual availability of each of the potential prey species/given unit of time; and the predictable patterns in the reactions of the ora and its immediate potential prey may indeed barely begin before deviations occur.

12

Coactions

Interspecific Coactions

Non-Parasitic

Varanus salvator togianus.—The water monitor, or *weti* (see Appendix 2 for local names used for the water and Komodo monitors), is known in this part of the Lesser Sunda Islands from Sumbawa, Sumba, and Flores. It is absent from Komodo, Padar, and Rintja. Hoogerwerf's (1954a) statement that it occurs on Padar was based on the identification of footprints he found in the sand. We found no evidence of it on that island, and I agree with de Jong (1937), who found no differences between the prints of small oras and weti. Pfeffer (1959) believed the absence of *V. salvator* on these islands was due both to the strong currents between them and Flores and to the aridity. I doubt that the former is important in view of both the wide distribution of *V. salvator* in the East Indies and the demonstrated ability of this species to cross significant water gaps, such as its early reinvasion of Krakatoa (Hoogerwerf, 1954a). Pfeffer's second reason is also invalid, for the water monitor inhabits many islands that lack surface water (see Hoogerwerf, 1954a). Hoogerwerf (1954a) and de Jong (1937) believed that *V. salvator* is not found on Komodo because *V. komodoensis* feeds on its eggs and young, and Dunn (1927b) simply ascribed its absence to the interaction between the two species. It was partly to study this interaction that I visited Flores in 1971, for only on that island are the two species sympatric. Thus, a study of several important parameters of a population of *V. salvator* was undertaken along Nanga Nggoer and Nanga Loök (see Auffenberg, 1980, for detailed results).

The major factor determining the distribution of *V. salvator* in this area is the presence of surface water. No other is as important. When water is present it is found from coastal brackish swamps well inland to an elevation of at least

1000 m. In the Flores study area the only permanent water is the river Nanga Loök which has a river bank ecology that is varied, including mangrove swamp, dense gallery forest, savanna, and agricultural land, but there is no correlation with weti distribution—all riverine habitats are exploited. In addition, occasional individuals are also seen in and near the small patch of paddy cultivation near Loök. They rarely move far from the water, and most are seen immediately adjacent to or in it, swimming or walking and sunning on the bank. During hunting forays they rarely move farther than about 40 m from the river. Though they move greater distances for carrion, they are only rarely drawn out into the savannas and monsoon forests on either side of the river.

This subspecies of water monitor is much smaller than the other forms. According to Chinese hide merchant records from Sumbawa (Mertens, 1942), specimens over 1.5 m are extremely rare. Our small sample of five specimens from Nanga Loök, Flores, average 115.8 cm (90.0–134.5 cm total length), and I doubt that any attain a length over 150 cm in that area.

Schmidt and Inger (1957) stated that the number of water monitors along rivers in Southeast Asia is sometimes impressive. On Nanga Loök we estimated 15 weti per kilometer of bank upriver from the highest tidal effects. The density within the mangrove area to the coast was estimated at 8/km.

On Flores the water monitor appeared nearly as opportunistic as the ora, feeding on live prey, carrion, and even human feces. Live prey includes snakes lizards, small monkeys, and freshwater crabs. Several species of birds are also eaten, including ground nesting types, and particularly several wading birds (primarily *Nothophyx* and *Bucoroides*). Two species of freshwater ducks (*Anas gibberifrons, A. querquedula*) are also taken. Chickens are commonly captured in the villages, and the villagers usually house chicks in cages hung from the trees to protect them from the weti.

Though weti densities were high along the river, very few of the animals came to the carrion we placed near the stream, but oras visited it often. However, several weti closely followed a dragged dead goat trail for a distance of 120 m. Though weti feed on carrion, they seem to do so much more irregularly and at a much slower rate than oras of equivalent size. This factor may be of considerable importance in areas where both species are found, but where a freshwater fauna is absent or of low density.

Of the individuals observed foraging for live prey, most remained within 10 m of the stream bank. In areas where the adjoining habitat is open, they usually stay even closer to the water's edge, while in extensive gallery forest they may often move as far as 50 m from the stream. The villagers stated that under unusual conditions the wetis may move as far as 0.5 km from the water, particularly during the wet monsoon season.

The usual habitats of these two monitors and their spatial activity patterns within them allow relatively little chance for regular direct interaction. While the weti apparently rarely enters the habitat of the ora in search of food, it is clear that in this area oras often come to the river to drink, and at least smaller oras must catch snakes and frogs there occasionally. On June 23 we found a 55-cm ora in a tree at Nggoer within the weti's usual activity range along the river's edge, and the oras regularly came to the river when we placed carrion there.

The Nggoer villagers told us that oras and wetis often feed side by side when carrion is near the water's edge. However, when the carrion is nearly all eaten, the *V. salvator* move away, because the oras will then attempt to catch them. To test this statement and assess the degree of spatial overlap in carrion-feeding areas, carcasses of several goats and a water buffalo were offered to monitors in several parts of the Nanga Loök valley. While oras came to each of the three sites, weti came only to carrion placed near (10 m) the river. Only two weti (about 110 and 140 cm total length) fed on this carrion bait, though several individuals lived within a few hundred meters of the site. At least two other weti were known to live within the immediate area. However, two oras (120 and 200 cm total length), which had been regularly seen at a carrion bait about 0.5 km from the river, fed on the carrion at the bank. On July 7 carrion similarly placed on the bank attracted a 160-cm ora and a 120-cm weti at the same time, both of which fed practically side by side and paid no attention to one another.

In every instance, the *V. salvator* approached the carrion slowly and circuitously, often moving only a few meters at a time. Their feeding lacked the enthusiasm seen in oras. No weti ever came to carrion placed in the *Zizyphus* savanna of the Nggoer padang 0.25 km south of the river, but oras visited the site regularly. Although a dead water buffalo lay in the uppermost reaches of a small tributary stream near Djarek, none of the weti that lived in its wider, deeper, lower reaches visited the carrion. Similar circumstances in the coastal mangrove swamps near Nggoer produced identical results. Thus, though additional data are desirable, it appears that in at least western Flores the ora and weti normally inhabit entirely different habitats and rarely interact except at carrion, and then only in the habitat of the weti. The weti rarely seeks food (carrion or live prey) far from the stream. Natives' observations suggest that when the two species interact at riverside carrion, the ora is entirely dominant. In spite of the suggestions of Hoogerwerf (1954a) and de Jong (1937), there is no evidence that oras eat weti eggs, though they would probably do so if offered the opportunity. However, their usual hunting area does not include that of the streamside habitats. Dead weti are eaten by both dogs and oras.

Dogs.—Feral dogs occur on all the islands on which the ora is found and most of the interaction between them is in the nature of competition for food, both live and carrion. Most food competition is between dogs and smaller oras (Fig. 11–6). In addition, oras smaller than 1.5 m are often chased from carrion by dogs, though we never saw one caught. The oras usually escape by climbing a nearby tree. Pfeffer (1959) reported the same, adding that he once saw a

V. salvator caught by dogs on Borneo. Larger oras have a strong advantage over the dogs; it is clearly seen when one or more dogs are feeding at carrion and a large ora approaches. The dogs nimbly move out of the ora's way and allow it to feed, but sometimes bark furiously from a respectful distance. Whenever the lizard leaves the carrion, the dogs approach and feed from it, retiring only when the ora returns. The number of dogs at carrion varies from one to seven, but usually only one or two are present. The larger numbers usually discourage even the largest oras from feeding at a carcass. The stress of feeding or remaining near carrion when oras are close is well illustrated by the usual subordinate postures of the dogs: the ears and head are down, the front part of the body is lowered, and the tail is tucked between the legs. However, there are instances when dogs and oras feed on carrion together. When the carrion is nearly gone, the dogs often press the lizards more closely, circle the ora(s) repeatedly, and bark continuously. The sound of this continuous barking is an indication of the presence of carrion and of oras feeding on it.

Agonistic ora behavior directed at dogs includes chasing, hissing, warning postures, tail blows, and biting. The usual ora reaction is simply to ignore the dogs. The following interactions are typical:

> *July 7, Flores.* "A 1.2-m ora is feeding at bait when a dog slowly approaches from behind, sees ora, then begins creeping toward ora, keeping a low profile as it moves forward. From a distance of 3 m behind the ora it barked and sprang forward as if to attack the lizard. The latter bolted, easily out-distancing the dog chasing it, finally climbing a sambe tree."
>
> *July 15, Flores.* "A single dog chased a 1.5-m ora from bait today, but from a respectful distance."
>
> *June 28, Flores.* "Three dogs barking and circling a 2-m ora, which paid no attention to them but kept looking at bait. The dogs finally moved off several hundred meters and waited under a clump of trees until the ora left the area."
>
> *June 28, Flores.* "A dog chased a 1-m ora from the bait and into a tree."
>
> *March 2, Komodo.* "28W walked toward bait through brush, but literally stumbled on two dogs lying in grass waiting for 19W to leave carrion. Dogs and ora ran in opposite directions for a few moments, then dogs turned and chased ora, barking all the while. After about 50 m ora stopped running, turned back toward carrion, and the dogs following it scattered as it passed between them."
>
> *Sept. 5, Komodo.* "Dog approached 1.2-m ora, who first ran away, but then turned and chased dog out of immediate area."
>
> *Sept. 5, Komodo.* "Carrion almost all gone, and when 19W carried what was left into brush, the dog started barking and excitedly moved

around all of the lizards, snapping at some. It chased 5W off, but couldn't budge 19W, who was trying to swallow entire chest of goat."

Sept. 5, Komodo. "1.6-m ora pays absolutely no attention to furiously barking dog snapping at its tail as ora ambles away from carrion site."

In addition to agonistic and competitive coactions, dogs and oras often utilize the same trails. Additionally, droppings of each species are closely investigated by the other, though for much shorter periods of time than scat of their own species. One can only surmise that at least interspecific density information is exchanged in this fashion.

Wild boar.—Competitive and predatory coactions among boar and oras are discussed in Chapters 10 and 11 and injuries sustained by oras as the result of encounters with boar in Chapter 6. Other coactions noted involve warning displays and a few commensal relationships.

The boar's warning display includes both auditory and visual signals. The former includes certain characteristic grunts, as well as the clicking noises of tusks striking together. Visual display includes raising the long bristles on the back of the neck and down the center of the back, holding the tail up, and turning the body sidewards to the adversary, with the back arched and the legs extended. All three are sometimes seen in encounters between boar and oras at carrion. Pfeffer (1959) briefly described some of these displays in a boar confronted by an ora on Rintja, and on this basis he wondered if the boar recognized a danger in the ora. The large number of boar remains found in ora droppings suggests that this is certainly the case. I saw no evidence that oras paid attention to the displays.

Much more important is that the wild boar wallows are usually the only sources of standing water at lower elevations throughout much of the year, and the oras regularly utilize them for drinking (Chap. 9). On Rintja and Flores, water buffalo wallows are important in the same way. In both cases they are often the only place where frogs can lay their eggs and are thus important in maintaining this resource. The wallows are regularly visited by at least the larger prey of the ora, and some attacks are certainly associated with such movements, the ora being concealed along trails leading to the wallows.

The trails over which oras tend to move, in at least the thick cover of their activity ranges, are originally produced by deer, boar, and water buffalo. Thus, the regular movements of game, for whatever reason, determine to a large extent the directions taken by moving oras, whether their purpose is hunting or not.

Wild boar are also important in determining the specific place where oras dig some of their burrows. Along stream courses on Komodo boar enlarge natural openings under overhanging banks in which they sometimes spend the hotter part of the day. These spots are often later further excavated and deepened by oras. At favorable sites on hillsides boar often clear organic debris and rocks from small resting sites. Sometimes they are used as dusting places. They are often located under a large overhanging rock or in an isolated clump of trees on an otherwise nearly treeless hillside. The critical factors seem to be shade and a breeze. These shallow pits, or terrace-like spots, are often used by oras to initiate their own burrows under the rocks, etc.

These activities suggest that boars may also use larger ora burrows on occasion. Boar prints have been found at the mouths of several ora burrows in the Loho Liang area. In addition, on January 23 a boar was seen running out of an ora burrow, with the original ora occupant immediately behind it.

Smaller animals.—Certain small animals, such as snakes, lizards, and rats, utilize ora burrows for shelter. The gecko *Crytodactylus d'armandvillei* lays its eggs in ora burrows.

The interaction of carrion-feeding birds and oras has been described in Chapter 11.

Pfeffer (1959) said the green jungle fowl (*Gallus varius*) announces the presence of an ora by a series of alarm calls, although we were not able to confirm this. However, on several occasions along Kali Besar we saw an ora approach a megapode (*Megapodius freycineti*) nest, and heard one or both parent birds give alarm calls as they alternately flew and ran away through the surrounding forest. Lincoln (1974) reported a megapode scraping at a nest mound at the same time that a 2.7-m ora was digging into the center. After the ora left, the bird covered the hole.

Although monkeys tend to become silent and disappear into the forest on the approach of a human, a troop often gives alarm calls when it sees an ora nearby. Pfeffer (1959), who noted this behavior on Rintja, likened it to that of langurs (*Entellus*) in the presence of a leopard. Father Schmutz told me that in addition to the alarm calls the monkeys sometimes throw small branches and fruits at the ora below, which we confirmed during our studies near Nggoer, Flores. The most vigorous calls occur when a monitor (either *V. salvator* or *V. komodoensis*) surprises a monkey on the ground and it runs to the trees.

Man.—No lizard plays, or has played, such an important role in human affairs as the monitors. In fact, Konigsberger (1911–1915) described *V. salvator* as part of the human cultural fauna of Java. It is remarkably common in even some of the most urbanized sections of the large cities of Asia (Mertens, 1942). Several other species are commonly found near human dwellings. In some areas they prey extensively on house rats and chickens, so that the association between monitors and humans is almost "domestic." Many species of monitors are eaten by rural folk through the Old World. Mertens (1942) was incorrect when he stated that the flesh of *V. salvator* is rarely eaten. While large numbers of people do not eat it because they consider the flesh poisonous (Ceylon), because the animal is of a low type (Thai), or for religious reasons (Muslim), it and its eggs are regularly eaten by many Chinese, Filipinos, and most tribal people throughout much of Southeast Asia. On Bali hunting pressure has nearly extirpated the lizard from areas where it was once common. It is eaten by the Dongo on Sumbawa, and the Ngadas on Flores. The eggs of both this and other species are also eaten.

The skin of several monitor species is extensively used in the leather industry. Most skins are used for handbags, small hand drums, and recently as completely stuffed items for the tourist trade.

As far as I am aware, the flesh and eggs of *V. komodoensis* are not utilized, at least at the present time. Lallemant (1929), quoting Chinese records for 1926, stated that 100 Komodo monitors were shot on Rintja and a similar number on Flores, primarily for their skins. The same author referred to 65 specimens obtained from Komodo for their skins at approximately the same time. The attempt to commercialize the skins of the oras was unsuccessful, apparently due to the large and numerous osteoderms in the skin. Other authors (Delsman, 1927; Lallemant, 1929; and Anon., 1930, 1937) stated that oras were obtained to produce medicine, both, according to Chinese merchants, because the oras were "champion swimmers" and because their rendered tail fat could be used as a medicine for burns.

Oras often enter villages, where they cause considerable consternation. Most often these visits are made by oras that seem to be hungry. Under such circumstances, livestock of various types, including dogs, are frequently chased; goats are sometimes taken by marauding oras. During periods when the villagers dry large numbers of fishes in baskets and nets on the ground or on the thatched roofs, oras often wander into the area, obviously drawn by the scent of the drying flesh. Years ago the graveyard on Komodo was in a sandy region, and oras occasionally attempted to exhume recently buried human bodies. On at least one occasion a large ora was successful in digging through the loose soil to the body itself, on which it started to feed before it was driven off by the villagers. Because of this, Komodans now often tamp clay into the grave above the body. The earlier, indigenous Neolithic inhabitants of Komodo buried their dead in a flexed position in a very shallow grave. The body was protected by a low cairn made of very large stones built over the grave site. Other monitors are also known to feed on human bodies, and *Varanus salvator* is the species with which the habit is most frequently associated. Laidlow (1901) described this species feeding on bodies placed in trees on the Malay peninsula, Anderson (1889) reported the same on the Mergui Archipelago in southeastern Burma, and Taylor (1922) mentioned the same habit on the Philippine Islands.

Varanus salvator was noted eating human excrement on both Bali and Flores. We have only one observation of an ora doing the same. On September 20, 1969, an unnumbered ora (approximately 1.5 m in length) excavated much of our old latrine area at the base camp, and apparently fed on some of the fecal matter.

During our year on Komdo, oras regularly attacked our chickens, pet wild boar, and goats kept in camp for food and bait. This experience is shared by most of the villagers living within the ora's range. Along the north coast of Flores

Table 12–1. Livestock lost to oras per year in villages of western Flores[a]

Livestock	Villages		
	Djarek	Loök	Nggoer
Water buffalo	1	3–5	1
Horses	5	1–5	0–1
Goats	10	"many"	5–10
Dogs	"several"	"same"	"few"

a. Estimates of village chiefs and elders.

livestock depredation is often stated as the major reason for killing most of the oras that the villagers find there. Both feral and domestic horses, a few cows, and (on Flores) even water buffalo are often attacked and killed by larger oras (see Chap. 10) (Table 12–1). As pointed out elsewhere, atttacks, particularly on large domestic prey animals, are uncommon and are apparently caused by local, conditioned, individual oras. The trait is apparently not characteristic of all large members of an ora population, though all large specimens have the potential to develop such habits (see Chap. 9).

Man's effect on the oras in the Lesser Sunda Islands includes direct predation, particularly in response to livestock killing. Oras are hunted with dogs and speared, caught in baited snares and traps, and killed by poisoning carrion with insecticides. Most of this predation is directed towards large adults, though smaller specimens are not entirely immune. However, much more significant is the extensive habitat modification through both agricultural and hunting practices. Of these, the most important is repeated burning of the habitat (Zollinger, 1850).

A search of all the old literature on tribal customs and beliefs of the inhabitants of Sumba, Sumbawa, and Flores did not disclose any legends or myths reporting the ora, nor did the missionaries with whom I talked on Flores know of any. However, in Mangarrai Province I heard one story that explained that many years ago the oras originally moved to Komodo through a long tunnel that connects the two islands under the surface of the sea.

Economically less important, but perhaps of more interest, are the direct relationships between oras and local human populations.

In most of our close-range encounters with oras on both Komodo and Flores, we found that they were very shy and almost always bolted away. However, two of our activities tended to engender something other than flight. One was when we sat quietly on the ground in our observation blind; the other was when we approached oras that were in a trap, against a cliff face, or in a similar situation from which they could not escape.

In over 400 encounters with oras at feeding sites, very few showed any interest in us. Most fled when we disturbed them feeding. However, as Oesman (1970) and others have stated, large oras, once they have begun to feed, are often easily approached. On several occasions we stood in the open, our feet barely 1 m from the head of a large feeding lizard. However, this was unusual, and we found it true only in Loho Liang, where lizards are apparently not persecuted by humans, and where visitors are common.

However, on several occasions certain oras showed a decided interest in us as we sat either in the open or in our blinds. Most of these instances seemed to represent curiosity, more than attack intention. Smaller specimens were more commonly "interested" than larger ones. Those that fell into an "attack" category were generally large individuals. Excerpts from the field journal that illustrate typical "interest" and "attack" behavior under conditions unsolicited by the observers are presented in Table 12–2.

Table 12–2. Typical ora behavior to observer seated close to carrion

Interest behavior		
Date	Ora size (m)	Remarks
Nov. 7	1.2	"Came into blind. I hit it on head 3 times with my pencil, but it was not frightened off. It remained and flicked my tape recorder and my knife with its tongue. It then left the blind and climbed a tree to a height of 2.5 m. Approximately 1 hr later it descended the tree and again came into the blind, but this time walked straight through, smelling my bare foot with its tongue as it passed by."
Nov. 20	1.2	"Hit smaller ora on top of head twice with the flat part of my *parang* as the ora tried to enter the blind. It simply stepped back, moved to the right and attempted to enter the blind again."
Nov. 20	1.0	"Ora came into blind containing Putra and Kurt (2 persons). It walked around within the blind, licking various items with its tongue."
Nov. 20	1.4	"This ora stood at the side of the blind, peering behind the woven bamboo front on several occasions. It began to enter blind several times this afternoon but was frightened off when we shook a notebook in its face."
Feb. 9	2.1	"28W rushes around right end of blind and enters it, though I was sitting there, minding my own business.

(Continued)

		Ora did not threaten, but his size and his apparent disinterest in leaving sent me scurrying out of blind. This happened immediately after an unsuccessful attack of his on a large goat immediately in front of blind."
Oct. 1	1.0	"...smelled my paper as I continued to write my notes and moved my pencil all about. His "curiosity" in the paper, the pencil, and my hand was truly surprising. However, when I moved my body in an attempt to stand, he ambled out of the blind. This same lizard returned to the blind two more times during the afternoon."
Oct. 1	1.4	"Green Blotch entered the blind on three different occasions this morning. On none of these did he seem particularly disturbed by my presence, though I could not move about a great deal without sending him scurrying off. At one point, he even stretched out in the shade of the blind, with his front leg draped over mine as he lay there half asleep."
Mar. 7	2.5	"Yesterday at 1700 hr 34W came into camp, no more than 15 m from our tent."
Mar. 8	2.5	"Number 34W came back into camp at 1000 hr. He entered the tent and rummaged about in our backpack. He left, but kept circling our camp. At noon he returned and took Ishmail's shirt, shaking and tearing it as he carried it into the surrounding brush. At 1300 hr he returned and took a handkerchief that had been lying on the ground less than 2 m from Kurt, who was working in an excavated hole."

Potential attack behavior

Date	Ora size (m)	Remarks
Jan. 19	2.0	"32W tried to enter blind and was sharply hit on the nose with a stick approximately 2 cm in diameter. It ran away approximately 5 m then turned around and ran back into blind with its mouth wide open. As it was entering it was pelted with pebbles, which stopped its advance. The mouth was still help open and Putra decided to leave the blind to the ora. It remained there for at least 1 hr, when Putra returned to camp."

(Continued)

Table 12–2. *Continued*

Oct. 1	2.7	"23W drove me out of the observation blind near trapping site A on two occasions this afternoon. Though its mouth was closed, it entered the blind with the head held low and the roach slightly raised. I did not wait to see whether it would attack, but slipped out of the blind, sat down in the grass approximately 10 m away until it left the blind."
Jan. 16	2.0	"...saw 34W coming around from Sugito's Path as I sat in the main trail. It apparently saw me when I was approximately 5 m away, and instead of veering off the path, continued to move in my direction, with mouth held open slightly, hissing all the time, head held low and with roach erected. When it was approximately 2 m from me I quickly got up and entered the adjacent brush. Though the lizard was no longer threatening it slowly continued on its way, stopping only a moment to touch its tongue tips to the place where I had been sitting in the volcanic ash."
Mar. 3	2.0	"...and I saw 34W coming towards me down the beach, approximately 60 m away. When he saw me he moved close to the edge of the surf but continued in my general direction. At the same time, he exhibited a strong threat display with tail bowed as if to strike. I stopped on the beach and watched him carefully as he walked by me, still threatening only 6 m away."
May 21	2.0	"When 34W was released from the trap this afternoon, it walked straight toward both Wayne and Brian. Its mouth was slightly open, head held low and with roach slightly raised; the tail was arched and bowed. When it was approximately 2 m from both men, Wayne stepped out of the way. It veered only slightly from its original path, but it was not particularly intimidated by Brian's presence. It was only after it passed him that the lizard changed his slow stiff-legged walk to a more leisurely trot."
Feb. 3	2.0	"Ibraham had told us that the oras at Poreng will come to the camp if one lit a fire. We arrived at 1100 hr and started cooking near noon... . At 1650 hr 34W approached our camp to within 30 m... . The same ora returned at 1730 hr and approached to within 15 m of our tent."

However, though Burden (1928) stated that there is no indication oras will attack man, some individuals are decidedly aggressive, and attacks on humans *have* occurred on several of the islands. Lallemant (1929) is apparently the first investigator to mention attacks on humans. Others have reported attacks, though Pfeffer (1959) believed most are caused by the ora defending itself. While this is certainly true, some assaults are clearly unprovoked. However, the periodicity of the attacks and their localized nature suggest that only specific oras are involved.

Oras were involved in two deaths on Komodo. The first attack apparently occurred in 1931 and was reported by de Jong (1937), de Voogd (1950), and Pryor (personal communication in 1969). We talked to the father of the young man who was killed and learned that these two, and another son, sat in a small clearing in the forest cutting and trimming "yellow wood" in equal lengths. Suddenly an ora about 2 m long stepped into the clearing and started to move toward the three seated men. When it was within 5 m or so, all three jumped to their feet and started to run away. At this moment, the ora charged after the 14-year-old boy, who almost immediately ran into a low-hanging vine. The vine stopped the youngster for just a moment, and the ora bit him very severely in the buttocks, tearing away much flesh. Bleeding was profuse and the young man apparently bled to death in less than one-half hour. Darevsky (1962) was apparently referring to this incident when he said a child had been killed on the island.

The other ora "attack" on Komodo is that referred to by Collins (1956). Details of this instance were provided me by the elders of Kampung Komodo, some of whom were involved in the incident. Four men from Sapi, a village on the neighboring island of Sumbawa, were hunting water buffalo at Poreng, the same place known to be the core activity area for 34W. One of these men apparently became very ill and was left at Poreng. The remaining three went to Kampung Komodo and returned with 30 villagers and the chief on the next day. However, on their arrival the man left at the camp was dead. All the flesh of the arms and legs, as well as the viscera, had been eaten by oras. The few remains were buried at Poreng. Though many of the elders insisted that the man had probably been killed by the ora, there is no proof of this. However, the body was eaten by one or more oras.

Two unconfirmed deaths have been reported on Komodo since this study was completed in 1973. Both attacks were supposedly mounted against tourists and resulted in their deaths; one of these is said to have occurred at Poreng—the foraging area of ora 34W.

According to Komodo villagers a policeman from Flores was bitten in 1947 while petting a 2-m ora that was completely tied. The ora apparently lunged upward and tore out the man's right biceps. The villagers heard he had died one week later in the Ruteng hospital from a severe infection.

One of a pair of Komodans was evidently attacked while carrying a dead deer

(Darevsky, 1962), and one of our workers was almost bitten while he slept on a savanna in the daytime. Another was attacked as he walked past an ora burrow and only drove the ora off after pelting it with large stones.

Rintja villagers with whom I spoke stated that they knew of no attacks on that island. However, on Flores several reports of attacks were related to us. On the southeast coast of that island, near Nangelele, F. G. Thompson was told that oras were once common at the village of Nisar, but that in 1957 a person was killed and eaten by a large monitor, and that oras were now killed whenever possible.

In extreme western Flores, Nggoer villagers told me of an attack several years earlier in which a man was bitten below the calf by a 1.5-m ora. The incident occurred in the man's garden area. The ora had been chasing his chickens and the villager, in turn, had chased the ora. It ran for some distance, but then suddenly turned and attacked the man, who is still living in the area.

Another incidence of an ora attacking an adult man occurred a number of years ago when a villager was cutting grass for his roof on a mountain near the village of Nggoer. He was apparently squatting and did not see or hear the approach of a "large" ora. The ora bit the villager in the calf of the right leg and ripped out most of that muscle. Villagers working nearby said the man screamed and fell on his right side, and the ora ran away. Weeks later the man became pallid; he remained sickly for two years, then died.

Several people have been attacked while sleeping on the savanna in the daytime. One, Abdullah by name, living in the village of Nggoer, still carries the scars of an attack that occurred many years ago by a 1.5-m ora which had grasped him twice by the neck during the struggle.

In January, 1970, Sugito Tirtowono, Wildlife Officer, was approached in a threatening manner by a large ora (34W?) in the savanna behind the base camp. It apparently advanced toward him deliberately and slowly, with mouth open and back arched. The ora ran off only after being pelted with stones. A Komodo villager, Mahamu, told me a similar story of an ora that rushed out of its burrow and tried to bite his legs. It turned and reentered the burrow when struck with the heavy, machete-like *parang* carried by most Malays.

Ectoparasitemia

Ticks and mites are apparently common on most species of *Varanus*, and several monitors are specific hosts for particular parasite species. Other ectoparasites are rare, though a leech (*Glossiphinia* sp.) attacks individuals of *V. niloticus* (Mertens, 1942).

The only ectoparasites we found on *V. komodoensis* were ticks. These were *Aponomma komodoense*, *Amblyomma robinsoni*, and *Amblyomma helvolum*. All three species have been described from *V. komodoensis* previously.

Aponomma komodoense Oudemans was formerly considered a poorly known species, represented in collections by a few specimens from only Komodo. During the study we collected additional series from *V. komodoensis* captured on Komodo and near Nggoer, Flores. This is the first record of this tick from a locality other than Komodo. The Komodo monitor is the only host reported (Anastos, 1950). A related species, *Aponomma gervasi*, is a parasite of *V. bengalensis* in Northern India (Hoogstraal and Rack, 1967).

Amblyomma robinsoni Warburton is also known only from the same host and island as *Aponomma komodoense*. Toumanoff's (1944) reference to this species from Indochina is apparently incorrect (Anastos, 1950).

Amblyomma helvolum Koch is known from a number of reptilian hosts from a wide area in Southeast Asia (Nicobar to Australia). These include several species of large snakes, a turtle, and representatives of the genus *Varanus* (*V. salvator, V. indicus, V. bengalensis,* and *V. komodoensis*). In addition to earlier records of *V. salvator* serving as hosts for *A. helvolum* (Anastos, 1950), we found it on all four specimens of *salvator* searched for ticks near Nggoer, Flores. *V. salvator* also serves as host for *Amblyomma lucasi* (Hoogstraal et al. 1968). *Amblyomma helvolum* occurs on *V. komodoensis* from Komodo, and we discovered specimens on both oras examined for ticks at Nggoer, Flores.

All these tick species are generally found on the back and sides of the ora's body, most of them above the lateral fold. Concentrations sometimes occur at the insertion of both the front and hind legs. Occasional specimens are found on the tail and head. The proportionate number of ticks occurring on the oras was established by counting all the ticks of each species between the posterior edge of the insertion of the front legs and the anterior edge of the insertion of the hind legs, down to the lateral fold.

In general, larger oras are more heavily parasitized by ticks than are smaller ones, a common pattern in most vertebrates (Fig. 12–1). The mean number of ticks found on each ora is 15.9 (lizard sample N = 58). Each of the three tick species has a significantly different density on the oras (Table 12–3). *Amblyomma robinsoni* is the most common, comprising 59.7 percent of the total of 687 ticks tabulated; *Aponomma komodoense* 30.8 percent, and *Amblyomma helvolum* 9.5 percent. The mean numbers of ticks of each species in the sample areas per lizard are: *A. komodoense,* 5.4; *A. robinsoni,* 9.2; and *A. helvolum,* 1.2.

There is no significant seasonal variation in tick density on oras from January to September (Jan.–March, $\overline{X}$ ticks = 14, Apr.–June, $\overline{X}$ = 16, July–Sept., $\overline{X}$ = 10). However, at the height of the dry period the number of ticks per ora increases sharply (Oct.–Dec., $\overline{X}$ = 24.2), apparently being reduced quickly to about 13 per lizard sample area with the advent of the rainy season. Unfortunately, few lizards were rechecked throughout the year for tick infestation level. However, of those that were, the data seem rather conclusive. Number 33W had

50 ticks in the sample area in November, 31 in March; 34W had 57 in November, 23 in February; 7W had 15 in early October and 24 at the end of the month; 1W had 14 in October, 5 in November; and 3W had 16 in January and 18 in October. The two species involved in both oras were reduced evenly (about 50 percent). Number 23W was checked before and after the rainy season and the number of ticks was almost identical (15, 16). The nymph to adult tick ratio in the dry season was 1:1.25, in the wet season 1:12.13. Males outnumbered females in all three tick species. Female to male ratios were *A. helvolum,* 1:18, *A. komodoense*, 1:13, and *A. robinsoni* 1:23.

In June and July 1971 attempts were made to locate resting ticks and nymphs by "flagging" likely areas with toweling attached to a staff. Game trails and ora holes were all checked with very poor results. Only one tick was found in this way: an active adult *A. komodoense* on the soil of the floor of an ora burrow on the beach near Nggoer, Flores.

The number, distribution, and species of ticks found on ora prey species is given in Table 12–4. *Haemaphysalis papuana* is a pig-parasitizing tick, known from the same species of suid in India and the Republic of Indonesia (Hoogstraal and Rack, 1967). *Haemaphysalis renschi* is a parasite of artiodactyls and perissodactyls and is known from Flores westward to and including Sumatra (Hoogstraal et al. 1968). Nymphs are recorded from a crested tern (*Thalasseus bergii*). *Boophilus microplus* is widely distributed in Southeast Asia and has been reported from sambar deer (*C. unicolor*) in several papers.

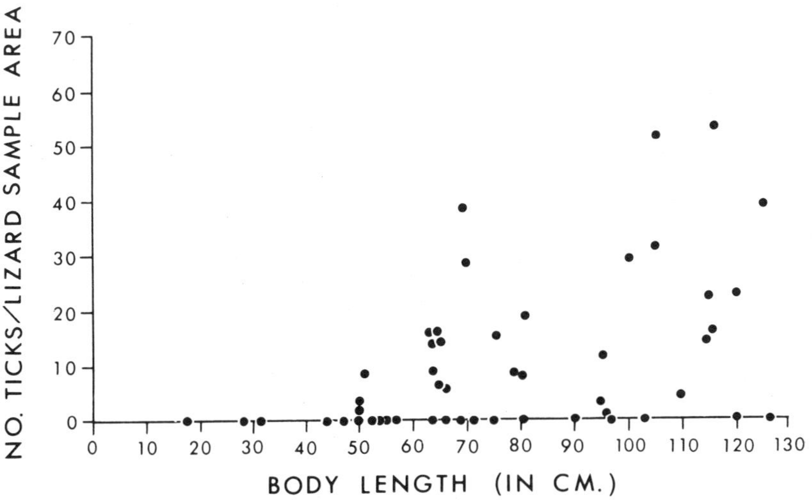

Fig. 12–1. Large oras tend to have a proportionately greater tick load than smaller specimens.

Table 12–3. Numbers of adult ticks of three species found on parasitized *V. komodoensis*

Specimen	Date	*Aponomma komodoense*	*Amblyomma robinsoni*	*Amblyomma helvolum*	Total
6W	Oct. 30	1	3	0	4
18W	Oct. 28	7	14	7	28
2W	Oct. 30	1	17	0	18
33W	March 2	14	17	0	31
	Nov. 5	29	21	0	50
34W	Feb. 2	11	12	0	23
	Nov. 21	27	30	0	57
23W	May 11	0	16	0	16
	Oct. 2	0	15	0	15
27W	Jan. 16	5	4	0	9
7W	Oct. 1	3	12	0	15
	Oct. 31	5	17	2	24
25W	Oct. 5	4	5	0	9
11Y	Oct. 6	3	5	3	11
2Y	Sept. 13	0	9	0	9
28Y	Oct. 19	0	4	2	6
30W	Jan. 12	4	2	0	6
31W	Sept. 18	0	9	2	11
36W	Nov. 17	6	2	0	8
32W	Nov. 23	9	26	5	40
4Y	Jan. 13	9	8	6	23
21W	Jan. 24	0	1	1	2
	Oct. 30	5	9	0	14
1W	Nov. 26	0	5	0	5
43W	Jan. 13	6	8	0	14
35W	Jan. 19	8	11	10	29
3Y	Jan. 25	4	12	0	16
	Oct. 30	10	8	0	18
37W	Feb. 26	0	4	0	4
47W	March 20	0	8	0	8
46W	March 20	5	9	0	14
41W	March 20	10	5	0	15
50W	March 28	0	1	1	2
5W	Oct. 30	14	10	9	33
9W	Oct. 30	16	7	14	37
8W	Oct. 31	1	34	0	35
9Y	Nov. 1	2	24	2	28
Totals		219(30.8%)	404(59.7%)	64(9.5%)	687

Table 12–4. Ticks found on two species of ora prey

Host species	Tick species	Total ticks
Sus scrofa (N = 3)	*Rhipicephalus pilans*	57
	Haemaphysalis papuana	1
Cervus (rusa) timorensis (N = 1)	*Rhipicephalus pilans*	4
	Haemaphysalis renschi	14
	Boophilus microplus	11

Endoparasitemia

Blood.—Samples of blood from female and male oras, 1.5 m long, were collected on Flores and later studied at the University of Florida by S. R. Telford. Though the presence of several species of habitual ectoparasites would suggest the presence of some type of blood parasite, none was found in the small sample we provided. However, a hemogregarine infection (Sporozoan) was discovered in one of three blood samples from *V. salvator* gathered at the same place on Flores. This as yet unidentified red blood cell parasite is a type most likely transmitted by a tick, possibly a leech, or even a mosquito.

Subcutaneous.—Larval cestodes of the order Pseudophyllidea (spargana type) were exceedingly abundant beneath the skin and in the fascia of the two oras dissected at Nggoer, plus all four of the *V. salvator* studied. The larval nature of the material precludes precise identification. The lizards are accidental hosts and are essentially intruders in the development cycle of the parasites, since the larvae will not mature. Infestation rates were very high. I estimated about 150 cestode larvae under the skin of an ora 1.5 m long. The larger larvae formed lumps under much of the skin of not only the varanid lizards examined, but of many of the frog-eating snakes of the area as well (*Trimeresurus albolabris* and *Dendrelaphis pictus*). Pseudophyllideans are often found in fish-eating vertebrates, and the varanids must have been infected by feeding on aquatic organisms. There is no evidence that the infection level caused the lizards any discomfort or loss of locomotor efficiency.

Fecal.—Feces from the two oras dissected and several samples of fecal pellets found in the field and preserved in polyvinyl alcohol were sent to Florida for study. In all *V. komodoensis* samples examined we found a small number of amoebae of the genus *Endolimax* (species indet.) and cestode ova. The former is not believed to be pathogenic and has been reported from captive oras previously (Gray et al. 1966).

Gastric.—The only gastric parasites found were nematodes of the genus *Tanqua* (Spiruroidea, Gnathostomatidae) taken from two of the three *V. salvator* examined at Nggoer, Flores; none was found in the *V. komodoensis* examined.

These nematodes have a dietary distribution, without phylogenetic implications. Other reptilian hosts from which they are known are *Varanus* sp. and *Xenochropus piscator.* The parasite probably uses an arthropod intermediate host; many spiruroids use ants.

Intestinal.—In *V. komodoensis* mature cestodes of several types were found, representing two orders: Pseudophyllidea and Protocephalidea. The former are represented by at least two species, *Duthiersia* cf. *expansa* and *D. venusta* (or *D. sarawakensis*). The genus is a small one, showing a phylogenetically correlated distribution. It is known only from varanids, but is widespread from Africa to Southeast Asia. *Duthiersia elegans* and *D. robusta* are reported as parasites of *V. niloticus; D. latissima* of *V. exanthematicus; D. sarawakensis, D. venusta,* and *D. expansa* of *V. salvator; D. expansa, D. crassa,* and *D. venusta* of *V. bengalensis;* and *D. expansa* of *V. indicus* (Woodland, 1938). In our sample of *V. salvator* from Flores we found a heavy infestation of *Duthiersia*, including three species in two specimens of *V. salvator* (*D. expansa, D. crassa,* and *D. venusta* or *sarawakensis* in one; *D. crassa* and *D. venusta* or *sarawakensis* and a probably undescribed species in the other), and two species in the third *V. salvator* (the presumed new species and *D. venusta* or *sarawakensis*). Though the life cycle of *Duthiersia* remains unknown, it surely involves an intermediate host eaten by the monitor.

Still another pseudophyllidean (*Scyphocephalus bisulcatus* or *secundus*) was found in *V. salvator.* This very small genus is apparently known only from varanid lizards in the Philippines and Java, probably *V. salvator.*

The second order of cestodes found in the intestine of *V. komodoensis* *(Proteocephalidea)* is represented by the species *Acanthotaenia varia*. The genus is widespread in the tropics and is a parasite of many species of lizards and snakes. Carter (1943) reported it as occurring in an ora originating from Komodo.[1]

Respiratory.—No pentastomids were found in *V. komodoensis* or *V. salvator,* though they have been reported from *V. indicus* (Self and Kuntz, 1957) and *V. salvator* from the Philippines (Self and Kuntz, 1966).

The endoparasitic data available from examination of even a few specimens of *V. komodoensis* and *V. salvator* (Table 12–5) suggest a rather extensive fauna, of which certain elements are shared; others are not. In general, *V. komodoensis* is not heavily parasitized in terms of parasitic variability or population density. However, the *V. salvator* examined showed a heavy parasitemia, both in faunal diversity and density of at least some of the elements (*Duthiersia* sp.). Whether the same low level of parasitism is found in the Komodo population of the ora is unknown. However, that the larval subcutaneous cestodes are absent (or rare) in

1. *Proteocephalus sandgroundi* Carter (1943) has been made a synonym of *Acanthotaenia varia* Yamaguti (1959).

Table 12–5. Endoparasites of *V. komodoensis* and western Flores populations of *V. salvator*.

Parasites	Place	*V. komodoensis*	*V. salvator*
Cestoda			
Pseudophyllidae			
larvae (spargana type)	Subcutaneous	+	+
Duthiersia expansa	Intestinal	+	+
Duthiersia venusta[a]	Intestinal	+	+
Duthiersia crassa	Intestinal	−	+
Duthiersia sp. indet.	Intestinal	−	+
Scyphocephalus bisulcatus[b]	Intestinal	−	+
Proteocephalidea			
Acanthotaenia varia	Intestinal	+	−
Nematoda			
Spiruroidea			
Tanqua sp.	Gastric	−	+
Sporozoa			
Hemogregarine	Blood	−	+
Amoeba			
Endolimax sp.	Fecal	+	?

a. Or *D. sarawakensis*.
b. Or *S. secundus*.

that population is obvious, for no evidence of them was seen or felt in the dozens of specimens handled during the work there.

Intraspecific Coactions

That social integration in some form or other occurs commonly among animal populations and is likely to be found virtually everywhere is well known. Socialization is a general phenomenon that has undergone progressive evolution, so that in the more advanced forms it has tended to become increasingly more conspicuous and complex. In general, this also seems to be true of the lizards where social adaptations in different groups vary widely. In every case, the behavior of the individuals is conditioned by the presence and actions of other members of an aggregation in the interests of facilitating the cooperation or dispersion of the group as a whole.

That social behavior can be of some value in various ways to the welfare of an aggregation of oras is here assumed; although when it comes to defining what precisely the advantages are, the results are not satisfying. In previous sections it has been shown that an aggregation of oras may be a factor in successful breeding, but it is obviously not a fundamental and universal necessity for successful matings in oras. However, ora aggregations are clearly important in establishing and reinforcing hierarchies, apparently maintained far beyond the immediate area of the carrion around which most of the aggregations form.

In fighting we find a widespread form of social activity in ora aggregations, with important implications. The benefits are undoubtedly successful attainment of social standing in the hierarchy, as well as feeding or territorial rights (Tinbergen, 1953, 1969).

No coordinated group behavior was witnessed. In all our encounters (N = 415), no walking oras were seen in pairs or in aggregations, except at kill or bait sites. Hunting is apparently a solitary activity. Meetings between hunting oras are probably rare and are perhaps avoided. Most encounters are probably brief and casual. However, even in large outdoor enclosures on Java, adults will sometimes lie adjacent to one another, even touching, or piled one on the other. DePrato (n.d.) described the same for captives in the National Zoo in Washington. Whether this ever happens in nature is unknown.

Feeding Aggregations

The high efficiency with which oras can locate either a fresh kill or carrion and their wide-ranging habits mean that groups of hungry individuals quickly gather around such food. Within several hours of a fresh kill, or one to two days at carrion, aggregations of three of four oras are common. The average number at carrion is 3.9 individuals and is rarely as high as 17 (in spite of Hoogerwerf's doubts of earlier records of large feeding groups by Broughton, 1936). Normally the aggregation is a fluid one, with individuals leaving and others joining at odd times during the day. Not all individuals in the local group feed at the same time or are even in the same small area. Members of the aggregation may be spread over several hundred meters, alternately feeding and walking about, or seeking shelter under vegetation. The group is comprised of individuals from approximately 1 m total length to the largest size classes and includes both residents and transients. This often large, fluid, mixed aggregation encourages a great deal of interaction, of which the most common is agonistic. It is apparently only during group feeding that the smaller oras are tolerated by the larger ones (see Kruuk, 1972, and Schaller, 1972, for similar data on hyenas and lions). Burden (1928), Broughton (1936), Hoogerwerf (1954a), Oesman (1970), and others have reported aggression in feeding groups of oras. Some of the rapid, voracious feeding of oras undoubtedly forestalls appropriation by conspecific competitors (see Schaller and Lowther, 1969).

Signals and Displays

In the Komodo monitor, displays related to spacing fall into four categories: auditory, chemical, tactile, and visual (Table 12–6, 12–7). A display may comprise signals from one or all four categories (Table 12–6). The auditory signal is a loud hiss, restricted to the threat display. Visual signals include color, position, and form; they are associated with a wide variety of displays, usually presented in open environments where they can be seen. Like auditory and tactile signals, visual ones are concerned entirely with communication in the present and lack any temporal perspective. Some chemical signals used by Komodo monitors (such as those providing sexual identification during courtship) are immediate. But others (such as fecal pellets) may indicate a past event (that is, the passage of an individual in that area). It may also indicate a future potentiality (possible aggression from another individual in the same area).

Chemical signals.—It is regrettable that our knowledge and understanding of the application of odors to social communication is so limited. This is particularly true in the case of the highly scent-oriented ora.

Chemical signals have the very characteristic quality of persistence, lingering long at the place where they were initially deposited. Of all the odors which might be produced by oras, the most obvious are those associated with the fecal droppings. Since these olfacients are deposited on the ground, it is possible for them to survive for a longer period than odors released into the air. This characteristic of ground-deposited scent tends to be much exploited, in particular by the terrestrial mammals that appear to specialize in chemical communication more than any other group. Among those that live a more or less solitary life like that of the ora, and at the same time at the low population densities that are generally incumbent on large animals such as the ora, it is undoubtedly possible to maintain the necessary social integration by fecal scent alone, even though any given individual ora may rarely come face to face with others, at least of his own sex.

Perhaps the single most important fact regarding the highly odoriferous fecal pellets of oras (when fresh) is that they are *not* hidden, but generally conspicuously placed in trails.

On the basis of field observations of small moving oras, and on the basis of tracks found in the volcanic ash during the dry season, oras of up to 1.5 m in length frequently stop and investigate ora feces along the game trails. The spoor around the droppings show that small oras usually spend considerable time at each of these sites, walking around the fecal pellets a number of times. Visual observations indicate that the fecal material is constantly touched with the tongue. Such investigative behavior may continue for as long as 10 minutes. Evidently information such as size and age of the depositor and freshness of the droppings is communicated. Leyhausen and Wolff (1959) suggest that olfactory

Table 12–6. Major communicatory signals in *V. komodoensis*

Signal category	Directed at	Sensory type	Behavioral signal	Presumed purposes	Remarks
I. Intraspecific					
A. Social (episematic)	Individuals of same species	1. Visual (positional)	*body*: "Schrägstellung," slightly flattened laterally, body raised with arched vertebral column; *head*: held straight out or up; *neck*: slightly arched, inflated, with hyoid apparatus expanded; *locomotion:* slow, stiff, deliberate, often parallel to or at an angle with recipient (*ritual walk*).	Assertion (Appeasement? Katasematic?)	Easily and quickly shifted to defensive behavior if necessary
		2. Olfactory	Scents associated with droppings and other spoor.	Provides information on distribution and density of oras with respect to size, sex, and number (?)	
B. Sexual (epigamic)	Other sex	1. Tactile	Males scratch females on neck and back with claws of one front foot.	Courtship	
		2. Olfactory	"Tonguing" of sides of head and body.	Sex identifications?	

II. Interspecific					
A. Warning (episematic)	Predators and other assailants	1. Visual	*body*: "Schrägstellung," with body sometimes slightly compressed laterally, bowed away from assailant; *head*: held low, with dorsal surface sometimes directed to assailant—in stronger reaction with mouth open, rarely with slight foam production; *neck*: neck often inflated, with hyoid apparatus expanded; *locomotion*: none; *tail*: bowed away from assailant, ready for sweeping tail blow. Sometimes tail switched back and forth, or vibrates (displacement?-parasematic).	Warning that defensive tactics (biting, tail lashing) imminent.	Sometimes difficult to distinguish from intraspecific visual signal (I. A above).
		2. Auditory	Hissing	Same as above	Sometimes directed to individuals of own species (episematic)
B. Passive predator deterrent	Predators and others	Chemical	Cloacal emptying	"Last ditch" deterrent effort	

Table 12–7. Communication types important in social and feeding behavior of Komodo monitors

Stimulus	Communication	Response	Probable communicatory function
Fecal odor	Chemical	Investigatory behavior	Sexual maturity, individual identification, density, dominance odors, territorial, and home range markers
Casual bodily contact	Tactile	Remaining in close, nonaggressive contact, or alarm behavior (stress and flight)	?
Body surface olfacients	Chemical	Investigatory behavior, leading to avoidance, courtship, or other behavior	Sexual maturity and individual identification
Stress reaction	Visual	Usually nil if not combined with other cues	Introduction to an important, perhaps aggressive display
Hiss	Auditory	Same	Often part of threat display
Weaponry exhibition or low-level threat	Visual	Usually avoidance behavior, sometimes threat	Prelude to a potential conflict in territorial dispute
Appeasement display	Visual	As above or nil	Nonaggression
Body scratching by male	Tactile	Nil, aggression or alarm behavior	May serve to quiet the female by reinforcing male intent to breed rather than fight
Mounting	Tactile	As above, sometimes leading to courtship	Same
Tail invitational movements	Tactile	Leads to hemipenial insertion attempts	Communicates female's intent to breed, rather than flee or fight
Neck biting	Tactile	Restrained movement on part of female	Perhaps not communicatory, except indirectly to indicate male's readiness to breed

marks of these types prevent unnecessary encounters in house cats. According to them, a fresh mark communicates to other cats that "This section is closed"; less fresh "Proceed with caution"; old mark "Go ahead." Whether oras use the scent of fecal pellets in the same way is, of course, unknown, but highly likely.

With the exception of the concentration of feces near the thermoregulatory sites (Chap. 5), oras do not establish special sites at which individuals deposit their scent. However, the scent of the fecal deposits at the thermoregulatory sites is extremely obvious. The fecal material is usually additionally conspicuous by its color and topographic position. There is no doubt that on ascending practically any major hill overlooking any large forest or valley on Komodo, a newly arrived large ora can quickly learn from these fecal deposits of the presence of other oras in the area and perhaps the sex of those individuals and some information regarding their maturity. As in wolves (Seton, 1909), the major information essential to the life of the adult ora may be obtained by this system of signal sites.

The adoption of regular dunging places is a widespread habit, particularly in mammals—rarely reptiles. The spotted hyena (*Crocuta crocuta*) feeds largely on bones, and its feces turn white on drying, as do those of the ora. These animals habitually defecate in regular latrine areas, which are spaces of limited extent. As in the ora, these dunging areas are conspicuous at a distance when the sun is shining on the chalky white feces (Kruuk, 1972).

I suggest that *most* members of the local ora population probably do not know one another individually, and that their rights as residents need constant reinforcement—not only through interaction at carrion, but also by means of scent signals associated with fecal pellets deposited in dunging places and along trails. This is a conventional system, quite common in mammals, but apparently little used by reptiles as far as we know. It may be very common in snakes. In oras such signals apparently safeguard the vested interest established by membership in a localized society. They are apparently important in restraining the admission of nonresidents who might be seeking similar privileges. To maintain their status of acceptability and the rights that go with it, oras can no more afford to neglect conventional observances such as greetings, social assemblies, and status-seeking exercises than can their mammalian solitary predator counterparts.

Other important chemical signals are those associated with the discharge of cloacal and stomach contents. Cloacal emptying as a fright reaction is common in many lizards and was noted on many occasions during manipulation of particularly our smaller captured Komodo monitors. It occurs most often when they are attacked or handled, suggesting the behavior may have some importance as a deterrent. Disgorging the stomach contents is also common, particularly during the first several days following a large meal. It occurs most often before an impending attack and seems to be either a preparation for action or a deterrent to attack.

Evidence presented in Chapter 8 suggests that the integumentary surface of females may be used to communicate their sexuality and/or receptivity to attentive males. That this is probably also true in gekkonid lizards has been shown by Greensberg (1943). Sex identification by chemical means is apparently common in snakes. Madison (1977) reviews this literature as well as other cases of chemical communication among lizards.

Visual signals.—Visual signals include color, pattern, position, and form. Color is not as important in this species as in others of the genus. Color and pattern of potential importance in this context include the vivid juvenile pattern, the bright yellow eyelid of adults, and the bright yellow tongue in oras of all sizes.

Form and position also play a role in communication between individuals within the normal visual distance. Those components believed to be important are: raising the roach, tail position (held straight and stiff, as opposed to lashing it or bowing it to one side), head position (at angle to ground, sometimes also turned to side), *Schrägstellung* (positioning body so that the largest possible surface is shown), lateral body compression, hyoid expansion, mouth open, and body raised high on legs.

The *Schrägstellung* is associated with defensive tactics and is part of the threat display. It is common in several species of *Varanus* (Mertens, 1946a; Honegger and Heusser, 1969; Murphy and Mitchell, 1974). Preceding and during an actual attack the head and anterior part of the body are usually held at a low level. A lateral sigmoid curve in the neck is most often associated with an impending attack (common in snakes).

Many *Varanus* species, including *V. komodoensis*, lower the head and bend up the back (Mertens, 1946a; Auffenberg, in press c), often during a threat display. The head is usually turned slightly to one side so that the dorsal surface is presented to the offending object. The eyes peer out from underneath the bony projections and an attack in either offense or defense is usually imminent. The arching of the back and tail is an obvious visual signal used particularly by smaller individuals when approaching other monitors, or by large individuals when approaching or being approached by humans. It is often accompanied by a slow, stiff-legged, stereotyped walk in which the body is slowly thrown from side to side in exaggerated lateral undulations (Fig. 6–2).

Often accompanying such movements, but sometimes independent of them, is a display in which the throat is visibly enlarged (Fig. 6–2). This is accomplished by posterolateral movement of the hyoid apparatus so that the bottom of the throat looks like a slightly inflated balloon, compressed laterally to a certain extent. It has been described in many *Varanus* species (e.g., Cowles, 1930; Mertens, 1946; Vogel, 1979; Auffenberg, in press c), and it is often associated with lateral body compression. The roach is often raised along the vertebral line of the neck and anterior half of the body, making the body appear even larger.

The multiplicity of contexts in which the roach is raised, the hyoid is expanded, and the head is at an angle suggests that this complex is probably best treated as a stress reaction (Auffenberg and Ganci, in press). The complex is apparently not intended to impart communicative specificity, but rather a general state of being—a behavioral framework to which other signals are added. Together the entire signal set then constitutes a specific display signal (Table 12–6; Fig. 6–2).

Varanus komodoensis exhibits several examples of behavioral scaling (= variations in the magnitude or in the qualitative state of a behavior), particularly noticeable in the visual display categories. One of these is agonistic display rate correlated with population density. Thus, the number of appeasement and threat displays are directly proportional to aggregation size. As an example, the average number of stress reactions per hour per individual 1.5 m or less at carrion was 1.6 at aggregations of from 2 to 5 individuals, 2.8 at aggregations of from 6 to 9, and 4.1 at aggregations of 10 or more. Another example is tongue flick rate per individual at different aggregation densities. It too is directly proportional to group size ($\overline{X} = 8.2$ flicks per minute at aggregations of from 2 to 5, 12.1 at aggregations of from 6 to 9, and 19.6 at 10 or more). Fighting probably falls into the same category, but the differences in the three density classes are statistically insignificant.

Tactile signals.—Many reptiles use parts of their body when fighting. Pushing with the body, tail, head, or even with a wide open mouth (jaw fencing) is common in many species of lizards (Mertens, 1946a). None of these was observed in interactions among Komodo monitors.

A common defensive tactic of many species of *Varanus*, including *V. komodoensis,* is the "tail blow." In the threat display preceding it, the tail is bowed away from the stimulus. When the stimulus comes within striking distance, the lizard lashes laterally with its tail (the force of the blow is often exaggerated in popular literature) (Fig. 6–2).

Biting is common in both offense and defense, producing severe wounds that often lead to the death of one of the individuals.

The various signals of *V. komodoensis* in respect to approximate function are shown in Table 12–7. It is obvious that the categories listed are not rigidly separated. While scent may be the most important integrating display in the core areas, this is not the case in the feeding aggregation where visual displays are most important. This communicative behavioral complex depends more on the characteristic general visual appearance of the monitors moving around the common food resource than on any particular marks or patterns. The movement itself is apparently the chief component of the system of social interaction. The constant view of the oras, particularly the larger ones around the carrion, apparently provides the entire group with a frequent sample of visual recognition contacts, which results in compensatory movements and density adjustments, especially on the part of the smaller individuals.

Only around large carrion can this recognition and response be maintained visually. It is undoubtedly for this reason that visual signals predominate. Tactile and chemical signals, though they are important in certain behavioral patterns, such as mating, are not important in group behavior. However, social interaction of the entire population within a large habitat cannot depend on visual signals. Therefore, this interaction depends on communication at the olfactory level, in which the scent of Komodo monitors, and particularly their droppings, is emphasized. The frequency of olfactory contacts throughout the entire habitat and the visual contacts around carrion depend on population density, speed of movement, maximum range of recognition, and persistence and frequency of chemical signals.

Hierarchies and Spacing

Varanus komodoensis possesses a strong hierarchical system that is maintained by frequent aggression. Included in this aggressive, agonistic behavior are threats, actual attacks, and displacement in space or rank (Table 12–8). Under submissive agonistic behavior are included those patterns resulting from an attack (often flight), displacement in either rank or space, reaction to threats, and appeasement. It might also include the behaviors of either a homosexual mounted individual or the recipient of certain courtship behavior, such as back scratching.

For purposes of analysis, an attack was considered any hostile encounter involving physical contact. Attacks were not common and were usually comprised of one individual biting or attempting to bite another. I saw no indication that attacks were ever redirected to smaller individuals lower in the hierarchical system. Most attacks were precipitated by the close approach of a subdominant individual to a dominant one, rarely reversed, and then only in courtship.

On April 25, ora 18Y (1.54 m) was in the base camp trap feeding on wild boar carrion bait. At 1750 hr 19W (2.5 m) entered the trap and immediately started to chase 18Y back and forth; the latter attempted to climb the wall and escape. Finally, the larger lizard managed to drive the smaller one into the narrow end near the release box. There it quickly seized 18Y by the body and killed it by repeatedly slamming the smaller animal against the ground and part of the trap. It ate most of the intestines and then dragged the dead ora out of the release box and into the thick brush 60 feet away. There I frightened off 19W and retrieved the carcass of 18Y. In addition to the large ventral wound, the left front limb was almost ripped from the body, and the vertebral column was severed in the region of the thorax. The actions of 19W made it clear that it would have eaten the carcass if it had been allowed to do so.

On January 9, ora 27W (1.44 m) was in the base camp trap eating goat carrion. Ora 19W (2.5 m) entered the trap at 1615 hr and started to feed on flesh. Ora 27W

Table 12–8. Aggressive behavioral categories in *V. komodoensis*

Type	Function	Remarks
Territorial or dominance	Repulse potential users of resources claimed by one individual. Most commonly the resource is carrion prey.	It is not clear whether *V. komodoensis* exhibits aggressive behavior to repulse trespassers. Appeasement displays are used largely by females, subadults, juveniles, and most transients to transmute aggressive displays into conciliation.
Sexual	Males threaten and attack females for mating purposes. This includes biting, chasing, and mounting.	Males sometimes engage in homosexual back scratching, neck biting, chasing, and mounting in connection with dominance aggression.
Predatory	Large specimens sometimes attack and kill smaller ones, apparently to feed on them.	

failed to move out of the way when 19W tried to get into a different position, whereupon 19W suddenly attacked the smaller ora. Grasping it by the body, he killed it by slamming it against the ground and shaking it back and forth like a dog with a slipper. Several of the wounds were very severe (up to 3.6 cm deep and 10.6 cm long), though death undoubtedly resulted when the spinal cord was severed.

Other aggressive behavior included a large female (23W) biting a small male in the side when it attempted to mount her, and a large number of short chases in which individuals were bitten on the hind legs, tail, and back. During the year we also noted several instances in which one ora struck another on the body with its tail, producing a loud thump that could be heard many meters away.

Male combat has been described in several varanids. Murphy and Mitchell (1974) described the ritualized combat of male *V. gilleni* in which the combatants grasp one another by both the front and the hind limbs, using the head and tail as supports. A somewhat similar position is sometimes assumed by male *V. bengalensis* during combat. A more commonly reported combat position is bipedal in which the combatants grasp one another with their front limbs (described by Deraniyagala, 1958, and Auffenberg, in press c, in *V. bengalensis;* by Waite, 1929, in *V. giganteus*; and by Honnegger and Heusser, 1969, and Vogel, 1979, in *V. salvator*). Though biting is not normally part of the bipedal combatant behavior, Honegger and Heusser (1969) reported it in *V. salvator*, and Worrell (1963) reported biting and scratching in a male *V. varius*. The only male combat seen in *V. komodoensis* during this study involved two large adults, 8W and 19W. In both cases 19W was the aggressor, lunging at 8W with no prior warning; 8W dashed off, with 19W in pursuit, obviously trying to rake his claws down the back of 8W, who never allowed him to do so. These sequences were identical to those in courtship (Chap. 6), and I can only interpret them as homosexual behavior on the part of the dominant 19W. The behavior of 8W can only be interpreted as submissive escape.

During the year we noted no bipedal combat in oras, but Oesman (1970) reported savage biting in two previously feeding, large oras that stood on their hind legs. Abdoessoeki (n.d.) reported a similar case in two individuals on Komodo, but provided no details as to their size or sex.

Displacement in space and rank were the most common social interactions observed. All occurred around carrion. The interactions on November 12, 1969, are typical. The bait was a goat that had been killed two days before. Six different-sized lizards came to the bait; their respective total lengths and interactions are summarized in Figures 12–2 and 12–3. All were residents except Whiteface. To a large extent their reactions to one another were related to their sizes. The largest (19W) was always allowed to feed on the carrion whenever it wished to; all others moved out of his way. During movement around carrion, smaller oras almost always gave ground (Fig. 12–2); the transient (Whiteface) gave ground more often than any of the residents. Ora 19W showed stress-related displays far fewer times than almost all others (Fig. 12–3) and never gave ground (Fig. 12–2). The highest number of stress displays was exhibited by the mature female 23W, though almost all the displays occurred during her interactions with the dominant male, 19W. Oras 7W and 5W presented appeasement displays to 13W, 23W, and 19W, but the larger lizards paid no attention. The smallest members of the group (7W and 5W) fed on the carrion only when 13W, 23W, and 19W had left the bait. When all the lizards were frightened away from the carrion, the smallest were usually the last to leave and the first to return. Ora 7W had been at the bait site longer than 5W and drove off the latter twice by attempting to bite it. On the other hand, neither 7W nor 5W attempted to bite the larger individuals. The following is a rather typical series of interactions at the carrion.

September 5, 1969

1030 hrs First lizard (A) arrives (1 m long).

1042 hrs Lizard D (1.5 m) walks straight to bait and walks in circles around bait. A in shade nearby.

1058 hrs A at bait, 19W (2.5 m) arrives and walks to bait, arches and inflates neck slightly. A moves in circle around bait, definitely avoiding large one. 19W eats, A circling. 19W moves to shade, A rapidly moves to bait and eats. 19W returns, eats intestine, then moves into shade. A returns to bait. 19W returns and A leaves, giving appeasement display. 19W lays in shade. A moves back and forth in arc upwind of bait. 19W leaves shade and moves to bait; A gives appeasement display but continues walking in arc upwind. A moves closer to bait, gives appeasement display and skitters off at a surprise approach by 19W. Latter eats intestines, using front feet to hold goat down.

1300 hrs D (1.5 m) approaches and moves upwind of bait; moves into brush. A approaches bait and nibbles on it.

1330 hrs All lizards near bait, but in shade.

1400 hrs A approaches bait, stays short time, crawls into shade, moves out again and finds small piece of intestine shaken loose by 19W earlier, rushes back into the shade and eats it.

1510 hrs 19W returns to vicinity of bait, resting in shade of brush.

STANDS GROUND

MOVES AWAY

	7W	13W	White Face	5W	23W	19W
7W		6	4	2	5	3
13W	4		2	3	7	8
White Face	6	1		3	6	13
5W			1		1	3
23W			1	4		4
19W						

Fig. 12–2. Results of close approach of ora pairs around carrion, Nov. 12, 1969.

STANDS GROUND

MOVES AWAY	7 W	5 W	13 W	White Face	23 W	19 W	Total
7 W (132)		2			2	3	7
5 W (167)			1			2	3
13 W (170)		2			7		9
White Face (185)		1					1
23 W (235)			2			12	15
19 W (250)			1				1

Fig. 12–3. Number of times a stress display was shown by ora on vertical axis when near oras on horizontal axis, Nov. 12, 1969. Total length in parentheses.

1520 hrs B (1 m) appears with lizard C (1.2 m). Both approach bait together, with C behind B.

1530 hrs D approaches bait, noses it; B, who is circling upwind from bait, runs off giving a weak threat to D. D, apparenty hot, moves into shade. 19W approaches, threatens A, who runs off. D and A eat goat together from same side. No aggression. 19W returns and eats, D runs off a short distance, 19W moves back to shade. A returns to bait, giving appeasement display to both 19W and D, both of whom are sitting in the shade. D feeds at bait. 19W leaves shade and approaches bait, D runs off and walks in circle around 19W. A approaches, giving appeasement display to both 19W and D. All three lizards move into shade. D approaches bait, eats; A approaches bait, appeasement to D, stays out of its way and makes an upwind arc. A continuing to appease D while moving back and forth. B moves back into shade and A moves to bait and begins to eat it without any display to B. D approaches bait and threatens no one. A appeases D as it moves away to make room for it. 19W moves out of brush, and joins D at bait. A appeases both D and 19W at bait, at the same time moving farther upwind and back and forth in an arc. 19W begins to eat more vigorously and both D and A are frightened off as 19W begins to swing the bait around. 19W continues to eat, while D steps on 19W's tail and licks his tail base. A still moving back and forth in arc, continually appeasing 19W and D. D appeases 19W, continuing to remain several meters away, moving back and forth in an arc. A continues to move in an arc approximately 10 m from bait and upwind. D appeases carcass when it swings on being released by 19W. D and 19W eating goat from same side. Wild dog appears barking at C, without any effect on lizards. 19W moves into brush, just as wild boar walks within 20 feet of bait and only a few feet from the resting 19W, who lunges at boar and misses completely. B approaches bait quickly and eats while no other lizard is there. A approaches, chases B into brush. They threaten one another with mouths open, standing side to side with heads turned toward one another. A lunges at B as latter runs off, A chases and tries to bite B.

1600 hrs 19W swallowed entire head, neck, and anterior part of chest and throat of goat. While so occupied, D and C tear flesh off back of goat for at least 10 minutes. Finally 19W severs rope around the goat's neck and carries carcass into brush. At this point the dog who had still been waiting in the brush worried all of the lizards as much as he could, chasing C off after barking and snapping at it. The dog also barked at 19W (he still had a large part of the chest sticking out of his mouth). 19W pays no attention to dog. The smaller lizards pay no attention to the dog, which stops to rest. The smaller oras crawl over 19W and seem excited by the rapidly disappearing food. The hierarchy has completely broken down, undoubtedly due to the fact that 19W was incapacitated with his mouth full. When the remaining large parts of the carcass were completely swallowed by 19W, the smaller lizards wandered about the bait site looking for scraps and finally walked off into the brush.

The following sequence illustrates a typical reaction between two large adult males:

February 21, 1970

1515 hrs 19W approaches bait and starts to eat boar.

1600 hrs 8W approaching to within 4 m, when 19W spins around, threatens 8W with inflated throat and open mouth, and attacks it immediately. 8W darts off about 3 m before 19W can catch it. Then it stops. 19W threatens with its back arched, tail held high, standing very high on legs with throat inflated, mouth slightly open. 8W remains still, looking at 19W out of the corner of its eye. 19W charges and 8W dashes off, staying about 1–3 m ahead of 19W. Once during the chase 19W tried to raise its left foot onto 8W's back as they were running. The mouth of 19W was open and it was inflating its neck. They ran in a circle of approximately 75 m. The total length over which 19W chased 8W was 125 m. 8W ambled back to the bait site, completing the circle they had begun and 19W went to the hog wallow where he lay in the water and drank for a long time. 8W slowly moved off into the brush in the same direction from which he had come.

March 2, 1970

1441 hrs 23W trying to swallow entire head of boar. 8W rushes up quickly from downwind. 19W threatens 8W and immediately dashes after him. 8W runs off. 19W returns to bait and 8W joins him. Both eat together, though both continue to threaten one another, with mouth open and hissing; 19W remains aggressive and continually threatens 8W, who finally moves away. 23W resting in shade. 19W also moves into shade. Both return to carrion. 19W continues to feed as 23W enters dense brush. Immediately there is a great thrashing and it sounds as though two lizards are running. Presumably 8W is trying to mount 23W.

Active displacement in which one monitor presented a threat display that caused another to move was uncommon compared to passive displacement in which no threat was involved. In general, smaller individuals simply moved out of the way when a larger one approached or avoided them entirely during their movements in the vicinity of carrion (Table 12–9). There was no correlation between displacement and sex (Table 12–10). Specimens of equal size and of either sex tended to ignore one another unless overtly threatened. Threat was normally communicated at fairly close distances ($\overline{X} = 2.2$ m).

Displacements were highly reliable indicators of rank. There were no altera-

Table 12–9. Percent of spatial displacement per interacting size classes (m) (N = 167)

Size classes displaced	Size classes not displaced: 1.1–1.5	1.6–2.0	2.1–2.5	2.6–3.0
1.1–1.5	52(48)[a]	61(38)	78(22)	93(7)
1.6–2.0	1(38)	54(46)	65(21)	73(24)
2.1–2.5	0(22)	14(21)	49(51)	51(37)
2.6–3.0	0(7)	3(24)	12(37)	46(54)

a. Parentheses show percent of no displacement in either member of pair.

Table 12–10. Percent of spatial displacement per interacting monitor pairs (N = 167)

Sex displaced	Sex not displaced	
	Males	Females
Males	29 (40)	18 (27)
Females	23 (27)	30 (33)

Table 12–11. Percent displacement at carrion during any interactions of transient and resident *V. komodoensis* in which displacement occurred (N = 103)

Displaced size classes (m)	Displacements in	
	Residents	Transients
1.1–1.5	72	98
1.6–2.0	13	71
2.1–2.5	7	32
2.6–3.0	2	10

tions of dominance in at least the resident monitors during the entire study period. However, the hierarchical structure around the carrion changed constantly as new individuals entered and earlier arriving individuals left the feeding areas (Table 12–11).

Appeasement displays were most often engaged in by specimens 1–1.5 m in length and were never exhibited by large individuals (dominant or not). They occurred most commonly as the displaying individuals approached the carrion with other individuals nearby (irrespective of their size). There were also frequent displays from distances of 3–10 m (Table 12–12), as the displayer moved back and forth beyond the attack distance of larger individuals feeding on the carrion. That the displaying animals were in stress is suggested by the fact they were easily frightened by nonagonistic movements of nearby monitors, by human observers, by boughs moving with the wind, or even by leaves blowing across the ground surface. When disturbed, these individuals dashed headlong into the surrounding underbrush. Displacement of low-ranking individuals around a carrion always occurred with the movement of high-ranking individuals (Table 12–12).

Within the hierarchy established at carrion sites agonistic behavior was considerably higher in monitors of lower rank (i.e., generally smaller size). Agonistic behavior was least often demonstrated by the dominant residents (Table 12–13). Aggressive agonistic display was no indication of social rank, since high-ranking monitors at a feeding site often ignored those of lower rank. However, the reverse was not true; appeasement agonistic display was most often directed to animals of high rank. This behavior was more reliable in determining social hierarchy by means of display than any other.

In general, large oras are relatively tolerant of intruders on feeding sites, but they frequently attack, and rarely even kill low-ranking individuals. Consequently, avoidance of large individuals becomes a behavior characteristic of low-ranking individuals. Large size confers special site and feeding privileges to certain individuals, and these are accepted and remain unchallenged by smaller ones.

The inflated throats, compressed body, and stately march of a small ora approaching a group of larger individuals is clearly an appeasement behavior, related to submission, since it fails to elicit antagonism or aggression on the part of the larger individuals (see Lorenz, 1966, and others for discussion of similar patterns). The function of the appeasement displays of these small oras is appar-

Table 12–12. Distance over which appeasement displays are presented by small to large *V. komodoensis*

Display distance (approx. m)	Frequency per monitor size class (m)				
	1.1–1.5	1.6–2.0	2.1–2.5	2.6–3.0	Total
1.0–1.9	8	8	6	3	25
2.0–2.9	15	11	9	1	36
3.0–3.9	38	18	15	0	71
4.0+	16	3	0	0	19
Total	77	40	30	4	151

Table 12–13. Demonstrations (N = 355) of agonistic behavior by resident and transient monitors

Size classes (m)	Number of agonistic displays	
	By residents	By transients
1.1–1.5	40	130
1.6–2.0	32	69
2.1–2.5	26	47
2.6–3.0	13	25

ently to block aggression (Auffenberg and Ganci, in press). Therefore it is a type of formalized nonaggression, where all functional movements of active defense are avoided. Carpenter (1961b) reported similar displays in juvenile desert iguanas when they approached or passed near larger specimens.

Large dominant male oras either threaten, flee from, or attempt to mate with intruding oras. Such unpredictable behavior suggests that these lizards are in "conflict," or in a behavioral deadlock. The behavior of oras around a bait often confirms this. They often fail to carry out any of the stimulated reactions fully, sometimes performing one or both of them partially, sometimes doing something else, and sometimes doing nothing; often they are simply restless.

Social spacing in feeding aggregations of oras is very obvious While feeding at carrion most large individuals are concentrated in an area within 1.5 m of the food source; smaller individuals are located more peripherally. Thus, there is a definite correlation between spacing around carrion and social rank (Table 12–14), but not with sex (Table 12–15). Between feeding periods large specimens (higher social rank) tend to retire to an area separated from other individuals.

Usually none of the activities of at least the larger oras is accompanied by displays, and there is little suggestion of a permanent hierarchy. Kaufmann (1962) has shown that there is no strong hierarchical system in the loosely organized mammalian social group of coatis, *Nasua marica,* apparently because they

Table 12–14. Size and social spacing of monitors around carrion

Size classes (m)	Distance from carrion (m)		
	0–2.0	2.1–4.0	4.1+
1.1–1.5	39	89	154
1.6–2.0	51	94	127
2.1–2.5	86	41	39
2.6–3.0	119	67	28

Table 12–15. Sex and social spacing around carrion when equal or larger monitors were feeding[a]

	Distance from carrion (m)		
Sex	0–2.0	2.1–4.0	4.1+
Male	216	205	275
Female	79	86	71
Total	295	291	346

a. Male to female ratio is normally 3.5:1.

have a limited repertory of social displays. Thus intraband disputes are commonly settled by fighting. The behavior of individuals within a feeding aggregation of Komodo monitors more closely approximates this situation than that reported for other lizards. However, because *V. komodoensis* is cannibalistic, space maintenance is practical, at least for smaller individuals. These are sometimes pursued and try desperately to escape. Carpenter (1961b) noted that chased desert iguanas often assumed a submissive pose, but this was never noticed in monitors. Had a small monitor stopped, the larger, chasing individual would have undoubtedly attacked it. During the present study two smaller monitors (1.44 m and 1.54 m total length) were attacked and killed without provocation by a resident male 2.5 m long. A 2.3 m female severely lacerated a 1.8 m nonresident male that had tried to mount her. In addition, we noted many instances in which smaller individuals were bitten on the hind legs, tail, and back. Spatial displacement is thus as important in the social role for small Komodo monitors as it is for other monitor species (Auffenberg and Ganci, in press).

The less antagonistic the higher level individuals are, the more active the feeding of the lower ranking ones. The high-ranking monitor of a feeding group is normally tolerant of the movement of lower ranking ones to the food itself. The dominant individual normally fails to display its superiority, except passively on the basis of its large mass. Even the presence of a large individual at a feeding site elicits signs of social stress in smaller ones. Thus the submissive appeasement behavior of smaller oras seeks tolerance from higher ranking adults and is similar to the pattern in dogs and wolves (Schenkel, 1967).

The open association of oras around carrion is thus based on a type of social tenancy in which individuals (particularly smaller ones) are anonymous as individuals, but categorized on the basis of their size as low-ranking individuals. Large, high-ranking individuals are also probably anonymous to the smaller ones. Large residents are not anonymous to one another, for they seem to recognize one another on the basis of both physical and chemical characteristics. This leads to a kind of personal relationship between resident individuals that is lacking between residents and nonresidents.

Territoriality

Territoriality is usually defined as that behavior by which an organism lays claim to a resource and defends it against exploitation by others, most often a delimited area in which the individual operates. In the case of the ora there is no evidence suggesting that individuals defend such an area from intrusion by other members of the same species. True space territoriality in at least males has been demonstrated in several families of lizards (see Milstead, 1961; Hunsaker and Burrage, 1969; and Brown and Oriens, 1970), but not in snakes, to which varanids are apparently also related.

If operative at all, the territorial system of oras is probably less rigid than it is in other lizards. As pointed out in Chapter 5, the foraging and scavenging ranges of oras are very large when compared to those of other lizards. Such large areas are not easily defensible, though oras often show hostility when together and mutual avoidance is common. However, the benefits of attempting to defend the entire foraging range are probably not sufficient to counteract the high cost of maintaining an area for exclusive use by running from one boundary to another. It is not surprising that there is strong overlap in at least scavenging ranges of the oras. The solitary foraging of oras for live prey can be achieved simply by mutual avoidance through chiefly chemical signals without the risk and energy expenditure involved in territorial defenses of foraging oras. There are, however, clear indications of a feeding territory around carrion—a pinpointed resource—largely based upon rank order established on the basis of size. As indicated above, this type of hierarchy tends to reduce fights over feeding rights to a minimum in ora feeding aggregations. A similar rank order may be operative at chance encounters in the field, though no data are available. The intense investigation of fecal pellets by smaller oras suggests that information in this general category is sought by these smaller individuals, and that it is related to rank order.

Data on activity patterns (Chap. 5) clearly show that oras of all sizes share foraging areas, a common phenomenon in many large predators (Guggisberg, 1961; Kruuk, 1972; Schaller, 1972; Curio, 1976). However, more often smaller areas are shared by larger individuals of the same sex than by individuals of mixed sex or smaller length. The clearest foraging area occurs among large resident males. The situation in tigers is apparently similar, for there is no evidence of territoriality in female tigers (they apparently share the range with one another and with males), though males are strongly territorial (Schaller, 1967).

The advantage of site familiarity often accompanies territorial behavior. However, familiarity of foraging range in the ora is facilitated in resident individuals through long occupancy of a rather restricted area; a period of from 25 to 40 years seems reasonable on the basis of available data. On the other hand, small, sexually immature individuals often lack individual long-term activity areas. Certain sexually mature oras also seem to belong in this category (see Schaller, 1967, for a discussion of a similar situation in the tiger). Thus, in the ora there are several area tenure patterns that are largely size and sex dependent. These patterns range from exclusive use for large, resident, sexually mature males to complete lack of established long-term ranges, particularly of the smallest oras.

Residency confers certain advantages to the individuals practicing it. The overlapping core areas of adult males and females surely are advantageous to reproduction. Long residence certainly enables the individuals to learn shelter location (often difficult to find for individuals this size), water holes, and sentinel sites, as well as the asscmbly and collation of data on the location and activity cycles of prey animals.

13

Epilogue

Before this study was begun, some biologists feared that the ora was becoming extinct. This study shows that both the number of oras and the area they inhabit are larger than supposed and that extinction is not imminent. But the geographic range of the ora is small, and history has shown that animals so situated are highly vulnerable. Therefore, vigilance and a policy of conservation and enforcement are necessary.

Some readers will criticize what follows on the basis that it suggests too much *management* and not enough *conservation*. I submit that, in view of the world's expanding human populations, successful conservation of wild animals species demands local involvement in effective management programs, not aloof, passive, purist policies of inviolate reserves. Inviolate areas are unrealistic in most of Asia at present. Much of Indonesia will soon experience extreme pressure for "unused" land. Human populations on Java are already pressing hard on available local resources, and schemes to colonize primitive areas are under way. A similar colonization program has been suggested for Komodo. Just as human population and culture must be manipulated, so the ora, its prey species, and its environment must also be manipulated, if we hope to obtain a compromise between an important wild animal resource and the rapidly expanding human population in the Lesser Sunda area.

This section contains various aspects of the management of oras, their prey species, and the natural habitats of both, as well as man's activities in those habitats. These objectives can be realized only through the application of scientific management techniques designed to control the environmental factors involved. The important factors have been grouped according to whether they directly affect the number of oras or the environmental conditions that affec oras.

Population Control

Historical background.—The earliest legislation protecting the ora was passed in 1915, just three years after the species was described as new by Ouwens. This remarkable accomplishment of the Sultan of Bima[1] was in effect and legal until 1924, when the Mangarrai District of the Sultanate (from central Flores to Komodo) was placed under the jurisdiction of Timor. It was not until 1926 that the new Self Goverment of Mangarrai took up the protection of the ora, canceling the sultan's order and passing an ordinance that made it illegal to catch or hunt oras, to take their skins or eggs, or to disturb their nests. The ordinance was approved by the Dutch Resident of Timor in January 1927. After another review, ordinances protecting the ora were passed by local decree in May 1930 and a fine of 250 florins for hunting or capturing oras was enacted. Essentially the same protective legislation was referred to in the 1931 central government's *Regulations for the Protection of Wild Animals* and again in the 1940 hunting regulations for areas outside Java and Madura (Anon., 1953).

Komodo was first made an official wilderness reserve area by the governor-general of the Netherlands Indies colony in November 1928, on the recommendations of both the *Senckenbergische Natuurforschungs Gesellschaft Frankfurt* and the *Frankfurter Verein fur Geographie und Statistik* after a joint expedition in the general area of Komodo. In December 1929 de Jong presented a report to the director of the *Lands Plantentuin* regarding the status of *Varanus komodoensis* on Flores, Komodo, and Rintja, in which he suggested that the southern part of Rintja and the northern part of Komodo would be the best areas for wilderness reserves. Nothing was done. Eight years later he pleaded again for the establishment of reserves on these two islands, and in 1935 van Hemn asked that Komodo be made a reserve. In May 1938 the Self Government of Mangarrai set up southern Rintja and Padar as reserve areas. A warden with a motorboat was to be located in Reo, Flores, but the position was not actually established until 1968 and then by the Conservation Office in Bogor. At that time the warden was housed in Reo without a boat, many kilometers from the closest oras, but in 1971 the office was moved to Labuanbadjo, Flores, and a motorboat was provided. An assistant warden is now occasionally stationed on Komodo, greatly improving the situation, so a conservation and management program is possible for the first time.

The island of Komodo is also now officially considered as part of the wilderness reserve tracts of Indonesia (Anon., 1970), though I know of no central governmental legislative action establishing it as such, only by local decree. In 1970 a new reserve was established in western Flores, just south of Labuanbadjo, called Wai Wuul.[2]

1. This interesting ruler is also said to have introduced feral horses on Rintja, as well as settling Komodo with criminal exiles at about the turn of the century.

2. Since this writing the central government has established Komodo as a national park, and a carefully considered land use plan is being developed.

Annual harvest.—Archeological data obtained during this study[3] show that humans have been in contact with oras for thousands of years. We know nothing regarding the effect of this early interaction on ora populations. However, it is reasonably sure that trade with Chinese merchants from at least the twelfth century A.D. probably involved oras and their skins. Until at least 1937 oras were used to produce a "swimming medicine" because, according to Chinese merchants, the oras were "champion swimmers" (Anon., 1937). Earlier the Chinese apparently used ora tails for "burn medicine" (Delsman, 1927). Lallemant (1929) stated that Chinese merchant records on Sumbawa for 1926 indicate that 100 oras were shot annually for their skins on Rintja and a similar number on Flores. Of an additional 100 oras said to have been captured on Komodo, only 65 skins were actually obtained due to various types of losses. In January 1939 the Dutch Resident of Timor estimated that "several dozen to several hundred" oras had been shot on Komodo, presumably during hunting expeditions for deer. Since 1926 approximately 80 oras have been collected for museums. An unknown, but probably low, number of individuals were crippled or died during various trapping operations (estimated at 5/year). Thus, in the last 50 years or so about 500 oras have been taken from Komodo, about 10/year. The Nature Conservation and Wildlife Management Office in Bogor has established a harvest limit of 5 oras every 5 years, though it was exceeded in several instances.

There are two main objectives of any limiting control on the number of oras removed from the wild population each year: to preserve sufficient breeding stock by "harvesting" oras on a sustained yield basis and to distribute the taking of oras over those portions of the species range in a proportion that reflects as much as possible differences in local density. Protection is imperative; it is highly likely that without it the species will become extinct.

As far as is known in the span of the historic period, oras were found in an exceedingly small part of the world, and their population density was always quite low. The establishment of wildlife reserves over most of its range and its interest to zoos as a desirable exhibit species, in addition to its demonstrated "unwillingness" to breed in zoos outside of Indonesia, suggest that regulation of native ora populations must be somewhat different than that of many other animals. Eventually it must be possible for a limited number of oras to leave protected areas in Indonesia for dispersal to zoos but only to those with approved facilities, and on this basis the IUCN allows and arranges for a limited number of oras to be so dispersed.

A major factor, heretofore ignored, in the capture and shipment of oras to zoos is the unequal sex ratio (Chapter 7). Until this study it was impossible for

3. Material is deposited in Airlangga University (Surabaja) and the National Museum (Jakarta) and is being studied by Adi Sukadani of the former.

zoos to determine with confidence the sex of those animals in their care. The normal practice of Indonesian suppliers was simply to say that any two animals shipped to zoos constituted a pair, and their stated sex was based on the assumption that larger specimens were male and smaller ones female. The fallacy of this approach is obvious. We have shown that the sex ratio of oras on Komodo is approximately 3.5 males for every female.

Unusually high ora densities (unlikely to occur under current conditions but possible with higher human densities) pose the probability of undernourished, heavily parasitized specimens. If such occurs, reduction of the ora population would be necessary to upgrade the nutritional level of the remaining animals and to allow a larger number of young oras to grow to adult size. It is obvious that permitting the capture of oras, particularly the females, is justified only if such reduction is clearly shown to be required. *Overpopulation is not now a problem*. Undernourishment has not been clearly demonstrated, in spite of the statements of several earlier writers. Animals in excess of the present carrying capacity are culled by factors related to population dynamics at the local level, thus obviating the need for a management program to reduce oras densities.

Since removal of female oras will undoubtedly alter population levels in a short period, it is doubtful that pairs of even potential breeding individuals should be sent to zoos. The number of male and female individuals leaving Indonesia should reflect approximately the sex ratio of animals in their native range, if the natural population structure is to be preserved. Thus if oras must be removed from the population, it should be at a ratio of about 4 males for every female.

Data on population structure in several parts of the ora's range show the most common size class to be 1.0–1.5m—the same class that is greatly reduced by starvation, disease, and predation. The small number of large adults found throughout the range makes it clear that these large breeding individuals, so sought after by zoos, should not be the ones transported from Indonesia. Rather, exported specimens should be between 1.0 m and 1.5 m in length.

The total collecting "take" of the ora has three main components: legal capture, illegal capture, and occasional trapping losses (total about 10/year). Through adequate enforcement of collecting restrictions by ora sex and size (stated above), game managers must funnel all collected animals through legal channels in Bogor. This practice is not always feasible under present procedures. But only such strict enforcement will make it possible to maintain an equitable distribution of the ora crop.

Perhaps my major concern is the increasing popularity and growth of zoological gardens throughout the world. The demand by zoos for interesting wild animals is bound to affect the oras. If zoo collecting is allowed without restriction, oras and conservationists will be paying handsomely by the time collecting begins to exceed optimum level. Increased trapping of oras will continue to produce returns. The yield will drop only slightly, initially in terms of the size of oras obtained. This may already be happening in Loho Liang, Komodo, but such effects can be masked for some time by expected annual differences in number of oras at carrion baits per day. Unfortunately, as usual, there will be a tendency to ignore the unwelcome signs of overcapture of oras of breeding size, and depletion could continue for a long time before any harmful effect is recognized or the cause of it admitted.

It is difficult to establish meaningful management policies in the face of local and international pressures to use the ora as capital. *All* participants have to reach an agreement on some workable method of limitation and then abide by it. A fundamental ingredient of conservation policies is that free enterprise, which creates the demand for sales of threatened animal species, must be exchanged for a common, cooperative policy. The IUCN and Nature Office in Bogor have worked together for some time to control both the number of expeditions to Komodo and the annual catch—a difficult task with often imperfect results.

Current regulations allow 5 oras to be taken from Indonesia each year. Of a population estimated at about 5,000 oras, this means a "harvest" of 1/10 of 1 percent each year. It is obvious that, under certain conditions, the capture and transmittal of oras could be higher, as demonstrated by the old, though nebulous, records of Chinese exploitation of ora skins and tails.

Field data for legally captured specimens are reasonably accurate but are difficult to interpret because of the problem of obtaining accurate estimates of the populations that produced them. Perhaps the most complete information is that for the Loho Liang area on Komodo, the valley in which originate most oras taken off that island and the area in which most of this study took place.

Calculated on a density/km^2/year basis, it is possible that the number of legally collected oras will have to vary as more accurate information is obtained on variations in density, particularly in reserve areas. A maximum sustained harvest should probably average no more than 10 animals/year in the size range of 1.0–1.5 m, or one subadult for every 780 ha. However, this number is not self-evident and can be determined and adjusted only in the light of experience.

Though the legal "harvest" of oras of small size on Komodo and Rintja could probably be doubled without damage to population levels, the capture of oras on Padar and Flores should be completely prohibited at present. Population densities on these islands are too low to allow any captures.

Arriving at sound management practices for oras is a problem of both law enforcement and public relations on the part of *all* interested persons and agencies, not just the Indonesian government. Most important is compliance by foreign agencies and governments with both Indonesian and international exportation laws and by zoological parks. Recently some progress has been made by the acceptance of the CITES agreement between a number of countries in the world.

Illegal capture, crippling, and transportation losses.—As indicated, one of the troublesome aspects of ora population control is illegal collection, which apparently takes place most commonly on Komodo, though illegal specimens from other islands are known. As late as 1974, individuals in several Southeast Asian cities suggested they could obtain illegal specimens for me. Local illegal capture is apparently arranged through villagers, whose inexperienced use of snares and other devices can harm the animals. Entrapped animals sometimes starve to death because native captors become frightened and fail to transfer the animal to the illegal buyer (or the buyer fails to arrive for similar reasons). This is apparently more common on Flores than the other islands. Sometimes illegal capture of oras is apparently combined with deer poaching in reserve areas, resulting in a wastage of both oras and deer. Though it is difficult to estimate these losses, I think they may equal the legally collected numbers of oras each year.

One potentially dangerous practice, apparently just beginning on Flores, is poisoning carrion bait with insecticides such as Aldrin. This is a simple and highly effective method of ridding an area of unwanted oras, and it could become one of the most difficult law enforcement problems in future ora management.

Controlling illegal capture of oras is basically a law enforcement problem, but it is not always practical to place the needed number of enforcement personnel on each of the islands on which the ora is found. Local villagers presently have reasonable respect for laws protecting the oras. Additional respect and compliance can be derived only through impartial enforcement of sound regulations, and impartial enforcement is not presently characteristic of local government in that part of the world. The efforts of enforcement personnel must be upheld by local administrators, and villagers must police their own ranks.

Transportation losses have been reported by several authors, usually from starvation and disease (Eshuis, 1937), though escape (Broughton, 1936; et al.) en route is also a factor.

Restocking and transplanting.—Wildlife biologists used to believe that wild animal stocks could be developed simply by releasing animals raised in captivity, and management programs were based on this premise. Later, releasing game farm animals was replaced by transplanting captured wild animals, and studies have proved that both practices usually fail to produce large populations of wild animals, since wild animals cannot survive on land lacking essential habitat requirements. Wildlife biologists no longer depend on restocking as their major program of game management.

It is conceivable that the future press of human population on available resources, particularly on Flores, could necessitate the relocation of certain troublesome lizards. This is a particularly cogent management possibility when one is dealing with an organism that can do, and does, intentional and directed damage to livestock. The practice of relocation is common in many other predator species. But at present it is important that ora populations be left alone as much as possible and that no stocking or transplanting programs be initiated unless needs and feasibility are clearly shown on the basis of sound ecological studies. I tend to disagree with Pfeffer's (1959) suggestion that large ora specimens be sent to zoos or released on Flores to ease presumed food shortages on Rintja.

Several other writers have also either stated or implied that the oras are starving (Darevsky and Kadarsan, 1964; Hoogerwerf, 1954b; et al.). There is no question that the young and subadults are faced with higher mortality ratios than the adults, but we do not know the primary cause of death. There is no clear evidence that any significant numbers of oras are starving anywhere. Piazzini's (1960) incorrect belief that oras are not permanent residents on Padar is based on the premise of less than minimal food reserves. Pfeffer (1959) believed cannibalism among oras was due to low food reserves caused by overhunting of deer. Hoogerwerf (1954b) proposed to supplement food reserves on Komodo by establishing feral herds of goats on that island, and Pfeffer (1959) suggested that monkeys and rats be introduced (the latter already occur on Komodo) and that 4 or 5 kerabau be slaughtered each year for the oras. Several forestry officers in Bogor have asked me if more water should be made available, by means of impoundments and wells, to all the wildlife on Komodo in order to increase ora populations. Water *is* available to the larger mammals, at least at higher elevations, through most of the year. The problem is not critical to oras, since they can go for long periods without it, and there is no evidence the populations on Flores are more dense in the vicinity of permanent water than in areas without water.

Environment Control

Refuges.—Providing oras and other wild animals with protection from disturbance by humans has long been considered an essential aspect of sound game management. Aside from specific legislation protecting the animals themselves, the establishment of reserve areas is the only overt evidence of management in this species.

There are many successful animal refuges in different parts of the world. In fact, some protected animals have become so numerous in certain areas that density reduction is often essential to preserve both the range and the animals themselves. One of the main purposes of any refuge is to provide a reservoir of animals for overflow into surrounding areas. In the ora's range, this is not possible since two of the refuges are entire islands.

It is doubtful whether the established ora refuges can be considered as a meaningful part of the ora management program. For many reasons, particularly financial ones, there is little or no enforcement of management policy. Thus during our study poachers came and went as they wished; villagers established gardens wherever they wished; uncontrolled fires burned until they ran out of fuel. However, the environment in the present refuges is suitable for maintaining oras. In fact, much of the range of the species is included in the reserve areas so that the potential for perpetual protection is extremely good.

In addition to maintaining ora populations, the established ora refuges (particularly Komodo) possess important recreational values. The encouragement of a low-level outdoor program improves the possibilities of enforcement of game regulations in some wildlife refuges. The ora refuges already selected are essential to perpetuating both recreational and esthetic values, and they are well suited to maintaining a stock of oras for the future.

Habitat Control

Wild animals necessarily depend directly on the natural environment for all of their requirements. Wildlife biologists often combine food and cover into a general concept of habitat—one to provide nutrients to sustain life and reproductive processes, the other to provide protection from the elements and from predatory animals, in this case, man.

Ideally the habitat should carry an optimum density, related to its capacity, without any area becoming so crowded as to subject individual oras to starvation or leaving other parts empty. It seems possible that an approach to these conditions may be realized within the ora's range. Therefore we must consider the factors likely to determine the optimum density of an ora population.

Specific habitat types occupied by oras vary within their activity range and also vary seasonally. One of the necessities of improving animal habitat is the presence of this environmental variety, to accommodate the animal's needs during varying weather conditions and different seasons. The idea embodied in the so-called Law of Dispersion, first forwarded by Leopold (1933), is important: "The potential density of game of low cruising radius requiring two or more types is, within ordinary limits, proportional to the sum of the type peripheries." While all the conditions set forth do not necessarily apply to oras, the essential principle holds true, that ora density is highest in ecotonal zones between monsoon forest and savanna. The amount of boundary area between the two habitats is increased with the subdivision of monsoon forest into relatively small area units. A tract of 65 ha of monsoon forest surrounded by *Zi-*

zyphus savanna has a total edge of approximately 3.7 km. However, if the same tract is subdivided into 16 units of 4 ha each, alternated between monsoon forest and savanna, the total edge is about 9.3 km. In view of the preferred ora environment (Chapter 4), the latter situation would provide a better habitat. Concomitantly, this type is also the preferred habitat of both of the major prey species of adult oras, deer and wild boar.

The size of the two interspersed cover types obviously influences the animals utilizing them. For species of relatively low absolute mobility, such as rats and megapode birds, the critical size of the cover types may be only a few m^2; for the larger, more mobile prey animals, such as deer and wild boar, critical size may be many hectares in extent. Successful management of ora habitat depends on a thorough understanding of the dispersion of both the ora and its prey, as well as the interspersion of cover types.

Controlled burning of ora habitats.—Perhaps the oldest means of controlling unwanted natural vegetation and at the same time providing forage for grazing and browsing animals (including deer and wild boar) is the use of fire. Many rural inhabitants throughout the world periodically burn forests to create openings, for differing reasons: agriculturists to produce field sites, livestock herders to rid the ground of dry vegetation from the previous year and to stimulate the production of new vegetative growth. Fires are also produced by lightning and volcanism. Plant ecologists universally recognize the importance of fire in perpetuating extensive savanna forests throughout the world (Chapter 4).

The savanna forest is common throughout most of the ora's range and is undoubtedly largely produced and maintained through the use of fire. It is illegal to set fire to reserve lands, partly because fires are set mainly in connection with deer poaching and partly because much of the burning is of a highly destructive, wildfire type.

While wildfire certainly has no place in any management program, controlled burning can be very beneficial. Fire carefully used under proper wind and humidity conditions has distinct advantages in controlling plant growth in environments originally developed largely as a result of fire conditioning. To allow the islands on which the oras exist to be without the benefit of fire for many years will lead to the deterioration of the habitat, from the ora's standpoint.

One of the chief benefits of burning natural ranges is that it is thought to increase wildlife (see Komarek, 1966, for review) and to enhance the nutritive value of native browse plants. On Komodo this would be most important in management of deer and wild boar. Though some studies have shown no major effect on forage value, a recent study (Dills, 1970) reported a decided improvement in controlled burn areas. Biswell (1962) and others have shown that both population density and ovulation rate were higher in deer living on burned habitat than on unburned areas. As in other parts of the world, there is a preferred season for burning the vegetation in the Lesser Sundas. On these islands, the best time is probably the early to middle part of the dry season (Chapter 4). However, comparative studies on the nutritional advantages of burning in different seasons and in different Lesser Sunda habitats should be conducted before any program is initiated.

It is important to point out that, although forage nutrition may be increased by burning, total forage production may not be altered. Many studies, such as those by Leigh (1956), show that, while deer browse is often reduced for several years following burning, herbaceous forage is increased for at least several years. This seems to be true on Komodo as well, though studies are needed.

Considering the ecological importance of fire, the lack of data on its effects on small mammals is surprising. Most available studies show that rodent populations were initially reduced and sometimes even destroyed (Rice, 1932; Cook, 1959; et al.), though not all species were equally affected. Arata (1959) found that on savanna lands of the southeastern United States there was no net change in population numbers but that some species moved into unburned areas, greatly increasing rodent densities in those areas.

One major consideration in management programs in which fire plays a role is the frequency of burning. Most wildlife biologists recommend burning every second or third year because some plants will not fruit on a shoot until its second growing season. Where deer and wild boar have primary consideration in a land management program (in this case because they are the major prey of adult oras), burning on a rotation of 4–6 years may be the optimal plan. Investigations of these factors in the Lesser Sunda Islands are needed.

In addition to stimulating food production for deer, boar, and other wild animals, the proper use of fire on these islands can aid in acquiring a high degree of cover interspersion. This is essentially the same mechanism responsible for the interspersion presently seen. The only difference is that in the future the burning should be purposeful and controlled, that is, not set by poachers but deliberately by trained conservation officers. The use of fire in wildlife management is becoming common throughout the world (Komarek, 1963). Almost all recent studies regarding fire as a management tool point to its necessity in maintaining populations in an abundant and healthy status. It is clear that we must continue to have fire in ora habitat.

Intensive area manipulation.—The usual concept of wildlife management is regarded as an extensive rather than an intensive form of land use. I agree with this approach in the case of the ora. It means that practices specifically implemented for the benefit of oras and their prey species must be broad in scope, not intensively applied to isolated, small portions of land except as experiments.

Food crop plots involving fertilization, cultivation, etc., are all possible but need not be considered presently in view of the high population densities of the ora prey species.

Providing deer and wild boar with minerals, salts, and water are other techniques that have been suggested to increase productivity of ora prey. In many parts of the world, deer populations close to the sea ignore salt placed within their range for their use. I expect this would also be true in the Sunda Islands.

Water is one of the obvious limiting factors to wildlife, particularly on the lower slopes of most of the islands on which deer, boar, and oras occur. When available, surface water is undoubtedly used by both deer and boar, but there is no evidence that the lack of water causes great concern in either species. Deer are known to exist without water during drought periods for a month or more (Marchinton, 1964). Subjective evidence obtained from this study suggests that deer and, particularly, wild boar are without surface water for periods as long as 3 to 4 months. Goats kept on the small rocky islets near Komodo exist for almost their entire lives without fresh water.

Several conservationists have suggested that the ora, deer, and boar populations in the xeric area might be increased by drilling wells or building impoundments for water storage during the dry season. One of the main reasons this is not feasible is that as some studies (Strode, 1954, et al.) have shown, there is no increase in use of normal water sources during prolonged dry spells in ordinarily arid lands. More important, it has been shown that artificial pools provided in an area known to possess high deer populations remained largely unused. Thus there is a possibility that, even if water were made available to oras and their prey species, it may not be used and may not result in an increase in local animal populations. As pointed out, there is no evidence that ora populations are higher near permanent water on Flores than in areas in which water was only seasonally available; if anything, it is the reverse.

Livestock vs. oras.—It has long been common practice in this area to graze water buffalo and, particularly, goats on wilderness tracts. Fortunately for the deer, upon which large oras commonly feed, this type of grazing will become less common as villagers discover the greater profit of raising higher quality animals on improved pastures closer to their villages. The competition between water buffalo and deer for grazing resources varies seasonally. Studies of competition between deer and bovines conducted in several parts of the world suggest that many tracts of land could theoretically support 50 percent more deer if bovines were absent. Even greater increases in deer density could be expected if goats were also kept out of the refuges. Studies such as those by McMahan (1964) show high competition levels for browse between deer and all types of livestock, particularly during extreme weather conditions. Thus the greatest competition in the Lesser Sundas must occur during the dry season.

Moreover, the elimination of choice foods by bovines and goats may have an influence on the reproductive processes of deer. Enclosure studies by Goodrum and Reid (1954) revealed a sharp drop in fawn production after depletion of only 5 major deer browse species. McMahan (1964) concluded that preferred deer foods are sometimes so reduced by livestock grazing that the deer reproductive rate is seriously reduced, in direct proportion to grazing pressure from domestic animals.

Sufficient information exists to make it clear that, in certain areas, competition between deer and domestic livestock exists, and this competition will increase. Where deer production is viewed primarily as a means of increasing food for the ora, a grazing policy should be initiated as early as possible, before a major problem develops. Domestic cows, water buffalo, and goats should be removed or placed on restricted ranges, at least during the dry season. This practice will increase yields of cattle as well as deer. Duvall and Halls (1962) show that cattle grazed on forest ranges throughout the year have a higher mortality, calf crops of 50 percent less, and a lower weaning weight at 7 months of age than those placed on improved pasture for part of the year. Supplemental feeding regimens and improved grazing ranges are not now utilized in Indonesia, but it is obvious that such practices are sure to be necessary, particularly in the arid habitats where the ora is found.

Competition with other predators.—Competition between different size oras and other animals within their range was discussed in Chapter 12. None of these predators is sufficiently important to be considered in an ora management program, with one exception—feral dogs. These animals, the result of poaching activities, are the scourge of the islands. Hunting in small packs of several animals, they relentlessly chase and bring down deer. On Komodo their numbers are not great; during our study we killed 12 individuals and noticed a remarkable reduction in spoor within the entire Liang valley.

As pointed out by Hoogerwerf (1954a, b), Padar is particularly plagued by these animals. Droppings and tracks are very common. The signs around one deer carcass we found suggested strongly that dogs rather than oras had killed and eaten it. Hoogerwerf believed that the deer population on Padar could be appreciably increased if the feral dogs were all killed. I agree, and I urge that an extermination program in which dogs are shot or trapped by the warden be initiated, at least on that island, as soon as possible.

Tourism

Tourism may seem an unusual topic in a work on the Komodo monitor, but it is a factor in the ecology of the ora, albeit still in a small part of the range. I see tourism playing a larger role in the future, so I feel this discussion to be not only appropriate but necessary, both for a full understanding of future relationships between man and the ora and to ensure that the interaction is appropriate from the standpoint of the ora's conservation.

There is currently agitation for programs to increase educational and economic levels in Indonesia. In most areas where family income is low, the first suggested panacea is often to acquire more industry and the second is often to develop greater recreation and tourist facilities. Indonesia is considering both these courses. Due to the immediate paucity of industrial possibilities in the Lesser Sundas, tourism is viewed as a possible boost to the low economic potential of this area compared to that of the Greater Sundas. Some tourist and government agencies view the Komodo-Flores area as of great potential economic importance—partly because of the ora. In view of the obvious short supply of Indonesian capital for development, it is important that it be invested wisely.

It is somewhat curious that the presence of tourists sometimes eases the problems of management and reserve personnel. Tourism runs counter to poaching, livestock grazing, and many other practices generally injurious to wildlife management (Gee, 1964), although large numbers of tourists may be injurious to reserve areas, as illustrated by the overcrowded conditions in national parks and forest preserves in North America, Europe, and East Africa.

It is assumed that the objective of a tourism and recreational program is the preservation of the natural scenery and native wildlife of the Komodo area. Thus it is important to assure the success of such a recreational program. Though the number of tourists wishing to go to Komodo is now low, because of its remoteness and reputation, substantial increases are expected. Projections indicate a great increase in outdoor recreational needs for Indonesia by the year 2000.

A review of the ecology, comparative ora densities, and various opportunities presented by the ora reserve areas suggests that if tourism is to be encouraged in the area, it should be controlled in both number of visitors and access. Loho Kalo, Komodo, and/or Wai Wuul, Flores, could be set up as multiple-purpose tourist areas without adverse effect on ora populations, but only if properly regulated. Each piece of ground would not need to serve tourists and conservationists simultaneously but should be used for optimum benefit both to the public and to the Indonesian government, while still protecting the environment.

Wilderness, or natural, areas, without trails, gardens, or campsites, must be set aside to preserve the primitive environment of at least the southern portion of Rintja and most of the rest of Komodo. These areas should have restricted use, and entry should be by special permit only. Padar might be used for management studies, with possible manipulation of cover and prey under controlled conditions.

A major factor in any plan for tourism must take into account the possible adverse effects of nearly continuous baiting of oras with carrion, resulting in major shifts in the activity ranges of local oras, increasing densities around baiting sites, and increasing interspecific interactions. So if carrion offering were used by increased tourism programs, some plan should be formulated for limiting the number of offerings per month and for variation in where the carrion is offered.

The last factor of importance regarding tourism relates to their safety. Humans have been attacked and killed by oras, and this will continue and possibly increase in frequency as tourist numbers rise. This possibility must be considered in forming policy, particularly in regard to the aged and very young visitors to the island.

The Uniqueness of the Ora as a Wildlife Resource

Probably the most significant factor related to the necessity of conserving the ora for future generations is its great size. Even average hatchlings, at a total length of 49 cm, are considerably longer than adults of most lizard species. The great size and weight of adult oras in particular, combined with highly specialized teeth for slicing flesh, modify in an important way the social interactions of members of the population and the feeding biology. Both intraspecific coactions and predation in this species are quite different from any described so far among reptiles.

The ora is also unique in regard to its high locomotor facility; it can run swiftly for a reptile. It is occasionally bipedal and clearly is an effective climber, at least when young. It is a reasonably good diver and swimmer and excavates large burrows in the soil.

The ora is provided with a highly sophisticated neural anatomy, particularly in respect to those parts of the brain associated with vision and olfaction. Closely related species have been shown to possess a high metabolic rate for reptiles, and it is suspected that the ora has a similar energetic pattern. Though optimal body temperatures do not differ significantly from those of most other reptiles, the great mass of adult oras particularly confers important thermal advantages—deep body temperatures tend to remain more uniform day and night throughout the year than they do for any other reptile in the world. The anatomy of its circulatory system suggests that there is less mixing of oxygenated and unoxygenated blood during its rounds through the body than there is in any other lizard or snake. Finally, animal keepers in zoos and animal behaviorists who have had an opportunity to study the ora seem convinced that it is one of the most intelligent reptiles in the world.

Together these physical, physiological, and behavioral characteristics provide the ora with the potential for complex behavioral patterns. The studies completed to date on this organism prove that its potential is realized, particularly in regard to its feeding strategies. It is clearly one of the most unique reptiles in the world today.

For this reason the preservation of the ora should be accepted as one of the highest priorities in conservation of the world's reptiles. The existence of a sound nature and wildlife conservation program in Indonesia is a good indication of the nation's progress and development. I am confident that Indonesia and the various international conservation agencies involved will not fail in their duty of preserving this unique, spectacular animal.

Appendixes

Appendix 1. Major expeditions to study Komodo monitors

Year	Principal worker(s)	Place	Approximate length	Publication
1910	van Steyn van Hensbroek	Komodo	(?) 1 week	None
1911	Bogor Zool. Museum	Komodo	(?) several weeks	Ouwens, 1912
1923	Baron von Mecklenberg	Komodo (or Flores?)	?	None
1924	Horst	Komodo and Rintja	?	1926
1926	Burden-Dunn	Komodo and Padar	4 weeks	Burden, 1927a, b, c; 1928; Dunn, 1927b
1929	Avé Lallemant	Komodo and Rintja	(?) several weeks	1929
1929	de Jong	Flores, Rintja, Padar, Komodo	several weeks	1929
1934	Surabaja Zoo–Griswold and Harkness	Komodo	(?) 8 weeks	Anon., 1934

(Continued)

Appendix 1—*Continued*

1935	Lord Moyne	Komodo	?	Broughton, 1936; Anon., 1935
1936	de Voogd	Komodo	several weeks	1950
1937	de Jong	Komodo and Flores	7.5 weeks	1937
1937	Fejos	Komodo	1 week	None, but letter quoted in Hoogerwerf, 1954a
1938	Collins	Komodo and Padar	20 weeks	1956
1952	Hortojo-Hompes	Komodo	1.5 weeks	None
1952	Pryor	Komodo	several days	MS
1953	Pareira	Komodo	several days	None
1955	Hoogerwerf	Komodo, Padar, and Rintja	6 weeks	MS, 1954
1956	Piazzini-Pfeffer	Flores, Rintja, and Komodo	10 weeks	Pfeffer, 1959, 1965; Piazzini, 1960
1962	Darevsky-Kadarsan	Komodo, Padar, and Rintja	4.5 weeks	Darevsky, 1963, 1964; Darevsky and Kadarsan, 1964
1963	Surabaja Zoo	Komodo	?	None
1964	Bishop Museum Expedition	Komodo	?	Darevsky and Stusak, 1968
1967	Kern-King (NYZS)	Komodo	1.5 weeks	Kern, 1968; King, 1968
1969–70	Auffenberg (NYZS)	Komodo	44 weeks	This publication
1971	Auffenberg (NYZS)	Flores	5 weeks	This publication
1973	Lincoln-Patterson-Guinness	Komodo	2 weeks	Lincoln, 1974
1973	Auffenberg–Don Meir Productions	Komodo	4.5 weeks	This publication

Appendix 2. Local names for Komodo and water monitors (largely after Verheijen, 1967)

Komodo monitor	
Ora (also *hora, lawora*)	Komodo, Rintja, and West Mangarrai District, Flores
= Malay *biawak raksasa* ("giant monitor")	
= Malay *buaja darat* ("land crocodile")	
Rugu (= *ora*)	Central Mangarrai District, Flores
Si (also *ugu*)	Central Mangarrai District, Flores
= Malay *tjejak* ("lizard")	
Lio (also *ugu*)	Central Mangarrai District, Flores
= Malay *biawak besar* ("large monitor")	
Pendugu ("grandfather ora")	Central Mangarrai District, Flores
Water monitor	
Weti (= Malay *biawak*)	Mangarrai District, Flores
Lawora	Sumba
Lagura	Savu
Nggora	Roti
Ghora (or *hora*)	Nggada District, Flores
Uti	Sumbawa
Alu	Bali

Appendix 3. Circuit diagram and harness arrangement for temperature transmitters

Circuits (Fig. A3-1) were modified from the blocking oscillator types described by Stebbins and Barwick (1968). Transmitters were mounted on a web harness in the sacral region (Fig. A3-2). The thermistor transducers of each transmitter were provided with 15-cm leads; one attached to the median body line in order to transmit ambient temperatures at the level of the monitor's dorsal surface (Tr 2); the other thermistor was placed 3 cm deep into the flesh of the axilla of the left hind leg (Tr 3). The ½ wave length antennae for both transmitters were passed around the tail to serve as directional loops, held in place with nylon strap tape. Still another transmitter (Tr 4) provided ambient temperature under an open sky, 10 cm above the ground surface. Temperature data from each transmitter were broadcast on a different wave length in the 27 Mhz CB range. Tone and click frequency are adjustable at R2 and C4 to distinguish data from several individuals if the same frequencies were being used. Appropriate

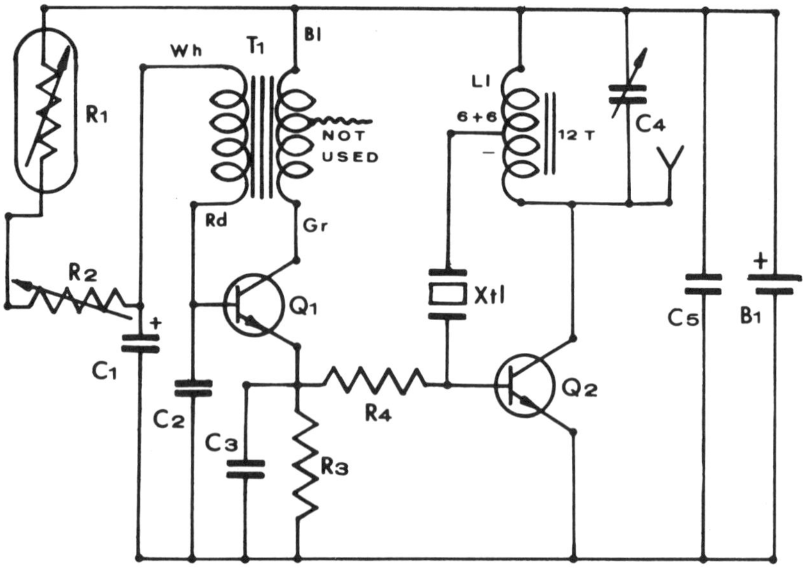

Fig. A3–1. Temperature sensing-transmitter diagram.

Transmitter Circuit Data

C_1 = 10 mf, electrolytic
C_2, C_3 = .01 mf, 50V, ceramic
C_4 = 40 pf, adjustable trimmer
C_5 = .05 mf, 50V, ceramic (not placed close to L_1)
R_1 = 1.0 M, bead thermistor
R_2 = 500 k, 1/10 W, subminiature potentiometer (may be replaced with 500 k fixed resistor)
R_3 = 1.5 k, 1/4 W, 5%, carbon (may be ± 100 ohms for improved transmission)
R_4 = 30 k, 1/4 W, 5%, carbon
Q_1 = Motorola Hep 54, silicon NPN
Q_2 = Motorola Hep 56, silicon NPN (high frequency)
T_1 = subminiature audio diver transformer
L_1 = slug tuned coil for printed circuits, wound with #27 magnet wire (do not place close to other components)
B_1 = 1.5 to 9V (not placed close to L_1)

modifications were made in a 3-channel, 3-watt transceiver in order to use one of the harness transmitters as a directional transmitter. A small near-field 80 Mhz temperature-controlled transmitter (Tr 1) (based on that described by

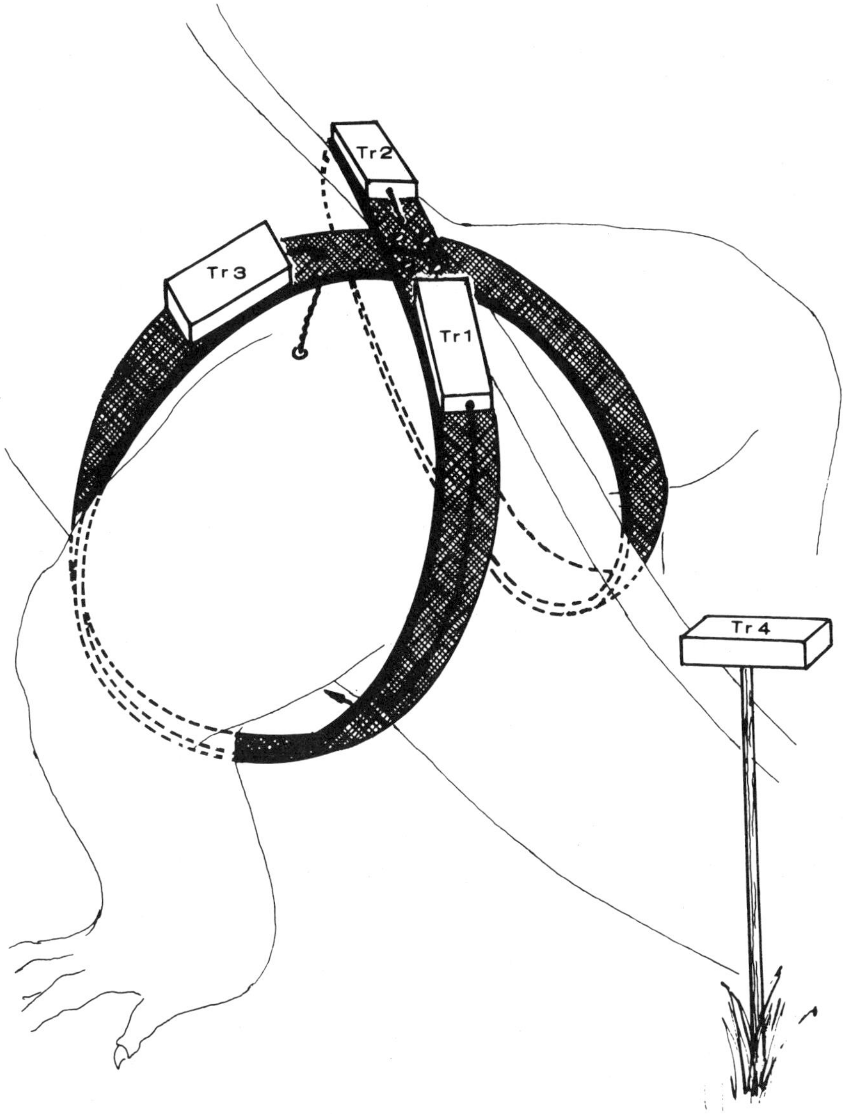

Fig. A3–2. Harness mounting of transmitters.

MacKay, 1964) was used to measure core temperatures by means of the thermistor lead being implanted 6 cm deep within the cloaca. When the animal was basking or otherwise inactive, a small transistorized AM receiver was held within 1 m of the transmitter and temperature data recorded.

Appendix 4. Common plants of the Komodo tropical savanna (*Zizyphus* and *Borassus* facies)

Taxa	Remarks
Grasses	
Eragrostis megadactyla	Not common, not in dense stands, scattered, sometimes eaten by deer
Setaria verticillata	Scattered, not dense stands, not common, to 2 m (>1 m) important cover only when combined with other grasses and forbs
Eulalia leschenaultiana	Most important deer forage when grass young, common, to 1 m (0.5 m), in large, dense stands, widely distributed
Chrysopogon tenniculmis	Thick stands, important cover at both high and low elevations; to 1 m (<1 m)
Leptochloa chinensis	Widely distributed, often patchy, stands usually not dense; to 1.7 m (0.5 m)
Perotis hordeiformis	Common, but rarely dense stands, often in waterlogged, usually bare soils; to 0.5 m (0.3 m)
Heteropogon contortus	Common, widely distributed, not in dense stands, not important cover, but often eaten by deer
Echinochloa colona	Not common, not in dense stands, not important cover, but sometimes eaten by deer
Themeda frondosa	Very common, in large, dense stands, very important cover grass for oras and prey; to 2 m (1.5 m), sometimes eaten by deer
Themeda triandra	A common and important cover grass for oras and prey on hillsides; to 1.5 m (1 m); patchy, but dense, a common deer food
Paspalidium flavidum	Uncommon, not in dense stands, not an important cover type; to 1 m (0.6 m)
Manisuris granularis	Not common anywhere, never in dense stands; not an important cover; to 1 m
Dactyloctenium aegyptium	Common, usually not in dense stands, patchy distribution; large stands on Padar, Nusa Mbarapu, and NE part of Komodo; not an important cover; less than 1 m

(Continued)

Appendix 4—*Continued*

Chloris barbata	Very common, not in dense stands, widely distributed; not an important cover, but sometimes eaten by deer; to 1.5 m (1 m)
Garnotia sp.	Dominant grass of hilltop savanna in dry zones; not an important cover; to 0.3 m
Dicanthium caricosum	Not common, not in dense stands; not an important cover; to 1.5 m (1 m)
Ferns	
Davillia solida	Not common, not at low elevations; not an important cover
Forbs	
Psoralea samborensis	Usually on hillside savannas, scattered, but common; not an important cover; to 2 m
Polanisia vicosa (N. Ketjah)	Locally abundant, often the dominant savanna forb, not in dense stands; not an important cover type
Shrubs and trees	
Morinda tinctoria (N. Nggudu)	Scattered tree of hillsides, not common
Schleichera oleosa (N. Sambi)	Scattered tree of savannas; characteristic of "tanah panas"; important shade producer
Tamarindus indicus (N. Asim)	Largest savanna tree; important shade producer; leaves and fruits eaten by deer
Cordia monoica (N. Dange)	Scattered shrub; not an important cover
Sterculia cf. *cinerea* (N. Niti)	Uncommon savanna tree; not important, except fruits eaten by deer and boar
Zizyphus jujuba (N. Lengkang)	Dominant tree of the savannas and important shade producer in areas where shade uncommon; fruits very commonly eaten by deer and boar
Borassus filiformis (N. Djantal)	Dominant shade producing palm tree locally
Avicennia alba (N. Haducia)	Common locally in savanna near shore

Appendix 5. Common plants of the *Pandanus-Spinifex* association

Taxa	Remarks
Grasses	
Spinifex littoreus	Common, in dense stands; an important cover; to 1 m
Digitaria adscendens	Common, forms dense stands to 1 m in optimum habitat; sometimes eaten by deer
Chrysopogon tenniculmis	Tufted grass in dense stands to 1 m (0.5 m); important cover
Sporobolus virginicus	Rarely in thick stands
Forbs	
Barleria prionitis (N. Lopoh)	Thorny, dense stands behind *Spinifex*; important cover in areas where stands are large; small, open stands occasionally occur inland
Calotropis gigantea (N. Kore)	Not important, scattered; to 2 m
Josephinia imperatricis	In scattered colonies, never dense, low
Crotalaria retusa	One of dominant forbs of this zone; to 1 m; important cover when in dense stands
Shrubs and trees	
Jatropha curcas	An important shade tree, forming dense thickets near coast and in damp soils
Tamarindus indicus (N. Asim)	An important shade producer wherever it occurs; fruit and leaves often eaten by deer and boar
Schleichera oleosa (N. Sambi)	An important shade producer wherever it occurs
Pandanus tectorius (N. Landes)	Scattered tree; not important, except that it characterizes this zone
Hibiscus tiliaseus (N. Peninta)	Widely scattered tree, but good shade producer
Thespesia populnea (N. Mbeloti)	Common tree of beach area; an important shade producer in areas where shade is often rare
Zizyphus jujuba (N. Lengkang)	Scattered, but common tree, sometimes extending into *Spinifex* zone; fruits often eaten by monkeys (on Flores), deer, and wild boar

Appendix 6. Common plants of the Loho Liang, Komodo, mangrove association

Taxa	Remarks
Grasses	
Cyperus javanicus	Common, forms thick, dense stands and provides important cover for oras
Shrubs and trees	
Avicennia alba (N. Haducia)	Common, often the largest tree of association, rarely forming thickets; not an important cover type
Clerodendron inerme	Shrub, forming dense thickets, common near creek mouths; important cover for wild boar and oras
Azyma sarmentosa (N. Vetok)	Evergreen, spreading, thicket-producing shrub, very important as cover for boar and deer—particularly in the dry season; often the only shade producer at this time of year; found on edges of association
Lumnitzera racemosa (N. Rumput Atu)	Forms dense stands, often the dominant shade producer of association; very important cover for wild boar and oras
Rhizophora mucronata (N. Banku)	Forms thickets, common, but not too important because stands too far out in water
Capparis sepiaris (N. Kerkoa)	Often forms thickets at creek mouths and on edge of mangrove associations

Appendix 7. Common plants of the deciduous forests of Komodo (below 500 m)

Taxa	Remarks
Grasses	
Cyperus zollingeri	Not common, except locally, under open canopy in damp places, not an important cover; to 0.5 m; sometimes eaten by deer
Panicum trichoides	Common under open canopy in damp places; forms dense stands; to 1.5 m high (0.7 m); often eaten by deer
Oplismenus burmanni	Common in sunlit parts of forests, often dense stands; to 1.5 m (1 m); important cover in March

(Continued)

Appendix 7—*Continued*

Brachiaria ramosa	Common, especially in more open forests, first grass to bloom and first to die; forms very thick stands; important cover; 0.5–1 m
Forbs	
Asystasia intrusa	Common understory forb, to 0.5 m, not in dense stands; not important
Acalypha boemerioides	Under dense canopy, not in dense stands; not important cover
Lepidagnathis sp.	Under dense canopy, to 0.7 m (0.3 m), often the only cover in the densest forest; not important cover, usually too scattered; sometimes eaten by deer
Shrubs and trees	
Flacourbia indica (N. Timur)	Shrub to small tree; very common understory evergreen; when large, important cover; fruits eaten by wild boar and deer
Murraya paniculata	Shrub to small tree, understory type, widely distributed, common, sometimes forming small thickets; this, plus evergreen feature, make it an important cover in dry season
Sterculia foetida (N. Vau)	Common, scattered, large tree; produces much shade; forms great deal of duff
Sterculia cf. *cinerea* (N. Niti)	Common, scattered, large tree; nuts eaten by many animals
Combretum latifolium	Common, scattered, large tree; leaves eaten by deer
Schoutenia ovata (N. Mbuhung)	Common, medium-size tree
Schleichera oleosa (N. Sambi)	Common, large tree, characteristic of "tanah panas"
Pterospermum diversifolium (N. Vadju)	Common, large tree; rare in "tanah dingan"
Bridelia ovata (N. Rau)	Common, large tree
Cladogynos orientalis	Common, medium tree, often in clusters
Tabernaemontana floribunda (N. Sepada)	Small tree, common, widely distributed, not important cover, except in ecotonal situations

(Continued)

Appendix 7—*Continued*

Zizyphus horsfieldii (N. Klanka)	Rather characteristic, but scattered, large tree
Zizyphus timorensis	Common, large tree
Randia patula (N. Dikik Ngalong)	Viney shrub, producing open thickets; not important unless associated with other shrubs, such as *Azyma*
Helicteres angustifolia	Tall shrub, scattered, not dominant except locally; not important cover
Melochia hospita (N. Hadju Ndungi)	Scattered, medium-size tree

Appendix 8. Common plants of the ecotonal zone between savanna and forest (below 500 m), Komodo

Taxa	Remarks
Grasses	
Setaria verticillata	Important only in combination with other grasses and forbs; to 2 m
Rottboehia escalbata	To 4 m (2.5 m) tallest grass on Komodo, in dense stands; very important cover for boar, deer, and oras
Brachiaria ramosa	Very common, dense in ecotone, to 1 m; important cover
Digitaria adscendens	Forms dense stands, to 1 m; important cover
Vines	
Mormordica charantia (Balsam apple) (pai)	Very common, covering many shrubs like a blanket from January to March or early April; an extremely important cover at this time of year for deer, boar, and oras; boar eat fruit
Muemerargia timorensis	Next in importance as vine cover to *Mormordica* in late February and all of March
Forbs	
Triumfetta indica	Important only when combined with other grasses and forbs; to 1 m; sometimes eaten by deer
Bidens pilosa	The most important forb for cover in the ecotones; to 1.5 m

(Continued)

Appendix 8—*Continued*

Hyphis suaveoleus	Sometimes forms dense stands locally; to 1 m (0.5 m)
Shrubs and trees	
Azyma sarmentosa (N. Vetok)	An armed evergreen shrub, of spreading habit, often in wet, well lighted areas, probably most important cover for deer, boar, and oras in the dry season; often the only shade producer at this time of year; forms dense thickets
Flacourbia indica (N. Timur)	Too scattered to be important cover
Callicarpa condicans	Shrub to 3 m, important cover locally, but produces little shade
Caesalpinia sappan	Habit too open to be important cover
Diospyros malabarica	More common at higher elevations than lower elevations; sometimes forms thickets, but too scattered to be important cover; sometimes leaves eaten by deer
Murraya paniculata (N. Kemunung)	Shrub to small tree, tending to form thickets, widely distributed; this plus evergreen features make it an important cover in dry season
Hypoestes malaccensis (N. Bai)	Shrubs to 2 m, forming extensive and dense stands in open, sunny areas; probably the most important ecotonal plant in terms of shade and thicket development; an important cover type
Harrisonia brownii (N. Mbelak)	An important light filter of ecotonal situations; area below plants not large, habitat open, not important cover, except in combination with other shrubs
Abutilon indicum (N. Bertai)	Very common and most important cover shrub after *Hypoestes*, but only in wet season; to 3.5 m (1.5 m)
Schoutenia ovata (N. Mbuhung)	Medium tree, scattered
Pterospermum diversifolium (N. Vadju)	Large tree, scattered

(Continued)

Appendix 8—*Continued*

Solanum junghulnii	Medium shrub, to 2 m, scattered, open habit; not important cover
Annona squamosa (sugar apple) (N. Garoso)	A small tree, common in open sunny places and early successional stages; fruit eaten by many animals
Acacia cf. *concinna* (N. Radah)	Common, but scattered small tree; not important as cover, open habit; leaves sometimes eaten by deer
Bauhinia sp. (N. Daling)	Same as above; leaves eaten by deer
Piliostigma malabaricum (N. Daling)	Same as above
Pithacolobium umbulatum (N. Hadju Lembur)	Same as above
Tabernaemontana floribunda (N. Sepada)	Small tree, common in ecotone, open habit, not important
Microcos paniculata (N. Soro)	Important shade producer, common in and on edges of forest; important successional plant; to 4 m
Grewia koordersiana (N. Nila)	Same as above
Helicteres isora (N. Tange)	Same as above, but does not form a great deal of shade

Appendix 9. Shrub and tree species in the Nggoer (Flores) transects (%)

Species	Transition monsoon forest (G. Wasedoo, el. 300 m)	Hillside monsoon forest (G. Insilung, el. 100 m)	Forest stand in Nggoer savanna (el. 20 m)
Murraya paniculata	23.8		
Diospyros malabarica	16.7		
Colona kostermansiana	14.3		
Microcos paniculata	12.2		12.5
Zizyphus timorensis	10.3		
Cladogynos orientalis	5.1		9.4
Bambusa blumeana	5.0		
Schoutenia ovata	4.7	30.6	

(Continued)

Appendix 9—*Continued*

Uvaria rufa	3.7		
Pterospermum diversifolium	2.3		
Abutilon latifolium		15.3	
Jatropha gossypiifolia		14.1	1.2
Hypoestes malaccensis	1.8	12.8	6.3
Tamarindus indicus		9.4	4.6
Zizyphus horsfieldii		7.6	3.1
Pithacolobium umbulatum		3.5	1.6
Zizyphus jujuba		2.2	1.1
Schleichera oleosa		2.2	9.4
Unident. (2 species)			29.7
Tabernaemontana floribunda		1.1	20.3

Appendix 10. Gazetteer of geographic names used in text

Adjung (Fig. 1-4), mountain, E part central ridge, Komodo.
Aimere, village, S central coast, Ende Province, Flores.
Amasabung (Fig. 1-4), mountain, E part, central ridge, Komodo.
Amasiri (Fig. 1-4), mountain, E part, central ridge, Komodo.
Angkor, village, Badjo District, SW Flores.
Banda, island in the Molucca group, NW of Komodo.
Banta (Fig. 3-1), island, NW of Komodo, and (Fig. 1-2), hill, Mangarrai Province, W Flores.
Banu Sambi, mountain area, above Kampung Kechil, Komodo.
Bara, village, near Nangelele, SW Mangarrai Province, SW Flores.
Base Camp (Figs. 1-11, 5-7), bait station and main trapping area, E Komodo.
Beliling Mountains, central Mangarrai Province, W Flores.
Boleng, village, NW coast Flores.
Buntal, village, coast, N Flores.
Dampek, village, NW coast Flores.
Djarek (Fig. 1-2), village, W Mangarrai Province, W Flores.
Ende, village, S central coast, Flores.
Flores (Fig. 1-1), large island, E of Komodo.
Galulai (Fig. 1-3), mountain ridge, central Komodo.
Golo (= hill)
Golo Barat (Fig. 1-2), hill, near Loök Village, Mangarrai Province, W Flores.
Golo Misi (Fig. 1-2), hill, near Loök Village, Mangarrai Province, W Flores.

(Continued)

Appendix 10—*Continued*

Gunung (= mountain)
Gunung Amasabung (Fig. 1-3), mountain, E Komodo.
Gunung Ara (Fig. 1-3), mountain, central Komodo.
Gunung Kalo (Fig. 1-7), low mountain, Loho Kalo, E Komodo.
Gunung Klinta (Fig. 1-3), mountain, E Komodo.
Gunung Kubur, mountain, E Komodo.
Gunung Lamas, mountain, NW coast, Mangarrai Province, W Flores.
Gunung Liseng (Fig. 1-4), mountain, S outlier, central ridge, Komodo.
Gunung Nadju (Fig. 1-3), mountain, S central Komodo.
Gunung Ntodo, mountain, near Nangelele Village, SW Flores.
Gunung Pahme (Fig. 1-3), mountain, N central Komodo.
Gunung Pronge (Fig. 1-7), mountain, separating Loho Kalo from Loho Liang, E Komodo.
Gunung Rankang (Fig. 1-4), mountain, S outlier, central ridge, Komodo.
Gunung Satilibo (Fig. 1-3), mountain, N central Komodo.
Gunung Soke (Fig. 1-7), mountain, Loho Kalo, E Komodo.
Gunung Vau (Fig. 1-3), mountain, S Komodo.
Gunung Wasedoo (Fig. 1-2), mountain, Mangarrai Province, W Flores.
Hill 19 (Fig. 1-4), Loho Liang, E Komodo.
Hill 23 (Fig. 1-4), Loho Liang, E Komodo.
Inaloah (Figs. 1-11, 5-7), bait station and trapping site, along Kali Inaloah, Loho Liang, E Komodo.
Indo-Australian Archipelago, island chain between Malaysia and New Guinea, including Komodo.
Jakarta, capital city of Java, Indonesia.
Jogjakarta, city, W central Java.
Kali (= creek)
Kali Adjung (Fig. 1-4), creek, Loho Kalo, E Komodo.
Kali Inaloah (Figs. 1-6, 1-11), creek, Loho Liang, E Komodo.
Kali Amasiri (Fig. 1-7), creek, Loho Kalo, E Komodo.
Kali Banu Adjung (Fig. 1-7), creek, Loho Kalo, E Komodo.
Kali Besar (Fig. 1-4), creek, Loho Liang, E Komodo.
Kali Inaloah (Fig. 1-4), creek, Loho Liang, E Komodo.
Kali Kechil (Fig. 1-4), creek, Loho Liang, E Komodo.
Kali Nikipolak (Fig. 1-4), creek, Loho Liang, E Komodo.
Kali Pronge (Fig. 1-7), creek, Loho Kalo, E Komodo.
Kalo (Figs. 1-11, 5-7), baiting station, Loho Kalo, E Komodo.
Kampung (= village)
Kampung Ara, village, N coast, NW Flores.
Kampung Djarek (Fig. 1-2), village, Mangarrai Province, W Flores.

(Continued)

Appendix 10—*Continued*

Kampung Kechil (Fig. 1-3), village, central Komodo.
Kampung Komodo (Fig. 1-3), village, central Komodo.
Kampung Lavi (Fig. 1-3), village, central Komodo.
Kandong, village, in mountains, Mata-wae District, W central Flores.
Kedindi, village, near, Reo, NW coast, Flores.
Komodo (Figs. 1-1, 1-3).
Krakatoa, volcano and island, between Java and Sumatra.
Kubur (Fig. 1-4), mountain, E Komodo.
Labuanbadjo, village, Mangarrai Province, W Flores.
Ladang (= overgrown garden).
Lanchang (Fig. 1-2), hill, near village of Loök, W Flores.
Larantuka, district, extreme NE Flores.
Lenteng, village, W Mangarrai Province, W Flores.
Lesser Sunda Islands, E part of Sunda Island chain, Bali to Aru Islands, including Komodo.
Lintja Strait (Fig. 1-3), W of Komodo, between Padar and Komodo islands.
Loho (= valley).
Loho Buaja, valley, N central coast, Rintja Island.
Loho Kalo (Figs. 1-3, 1-4), valley, E Komodo.
Loho Liang (Fig. 1-3), valley, E. Komodo.
Lombok (Fig. 1-1), island, between Bali and Sumbawa, W of Komodo.
Loök (Fig. 1-2), village and river, Mangarrai Province, W Flores.
Mangarrai (Fig. 1-2), province, extreme W Flores.
Maumere, city, coastal NE Flores.
Mbarat Mangge (Fig. 1-2), hill, Mangarrai Province, W Flores.
Mbuhung (Fig. 1-2), hill, Mangarrai Province, W Flores.
Mburak, village, coastal W Flores, S of Labuanbadjo, and wildlife reserve.
Missi (= mission).
Missi Nunang, mission in mountains, Mangarrai Province, W Flores.
Missi Ruteng, mission in mountains, Ruteng District, W central Flores.
Nanenepi (Figs. 1-6, 1-11), mangrove area, E Komodo and baiting station nearby.
Nangelele, village, Mangarrai Province, SW Flores.
Nanga (= river).
Nanga Loök (Fig. 1-2), river, Mangarrai Province, W Flores, and river, near border of Mangarrai and Riung provinces.
Nggoer (Fig. 1-2), village, Mangarrai Province, W Flores.
Nggokodjarung, mountain, Loho Kalo, E Komodo.
Nggulung (Fig. 1-11), baiting station, along Vai Nggulung, E Komodo.

(Continued)

Appendix 10—*Continued*

Ngondo, village, Ende District, W central Flores.
Nikipolak (Fig. 1-4), mountain, S outlier, central ridge, Komodo.
Nisar, village, Mata-wae District, SW Flores.
Ntodo (= mountain).
Ntodo Adjung, mountain, Loho Kalo, E Komodo.
Ntodo Klea (Fig. 1-3), volcanic mountain core, E Komodo.
Nusa (= small island).
Nusa Mbarapu (Fig. 1-3), islet, E Komodo.
Nusa Mules, islet, SW coast, Flores.
Overflow area (Fig. 1-4), delta area of Kali Besar, Loho Liang, E Komodo.
Padang (= steppe or savanna produced by repeated fires).
Padang Lanchung (Fig. 1-2), grassland patch, near Loök, W Flores.
Padang Nggoer (Fig. 1-2), grassland patch, near Nggoer, W Flores.
Padang Watumisi (Fig. 1-2), grassland patch, near Djarek, W Flores.
Padar (Fig. 3-1), island, E of Komodo.
Perangka (Fig. 1-3), low hill, NE Komodo.
Poreng (Fig. 1-3), long, open slope, N central Komodo.
Pota, village, NW coast, Flores.
Pulau (= island).
Pulau Longo (Fig. 1-3), islet, SE Komodo.
Pulau Seraja Besar, island, extreme NW Flores.
Randong, village, in mountains, Mata-wae District, central W Flores.
Reo, city, NW coast, Flores.
Rintja, island, E of Komodo, between Padar and Flores.
Riung, district, N central Flores.
Ruteng, city, inland, W central Flores.
Sabu, island, between Sumba and Timor, Lesser Sunda Islands.
Sapi, village, extreme E Sumbawa.
Sapi Strait (Fig. 1-3), E of Komodo, between Komodo and Sumbawa islands.
Selat (= strait).
Selat Molo (Fig. 1-2), strait, between Flores and Rintja.
Siru, village, Badjo District, SW Flores.
So (= bay).
So Kodang (Fig. 1-3), bay, NW Komodo.
So Vaso (Fig. 1-3), bay, NW Komodo.
So Vintji (Fig. 1-3), bay, NW Komodo.
Sokenor, village, Mangarrai Province, W Flores.
Sumba (Fig. 1-1), large island, S of Komodo.
Sumbawa (Fig. 1-1), large island, W of Komodo.
Sumbawa Besar, city, N central Sumbawa Island.

(Continued)

Appendix 10—*Continued*

Sunda Islands, Sumatra E through Aru Islands, including Komodo.
Surabaja, city, SE Java.
Takutega (Fig. 1-3), mountain ridge, central Komodo.
Tamboro, volcano, N central Sumbawa Island.
Tandjung (= point).
Tandjung Tongkir, point, E coast Rintja.
Tanah Dingan (= "cool ground"), general forested mountain area above 500–700 m.
Tanah Panas (= "hot ground"), generally savanna, below 500 m.
Telok (= bay).
Telok Gili Lawa (Fig. 1-3), bay, N coast, Komodo.
Telok Longo (Fig. 1-3), bay, SE coast, Komodo.
Telok Slawi (Figs. 1-3, 1-4), bay, E central Komodo.
Tereng, village, NW coast, Flores.
Termeda, village, in mountains, Mata-wae District, W central Flores.
Ternate, island, between Halmahera and the Philippine Islands.
Timor (Fig. 1-1), large island, near E end of Sunda Island chain.
Timor Laut, island, near Timor, Lesser Sunda Islands.
Timor Roti, island, between Sumba and Timor, Lesser Sunda Islands.
Tjereng, village, inland from Loök, Mangarrai Province, SW Flores.
Vai (= creek).
Vai Djamal, creek, near Nangelele, Mangarrai Province, W Flores.
Vai Nggulung, creek, central Komodo.
Valupaku (Fig. 1-4), mountain, S outlier, central ridge, E Komodo.
Wai (= creek).
Wai Ara, creek, S coast, W Flores.
Wai Keloa (Fig. 1-2), creek, Mangarrai Province, W Flores.
Wai Sadrap, creek, near Kampung Komodo, central Komodo.
Wai Sano, creek, in mountains, Mata-wae District, W central Flores.
Wai Wake, creek, Badjo District, SW Flores.
Watu Motang (Fig. 1-2), hill, Mangarrai Province, W Flores.
Wesang Ndalak (Fig. 1-2), hill, Mangarrai Province, W Flores.
Wetar (Fig. 1-1), island, NW of Timor Island, Lesser Sunda Island chain.

Literature Cited

Abdoessoeki, E. n.d. *Varanus komodoensis* Ouwens pada habitatnya di Pulau Komodo. Mimeographed. p. 1 to 9.

Ali, S. 1964. The Book of Indian Birds. Bombay Nat. Hist. Soc. 158 p.

Allard, P. 1947. Light intensity studies in Canaan Valley, West Virginia. Castanea 12:63–74.

Allee, C. 1951. Cooperation among Animals. Abelard-Schuman Ltd., London. 233 p.

Anastos, G. 1950. The scutate ticks or Ixodidae of Indonesia. Entomol. Amer., Brooklyn Entomol. Soc. 20:1–144.

Anderson, F. S. 1953. Analysis of a Danish roe-deer population. Danish Rev. Game Biol. 2:127–55.

———. 1961. Effective density on animal sex ratios. Oikos 12: 1–16.

Anderson, J. 1889. Report on the mammals, reptiles, and batrachians, chiefly from the Mergui Archipelago, collected for the Trustees of the Indian Museum. J. Linn. Soc. Zool. 21:331–50.

Andres, A. 1904. Intelligenz eines Wüstenwarans. Blat. Aquar. Terrar. Kunde 15:269.

Annandale, T. 1903. *In* Boulenger, 1903.

Anon. 1926. Das Drachenparadies von Komodo. Berliner Tageblatt (Berlin), February 24.

———. 1930. Ein nieuwe zeldzaam Beest in den Dierentuin. De Nieuwe Gazat (Antwerp), July 5.

———. 1934. Giant Komodo dragons come to America. The Literary Digest (New York), May 26.

———. 1935. Der Drache. De Nieuwes v.d. Dag (Buitenzorg), January 31, p. 12.

Anon. 1937. De Locomotif. July 3, 1939. p. 3.

Anon. 1953. Semua Undang dan Peraturan Pemerintah Mengenai Perlindungan alam dan Pemburan di Indonesia. Van Hoeve: Bandung. pp. 1–104.

Anon. 1970. The Nature Reserves of Indonesia. Office of Parks and Wildlife, pp. 1–36.

Arata, A. 1959. Effects of burning on vegetation and rodent populations in a longleaf pine–turkey oak association in north central Florida. Quart. J. Fla. Acad. Sci. 22: 94–104.

Attenborough, D. 1957. Zoo Quest for a Dragon. Lutterworth Press, London. 157 p.

Auffenberg, W. 1964. Notes on the courtship of the land tortoise *Geochelone travancorica* (Boulenger). J. Bombay Nat. Hist. Soc. 61(2):1–7.

———. 1972. Komodo dragons. Nat. Hist. 81(4):52–59.

———. 1978. Social and feeding behavior in *Varanus komodoensis*, p. 301–31. *In* N. Greenberg and P. D. McLean (eds.) Behavior and Neurology of Lizards. Government Printing Office, Washington, D.C.

———. 1979a. Intersexual differences in behavior of captive *Varanus bengalensis* (Reptilia, Lacertilia, Varanidae). J. Herp. 13(3):313–15.

———. 1979b. A monitor lizard in the Philippines. Oryx 15(1):39–46.

———. 1980. The herpetofauna of Komodo, with notes on adjacent areas. Bull. Fla. State Mus., Biol. Sci. 25(2):39–156.

———. In press(a). Courtship behavior in *Varanus bengalensis* (Sauria:Varanidae). Bull. Mus. Comp. Zool.

———. In press(b). Comparison of density estimating techniques in *Varanus bengalensis*. Rec. Zool. Surv. India.

———. In press(c). Combat behaviour in *Varanus bengalensis*. J. Bombay Nat. Hist. Soc.

———, and S. Ganci. In press. Agonistic behavior in adolescent *Varanus bengalensis*. J. Herp.

———, and W. G. Weaver, 1969. *Gopherus berlandieri* in southeastern Texas. Bull. Fla. State Mus., Biol. Sci. 13(3):141–203.

Barbour, T. 1926. Reptiles and Amphibians, Their Habits and Adaptations. Houghton-Mifflin and Co., Boston. 125 p.

Bartholomew, G. A., and V. A. Tucker. 1964. Size, body temperature, thermal conductance, oxygen consumption, and heart rate in Australian varanid lizards. Physiol. Zool. 37:341–54.

Bartlett, H. H. 1956. Fire, primitive agriculture, and grazing in the tropics. *In* Man's Role in Changing the Face of the South. Univ. Chicago Press, Chicago. p. 692–720.

Bastock, M. 1967. Courtship and Ethological Studies. Aldine Publishing Co., Chicago. 220 p.

Beddard, F. E. 1888. On certain points in the visceral anatomy of the Lacertilia, particularly of the monitor. Proc. Zool. Soc. (London), 312–16.

Bellairs, A. 1970. The Life of Reptiles. Universe Books, New York. 1, 2:590 p.

Bentley, P. J., and K. Schmidt-Nielsen. 1966. Cutaneous water loss in reptiles. Science 151:1547–49.

Berg, A. 1913. Unser Buurtwaran (*Varanus varius*). Blat. Aquar. Terrar. Kunde 24:717–19.

Bickmore, A. S. 1868. Travels in the East Indian Archipelago. John Murray, London. 542 p.

Biswell, H. H. 1961. Wildlife fire ecology in California. Tall Timbers Fire Ecology Conf. (Tallahassee, Florida), pp. 63–98.

Blair, W. F. 1960. The Rusty Lizard: A Population Study. Univ. Texas Press, Austin. 187 p.

Blower, J. H., A. P. M. van der Zon, and Y. Mulyana. 1977. Proposed Komodo National Park. Field Rept. No. 3, U.N.D.P./F.A.D. Nat. Conserv. Wildlife Mgt. Proj. INS/73/013. FAD (Bogor) October 1977. 105 p.

Bogert, C. M. 1949. Thermoregulation in reptiles, a factor in evolution. Evolution 3:195–211.

———. 1959. How reptiles regulate their body temperature. Sci. Amer. 200(4):105–20.

———, and R. M. del Campo. 1956. The gila monster and its allies. Bull. Amer. Mus. Nat. Hist. 109(1):1–142.

Boulenger, G. A. 1899. Third report on additions to the lizard collections in the Natural History Museum. Proc. Zool. Soc. (London), 912–23.

———. 1903. Report on the batrachians and reptiles. Fasc. Malay. Zool. 1:131–76.

Bourliere, F. 1961. Les populations d'angules sauvages Africains: characteriques ecologiques et implications economiques. CCTA:IUCN Symp. Conserv. Nature Nat. Res. Modern African States. Mimeographed. p. 1–4.

———. 1963. Specific feeding habits of African carnivores. African Wildlife 17(1):21–29.

———. 1965. Densities and biomasses of some ungulate populations in eastern Congo and Ruwanda, with notes on population structure and lion/ungulate ratios. Zool. Africana 1(1):199–207.

Brattstrom, B. H. 1968. Heat retention by large Australian monitor lizards. Amer. Zool. 8(4):144.

———. 1971. Social and thermoregulatory behavior of the bearded dragon, *Amphibolurus barbatus*. Copeia (3):484–97.

———. 1973. Rate of heat loss by large Australian monitor lizards. Bull. So. Calif. Acad. Sci. 72(1):52–54.

Brock, O. G. 1977. Preliminary observations on the prey trailing behavior of the eastern diamond-backed rattlesnake, *Crotalus adamanteus*. 1977 Ann. Meeting ASIH (Gainesville), Abstr., n.p.

Brodkorb, P. 1964. Catalogue of fossil birds: Part 2. Bull. Fla. State Mus., Biol. Sci. 8(3):195–335.

Broman, I. 1920. Das organan vomers-nasale Jacobsoni. Anat. Hefte, Abt. 1(58):137–91.

Brongersma, L. D. 1932. Uber die Eiablage und die Eier von *Varanus komodoensis* Ouwens. Der Zool. Garten (Leipzig), (n.s.) 5(1/3):45–46.

Broughton, M., Lady. 1936. A modern dragon hunt on Komodo. Natl. Geogr. Mag. 70:321–31.

Brouwer, H. A. 1942a. Granodioritic intrusions ... in the young Tertiary of Central Flores. Geol. Exp. Lesser Sunda Islands (Amsterdam), 4:291–317.

———. 1942b. Summary of the results of the expedition. Geol. Exp. Lesser Sunda Islands (Amsterdam), 4:345–402.

Brown, J. L., and G. H. Oriens. 1970. Spacing patterns in mobile animals. Ann. Rev. Ecol. and Syst. 1:1–206.
Burden, W. D. 1927a. The quest for the dragon of Komodo. Nat. Hist. 27:3–18.
———. 1927b. Dragon Lizards of Komodo. Putnam Sons, New York. 221 p.
———. 1927c. Stalking the dragon lizard on the island of Komodo. Natl. Geogr. 52(2):216–33.
———. 1928. Observations on the habits and distributions of *Varanus komodoensis* Ouwens. Amer. Mus. Nov. 316:1–10.
Burghardt, G. M. 1970. Chemical perception in reptiles. *In* C. J. Johnston, W. Moulton, and S. Turk (eds.). Communication by Chemical Signals. Appleton Century Crofts, New York. p. 241 to 308.
Busono, P. n.d. Facts about the *Varanus komodoensis* at the Gembiro Loka Zoo. Mimeographed. p. 1–8.
Cagle, F. 1946. A lizard population on Tinian. Copeia (1):4–9.
Campbell, H. W. 1970. Prey selection in naive *Elaphe obsoleta*—a reappraisal. Psychon. Sci. 21:300–1.
———, and W. E. Evans. 1967. Sound production in two species of tortoises. Herpetologica 23(3):204–9.
Carpenter, C. C. 1961a. Patterns of social behavior of Merriam's canyon lizard (*Sceloporus m. merriami*; Iguanidae). Southwest. Nat. 6:138–48.
———. 1961b. Patterns of social behavior in the desert iguana, *Dipsosaurus dorsalis*. Copeia (4):396–405.
———. 1962a. A comparison of the patterns of display of *Urosaurus*, *Uta* and *Streptosaurus*. Herpetologica 18(2):145–52.
———. 1962b. Patterns of behavior in two Oklahoma lizards. Am. Midl. Natl. 67:132–51.
———. 1967. Aggression and social structure in iguanid lizards. *In* W. Milstead (ed.) Lizard Ecology, a Symposium. Univ. Missouri Press, Columbia. p. 87–105.
Carr, A. F. 1962. Guideposts of animal navigation. BSCS Pamphlets (1): 1–36. D. C. Heath, Boston.
———, H. Hirth, and L. Ogren. 1966. Ecology and migration of sea turtles. 6. The hawksbill turtle in the Caribbean Sea. Amer. Mus. Nov. (2248):1–29.
———, and L. Ogren. 1960. The ecology and migration of sea turtles. 4. The green turtle in the Caribbean Sea. Bull. Amer. Mus. 121(1):1–48.
Carter, W. J. 1943. *Proteocephalus sandgroundi*, a new tetraphyllidean cestode from an East Indies monitor lizard. Trans. Am. Microsc. Soc. 62:301–5.
Champion, H. G. 1936. A preliminary survey of the forest types of India and Burma. Indian Forestry Rec. (n.s.) 1:1–286.
Chapman, V. J. 1970. Mangrove phytosociology. Trop. Ecol. 2(1):1–19.
Chew, R. M., and A. E. Chew. 1970. Energy relationships of the mammals of a desert shrub (*Larrea tridentata*) community. Ecol. Monogr. 40(1):1–21.
Clarke, R. F. 1965. An ethological study of the iguanid lizard genera *Callisaurus*, *Cophosaurus* and *Holbrookia*. Emporia State Res. Stud. 13:1–66.
Cloudsley-Thompson, J. L. 1965. Water relations and diurnal activity in the young Nile monitor. British J. Herp. 3(9):296–300.
Cogger, H. G. 1959. Australian goannas. Aust. Mus. Mag. (Nat. Hist.) 13:71–75.
———. 1967. Australian reptiles in colour. East–West Center Press, Honolulu. 112 p.
Cole, C. J. 1966. Femoral glands in lizards: A review. Herpetologica 22(3):199–206.
Cole, G. F. 1972. Grizzly bear-elk relationships in Yellowstone National Park. J. Wildl. Mgt. 26(2):556–61.
Collins, P. 1956. My "Komodo dragon" circus. Animal Kingdom, N.Y. Zool. Soc. 59(2):34–42.
Cook, S. F., Jr. 1959. The effect of fire on a population of small rodents. Ecology 40:102–8.
Cott, H. B. 1960. Scientific results of an inquiry into the ecology and economic status of the Nile crocodile (*Crocodylus niloticus*) in Uganda and northern Rhodesia. Trans. Zool. Soc. (London), 29:212–356.
Covich, A. 1971. Ecological economics of seed consumption by *Peromyscus:* A geographical model of resource substitution. Mem. Conn. Acad. Sci. (3):201–12.

Cowan, I. 1947. The timber wolf in the Rocky Mountain National Park of Canada. Can. J. Res. 25:139–74.
Cowles, R. B. 1928. The life history of *Varanus niloticus*. Science (n.s.) 67:317–18.
———. 1930. The life history of *Varanus niloticus* (Linn.) as observed in Natal, South Africa. J. Entomol. Zool. 22:1–31.
———, and C. M. Bogert. 1944. A preliminary study of the thermal requirements of desert reptiles. Bull. Am. Mus. Nat. Hist. 83:261–96.
Craighead, J. J., and F. C. Craighead. 1956. Hawks, Owls and Wildlife. Stackpole Co., Harrisburg. 443 p.
Crawford, E. C., and G. Kampe. 1970. Responses of the lizard *Sauromalus obesus* to increasing ambient temperatures. Fed. Proc. 29:1546.
Crisler, L. 1956. Observations of wolves hunting caribou. J. Mammal. 37:337–46.
Cuellar, O. 1968. Additional evidence for true parthenogenesis in lizards of the genus *Cnemidophorus*. Herpetologica 24(2):146–50.
Curio. E. 1976. The Ethology of Predation. Springer-Verlag, Berlin. 250 p.
Darevsky, I. S. 1962. Dragons are not necessarily cannibals. Tier. 11:31–33.
———. 1963. Zu Besuch bei den Komodo-Waranen. Aquar. u. Terrar. 10(1):74–80.
———. 1964. Die reptilien der Inseln Komodo,Padar und Rintja in Kleinen Sunda-Archipelago, Indonesia. Senckenbergiana Biol. 45(3/5):563–76.
———, and S. Kadarsan. 1964. On the biology of the giant Indonesian monitor lizard (*Varanus komodoensis*) Ouwens. Zool. J. 43(9):1355–60.
———, and I. M. Stusak. 1968. A recapture and growth of a *Varanus komodoensis*. Akad. Nauk. 47(7):1106–7.
———, and W. Auffenberg. 1973. Growth records of wild, recaptured Komodo monitors (*Varanus komodoensis*). HISS News J. 1(2):41–42.
Darlington, P. J. 1957. Zoogeography: The Geographical Distribution of Animals. 2d ed. John Wiley and Sons, New York. 405 p.
Dasman, R. F. 1964. Wildlife biology. John Wiley and Sons, New York. 231 p.
de Jong, J. K. 1927. *Varanus komodoensis* Ouwens. Ann. Mag. Nat. Hist. (London), (ser. 9) 19:589–91.
———. 1929. Rapport over het onderzoek naar de stand van *Varanus komodoensis* op de eilanden Flores, Komodo en Rintja en naar de megalijhheid van een natuurreservaat. Typed report in Leiden Museum. 13 p.
———. 1937. Ein en onder over *Varanus komodoensis* Ouwens. Natuurwet. Tijdschr. Ned. Indie (Batavia), 97(8):173–208.
———. 1944. Newly hatched *Varanus komodoensis*. Treubia 18:143–55. (published late in 1946, during Japanese occupation).
———, and L. D. Brongersma. 1927. Anatomische Notizen über *Varanus komodoensis* Ouwens. 1. Die Bewegungen im Schadel von *Varanus komodoensis*. Zool. Anz. Jahrb. 70:65–69.
Delacour, J. 1947. Birds of Malaysia. Macmillan Co., New York. 382 p.
Delsman, H. C. 1927. Nog este over de varanen van Komodo. Tropical Natur. 16:26–28.
———. 1951. Dierleven en Indonesie. Djakarta. 201 p.
DePrato, J. n.d. Komodo monitors in the National Zoological Gardens. Typed report in U.S.N. Zoological Park files. 58 p.
Deraniyagala, P. E. P. 1931. Some Ceylon lizards. Spolia Zeylanica 16:139–80.
———. 1957. Reproduction in the monitor lizard *Varanus bengalensis* (Daudin). Spolia Zeylanica 28(2):161–66.
———. 1958. Pseudocombat of the monitor lizard *Varanus bengalensis* (Daudin). Spolia Zeylanica 29:159.
de Rooij, N. 1915, 1917. The Reptiles of the Indo-Australian Archipelago. E. J. Brill, Leiden, Vol. 1 (1915) 305 p., vol. 2 (1917) 251 p.
de Voogd, C. N. A. 1939. Naar de Varanen van Komodo. 3 jaren Indisch Natuur leven, Nederlandsch Indosche Voreniging tat Natuurbescheming. Batavia. p. 304 to 309. In Leiden Museum.
———. 1950. Naar de Varanen van Komodo. Tropical Natur 30:128.
de Vos, A., and H. S. Mosby. 1963. Evaluation of habitat, Chap. 4, *In* Wildlife Investigational Techniques. Edwards Bros., Ann Arbor. 449 p.

DeWitt, C. B. 1967. Precision of thermoregulation and its relation to environmental factors in the desert iguana *Dipsosaurus dorsalis*. Physiol. Zool. 40:49–66.

———. 1971. Postural mechanisms in the behavioral thermoregulation of a desert lizard, *Dipsosaurus dorsalis*. J. Physiol. (Paris), 63: 242–45.

Dills, S. M. 1970. Nutritive value of forage plants in burned and unburned experimental plots. J. Wildlife Management 34(4):205–10.

Ditmars, R. L. 1910. Reptiles of the World. Macmillan Co., New York. 321 p.

Dryden, G. L. 1965. The food and feeding of *Varanus indicus* on Guam. Micronesica 2:72–76.

Dugmore, A. R. 1913. A Romance of the Newfoundland Caribou. J. B. Lippencott Co., Philadelphia. 191 p.

Dunn, E. R. 1927a. Lizards from the East Indies. Contrib. Dept. Zool. Smith College (145):113–27.

———. 1927b. Notes on *Varanus komodoensis*. Am. Mus. Nov. (286):1–10.

Duvall, V. L. , and L. B. Halls. 1962. Rotation burning. J. Range Management 17:322–26.

Edmund, A. G. 1960. Tooth replacement phenomena in the lower vertebrates. R. Ont. Mus. Life Sci. Contrib. 52:1–190.

———. 1969. Dentition. *In* A. C. Gans, A. d'A. Bellair and T. Parsons (eds.) Biology of the reptilia. Vol. 1, Morphology. p. 117–200.

Ehrat, H. 1928a. Geologisch-mijnbouwkundige onderzoekingen op Flores. Jahrb. Mijnwerk, 1925. 2:221–315.

———. 1928b. Die Tiefengesteine der Kleinen Sunda Inseln. Neues Jahrb. Mineral Geol. Paleontol. Abh. Abt. 58A:433–52.

Emlen, J. M. 1966. The role of time and energy in food preference. Am. Nat. 100:611–17.

Enders, F. 1975. The influence of hunting manner on prey size...in spiders...Am. Nat. 109(970):737–63.

Enlow, D. H. 1969. The bone of reptiles. *In* A. C. Gans, A. d'A. Bellairs and T. Parsons (eds.). Biology of the reptilia. Vol. 1, Morphology. p. 63–71.

Erikson, J. A., A. Jones, and K. Schultz. 1970. Estimating ages of mule deer—an evaluation of technique accuracy. J. Wildl. Manage. 34(3):523–31.

Errington, P. L. 1946. Predation and vertebrate populations. Q. Rev. Biol. 21:144–77, 221–45.

———. 1963. Muskrat Populations. Iowa State Univ. Press, Ames. 56 p.

Estes, R. D., and J. Goddard. 1967. Prey selection and hunting behavior of the African wild dog. J. Wildl. Mgt. 31(1):52–70.

Faussek, S. 1906. Biologische Untersuchungen in Transkaspien. Verh. Geogr. Ges. Abt. Allg. Geogr. 27:1–192.

Falk, K. 1921. Sudwestafrikanische Reptilien und ihre Heimat, VII Reptilien bei Omaruru. Wiss. Aquar. Terrar. Kunde 18:231–32.

Fejevary, S. J. 1918. Contributions to a monograph on fossil Varanidae and on Megalanidae. Ann. Mus. Nat. Hungarici 16:341–467.

———. 1927. Agro Kozlemenyeh mit tudunk Z. Komodo ories gyi Krob. Lermosgettindomanyi Kozlony. 59(3):1–176, 283–85.

Fitch, H. S. 1954. Life History and ecology of the five-lined skink, *Eumeces fasciatus*. Univ. Kans. Mus. Nat. Hist. Publ. 8:1–156.

———. 1956. An ecological study of the collared lizard (*Crotophytus collaris*). Univ. Kans. Mus. Nat. Hist. Publ. 8:213–74.

———. 1970. Reproductive cycles of lizards and snakes. Univ. Kans. Mus. Nat. Hist. Misc. Publ. 53:1–247.

Flower, S. S. 1933. Notes on the recent reptiles and amphibians of Egypt .... Proc. Zool. Soc. (London), 27:735–851.

———. 1937. The duration of life in animals, further notes, III–Reptiles. Proc. Zool. Soc. (London), 31:74.

Fraenkel, G. S., and D. L. Gunn. 1940. The Orientation of Animals. Clarendon Press, Oxford. 352 p.

Frazetta, T. H. 1962. A functional consideration of cranial kinesis in lizards. J. Morph. III (3):287–320.

Fritts, H. C. 1965. Dendrochronology. *In* The Quaternary of the United States. 7th Congress Internat. Assoc. Quat. Res., Princeton Univ. Press Princeton, p. 871 to 879.

Fukaka, H. 1977. Amount of food ingested by a snake *Rhabdophis tigrinus*. *In* M. Morista. (ed.) Methods of Estimating Population Density, Biomass and Productivity in Terrestrial Animals. Univ. Tokyo Press, Tokyo. p. 178–181.
Gans, C. 1961. The feeding mechanism of snakes and its possible evolution. Am. Zool. 12:217–27.
———. 1971. Biomechanics. J. B. Lippencott Co., Philadelphia. 261 p.
Galstaun, B. 1973. Eiablagen des Komodowarens (*Varanus komodoensis*) im Zoologischen und Botanischen Garten Jakarta. Zool. Garten 43:136–39.
Gee, E. P. 1964. The Wildlife of India. Collins, London. 192 p.
Geiger, R. 1950. The Climate Near the Ground. Harvard Univ. Press, Cambridge . 482 p.
Gibbons, J. W. 1970. Sex ratio in turtles. Res. Popul. Ecol. Kyoto 12:252–54.
———. 1977. Aging phenomena in reptiles. *In* Experimental Aging Research (Bar Harbor), p. 454 to 475.
Giltay, L. 1931. Observations teratologigues sur deux Squellettes de *Varanus komodoensis* Ouwens. Bull. Mus. Royal d'Hist. Nat. Belg. 7(29):1–7.
Golley, F. P. 1965. Food intake and assimilation by bobcats under laboratory conditions. J. Wildl. Mgt. 29:442–47.
Goodman, D. E. 1971. Thermoregulation in the brown water snake, *Natrix taxispilota*, with discussion of the ecological significance of thermal preferencia in the Order Squamata. Ph.D. dissertation. Univ. Florida, 85 p.
Gray, C. W., L. Marcus, W. C. McCarten, and T. Sappington. 1966. Amoebiasis in the Komodo dragon. Int. Zoo Yearb. 6:279–83.
Green, B., and D. King. 1978. Home range and activity patterns of the sand goanna, *Varanus gouldii* (Reptilia: Varanidae). Aust. J. Wildl. Res. 5:417–25.
Green, E. E. 1909. Arboreal habit of the "Kabaragoya" (*Varanus salvator* Laur.). Spolia Zeylanica 6:151.
Greenberg, N. 1977. An ethogram of the blue spiny lizard, *Sceloporus cyanogenys* (Sauria, Iguanidae). J. Herpetol. 11(2):177–95.
Greene, H. W. 1969. Brief notes on the Komodo dragon, *Varanus komodoensis*. Mimeographed. 2 p.
Greensberg, B. 1943. Social behavior in the western banded gecko, *Coleonyx variagatus* Baird. Physiol. Zool. 16:110–22.
Griffith, I. 1961. Skeletal lamellae as an index of age in heterothermous tetrapods. Ann. Mag. Nat. Hist. 4:449–65.
Grijs, P. 1899. Beobachtungen an Reptilien in der Gefangenschaft. Zool. Gart. 40:175–91, 210–26, 236–48, 267–77, 302–15.
Grote, H. 1911. Kurze biologische Notizen über einige Sänger und Reptilien Ostafrikas. Zool. Beob. (Zool. Gart.) 52:346–49.
Guggisberg, C. A. W. 1961. Simba, the Life of the Lion. Howard Timmins, Capetown. 304 p.
Hagen, B. 1890. Die Pflanzen und Thierwelt von Deli auf der Ostkuste Sumatra's. Natuurw. Skizzen Beiträge, Tijd. Kon. Ned. Arrdrijksk. Genootsch. (2):1–240.
Hamilton, W. J., and K. E. Watt. 1970. Refuging. Ann. Rev. Ecol. Syst. 1:263–97.
Harlow, R. F., and F. K. Jones. 1965. The white-tailed deer in Florida. Fla. Game and Fresh Water Fish Comm., Tech. Bull. 9. 239 p.
Harris, V. A. 1964. The Life of the Rainbow Lizard. Hutchinson and Co., Ltd., London, Hutchison Trop. Monogr. 174 p.
Heath, J. E. 1965. Temperature regulation and diurnal activity in horned lizards. Univ. Calif. Berkeley Publ. Zool. 64:97–136.
Hediger, H. 1934. Beitrag auf Herpetologie und Zoogeographie Neu Britanniens und einiger umliegender Gebiete. Zool. Jahrb. Syst. 65:441–582.
Henderson, G. R. 1950. Rumeruti grazing trials. Dept. Agric. (Kenya), Ann. Rev. 82 p.
Hendrickson, J. R. 1958. The green sea turtle, *Chelonia mydas* (Linn.), in Malaya and Sarawak. Proc. Zool. Soc. (London), 130:455–535.
Henne am Rhyne, R. 1903. Einege merkwürdige Kreichtiere der Sunda-Inseln. Zool. Anz. 26:167–72.
Hesse, P. 1889. Uber einige Reptilien des unteren Congogebietes. Zool. Gart. 30:257–67.
Heuvelmans, B. 1958. On the Track of Unknown Animals. Hill and Wang Co., New York. 341 p.

———. 1965. In the Wake of the Sea-Serpents. Hill and Wang Co., New York. 401 p.
Hirth, H. F. 1963. The ecology of two lizards on a tropical beach. Ecol. Monogr. 33:83–112.
Hisada, H. 1966. Komodo dragon lays eggs. Animals in Zoos 18:259.
Holling, C. S. 1959. The components of predation as revealed by a study of small mammal predation of the European pine sawfly, *Neodiprion sertifer*. Canad. J. Zool. 33:404–19.
———. 1969. The tactics of a predator. *In* T. Southwood (ed.) Insect Abundance. R. Entomol. Soc. (London). p. 47–58.
Hompes, W. 1952. *In* Hoogerwerf, 1954a.
Honegger, R. E., and H. Heusser. 1969. Beiträge zum Verhaltens—investar des Binder warans (*Varanus salvator*). Zool. Gart. 36(3/4):251–60.
Hoogerwerf, A. 1948. Enkele waarnemingen bij jonge Komodo-Varanen (*Varanus komodoensis* Ouwens) in der gevangen staat van eistadium bij dize Reptilien. Chron. Nat. 104(2):33–42.
———. 1954a. Rapport over nåar Komodo, Padar, en Rintja (Kleine Soenda eilanden) gemachte dienstreis van 26 Mei-6 Juli, 1953 (Unpubl. M.S.) Museum Bogoriense (Bogor, Java).
———. 1954b. Nature protection in Indonesia. Oryx 2:221–27.
———. 1955. *Varanus komodoensis*. Neth. Commissie voor Internatl. Natuurbescherming (15):47–54.
Hoogstraal, H., G. Anastos, and G. Rach. 1968. Studies on southeast Asian *Haemaphysalis* ticks (Ixodoidea, Ixodidae).... J. Parasitol. 54(6):1214–22.
———, and G. Rack. 1967. Ticks (Ixodidae) collected by Deutsche Indien-expedition, 1955-58. J. Med. Entomol. 4(3):284–88.
Hooijer, D. H. 1967. Indo-Australian insular elephants. Genetics 38:143–62.
———. 1972a. *Stegodon trigonocephalus florensis* Hooijer and *Stegodon timorensis* Sartono, from the Pleistocene of Flores and Timor. I, II. K. Ned. Akad. Wet. 75(1):12–32.
———. 1972b. *Varanus* (Reptilia, Sauria) from the Pleistocene of Timor. Zool. Meded. Rijks Mus. Nat. Hist. (Leiden), 47:445–48.
Hornocker, M. 1969. Winter territoriality in mountain lions. J. Wildl. Mgt. 33:457–64.
———. 1970. An analysis of mountain lion predation upon mule deer and elk in the Idaho primitive area. Wildl. Monogr. (21):1–56.
Horst, O. 1926. *Varanus komodoensis*. Tropical Natur 15:118–21.
Hunsaker, D. 1962. The significance of interspecific social dominance in inguanid lizards. Am. Midl. Nat. 81:500–11.
Impey, O. R. 1967. Functional aspects of cranial kinetism in the Lacertilia. Ph.D. Dissertation, Univ. Oxford. 67 p.
Jenssen, T. A. 1970. The ethoecology of *Anolis nebulosis* (Sauria, Iguanidae). J. Herpetol. 4(1-2):1–38.
Jolly, G. M. 1965. Explicit estimates from capture-recapture data with both death and immigration-stochastic models. Biometrika 52:225–47.
Kahmann, H. 1939. Von der Leistung des Jacobsonschen Organs bei den Wirbeltieren. Ergeb. Biol. 16:292–336.
Karawier, W. 1905. Recherches zoologiques des materiaux recueillis en hiver 1898–1899 sur l'Ile de Java. 2 Mem. Soc. Nat. Kiew, 19:1–17.
Kaufmann, J. H. 1962. Ecology and social behavior of the coati, *Nasua narica*, on Barro Colorado Isl., Panama. Univ. Calif. Berkeley. Publ. Zool. 60(3):95–222.
Kern, J. A. 1968. Dragon lizards of Komodo. Natl. Geogr. December: 874–80.
King, D., and B. Green. 1979. Notes on diet and reproduction of the sand goanna, *Varanus gouldii rosenbergi*. Copeia (1):64–70.
King. F. W. 1968. Ora-giants of Komodo. Anim. Kingdom. August: 2–9.
King, M., and D. King. 1975. Chromosomal evolution in the lizard genus *Varanus*. Aust. J. Biol. Sci. 28:89–108.
Klomp, H. 1956. On the theories on host-parasite interactions. 12 Cong. Internatl. Union Res. Organiz. (Oxford), 1956:1–32.
Knauer, F. 1911. Das psychische Leben der Kreichtiere und Lurche nach Beobachtungen im Terrarium. Wiss. Aquar. Terr. Kunde 8, Lacerta, 69–71, 73–4, 78–80, 82–4.
Koford, C. B. 1953. The California Condor, Dover Publ. Inc., New York. 362 p.
Koppen, W., and A. Wegener, 1924. Die Klimate der geologischen Vorzeit. Berlin. 128 p.

Konigsberger, J. C. 1911–15. Javanische Zoologisch an Biologisch. Batavia. 218 p.

Komarek, E. V., Sr. 1966. The meteorological basis for fire ecology. Proc. 5th Ann. Tall Timbers Fire Ecol. Conf. (Tallahassee), 85-125.

———. 1967. The nature of lightning fires. Proc. Calif. Tall Timbers Fire Ecol. Conf. 5–42.

———. 1969. A discussion of wildlife management, fire and the wildlife landscape. Proc. 8th Ann. Tall Timbers Fire Ecol. Conf. (Tallahassee), 177–94.

Komarek, R. 1963. Fire and the changing wildlife habitat. Proc. Second Ann. Tall Timbers Fire Ecology Conf. (Tallahassee, Florida), pp. 35–43.

Kourtz, P. 1967. Lightning behavior and lightning fires in Canadian forests. Canada Dept. For. Rural Devel. (1179):1–33.

Kratzer, H. 1973. Beobachtungen über die Zeitigungsdauer eines Eigeleges von *Varanus salvator.* Salamandra 9(1):27–33.

Kruuk, H. 1972. The Spotted Hyena, Univ. Chicago Press, Chicago. 572 p.

Laidlow, F. F. 1901. On a collection of lizards from the Malay Peninsula made by members of the Sweat Expedition, 1899-1900. Proc. Zool. Soc. (London), 301–11.

Lallemant, G. F. M. A. 1929. De Reuzen leguaan (*Varanus komodoensis)* op de eilanden Komodo en Rintja. Tropical Natur 18(8):125–28.

Lederer, G. 1942. Der Drachenwaren (*Varanus komodoensis* Ouwens). Zool. Gart. (Leipzig), (n.s.) 14(5/6):227–44.

Ledger, H. P., J. Schwartz, and B. Hillier. 1961. A preliminary investigation of the relationship between body composition and productive efficiency of meat processing animals in the dry tropics. Trans. 8th Internatl. Cong. Animal Production (Hamburg), 61p.

Leigh, T. 1956. Long-term studies of the effects of fire on deer browse. Proc. Georgia Wildlife Soc. 13(1):1–58.

Leopold, A. 1933. Game Management. C. Scribner's Sons, New York. 481 p.

———. 1947. A survey of overpopulated deer ranges in the United States. J. Wildl. Mgt. 11:162–77.

Leyhausen, P. 1956. Gerhaltensstudien an Katzen. Tierpsychol. Beih. 2:1–15.

———, and R. Wolff. 1959. Das Revier einer Hanskatze. Z. Tierpsychol. 16(6):666–70.

Licht, P. 1965. The relation between preferred body temperature and testicular heat sensitivity in lizards. Copeia (4):428–36.

———, W. R. Dawson, V. H. Shoemaker, and A. R. Main. 1966. Observations on the thermal relations of West Australian lizards. Copeia (1):17–110.

Lincoln, G. A. 1974. Predation of incubator birds (*Megapodius freycineti*) by Komodo dragons (*Varanus komodoensis*). J. Zool. London 174:419–28.

Locke, A. 1954. The Tigers of Trengganu. London. 321 p.

Löf, G., and G. Brown. 1966. World distribution of solar radiation. Univ. Wisc. Solar Energy Lab., July, Rept. 21. 21 p.

Longhurst, W. M. 1952. A survey of California deer herds. California Dept. Fish and Game Bull. (6). 136 p.

Lonnberg, E. 1928. Notes on *Varnus komodoensis* Ouwens and its affinities. Ark. Zool. (Upsala), 19A(20-27):1–11.

Loop, M. S. 1974. The effect of relative prey size on the ingestion behavior of the Bengal monitor, *Varanus bengalensis* (Sauria: Varanidae). Herpetologica 30:123–27.

Lorenz, K. 1966. The Nature of Aggression. Methune Co., London. 358 p.

Loveridge, A. 1923. Notes on East African lizards collected 1920–1923, with a description of two new races of *Agama lionotus*. Proc. Zool. Soc. (London), p. 935–69.

———. 1946. Reptiles of the Pacific World. Macmillan Co., New York. 314 p.

Lowe, V. P. W. 1961. A discussion on the history, present status and future conservation of roe deer (*Cervus elaphus* L.) in Scotland. Terre Vie 1961:9–40.

Lucas, A. H., and C. Frost. 1896. Reptilia. *In* Report Horn Scientific Expedition to Central Australia. Part 2, Zoology. Govn. Printer (Adelaide), p. 112–151.

MacArthur, R. H., and E. O. Wilson. 1967. The Theory of Island Biogeography. Princeton Univ. Press, Princeton. 203 p.

McCoy, F. 1878. Prodromus of the Zoology of Victoria. 2 vols. Gray and Co., Melbourne. 432 p.

McMahan, J. S. 1964. Competition between livestock and wildlife. J. Mammalogy 21(1):13–33.
McNab, B. K. 1963. Bioenergetics and the determination of home range size. Am. Nat. 97:133–40.
———, and W. Auffenberg. 1976. The body temperature of the Komodo monitor, *Varanus komodoensis*. Comp. Biochem. Physiol. 55:45–50.
Madison, D. M. 1977. Chemical communication in amphibians and reptiles. *In* Chemical Signals in Vertebrates. Plenum Publ. Corp., New York. p. 135–68.
Mahendra, B. Ch. 1931. How the monitor lizard sits in its burrow. J. Bombay Nat.Hist. Soc. 34:255–56.
Maleyev, Y. 1962. The Komodo dragon. Bull. Philadelphia Herpetol. Soc. 10(1):32–34. (reprinted from the New Times, Moscow, October 17, 1962:24–26).
———, and I. S. Darevsky. 1963. Dragons of Komodo. Priroda (Moscow), (3):24–25 (in Russian).
Marchinton, R. L. 1964. Activity cycles and mobility of ... deer, based on telemetric and observational data. M.S.thesis. Univ. Florida. 100 p.
Mech, L. D. 1966. The wolves of Isle Royale. U.S. Natl. Park Serv., Faunal Series (7). 56 p.
Meer-Mohr, J. C. v.d. 1930. Over eieren von *Varanus salvator* en *Python curtus*. Tropical Natur 19:156–57.
Meinertz, T. 1952. Das hertz und die grossen Blutgefasse der Komodoeidechse, *Placovaranus komodoensis* (Ouwens). Z. Anat. Entwicklungsgesch. (Berlin) 116:315–19.
Mertens, R. 1929. Über einege neu neu oder selten eingefuhrte Amphibien und Reptilien aus Sudost-Asien. III Eideschsen. *Varanus komodoensis* Ouwens. Blat. Aquar. Terrar. 40(23):410–12.
———. 1930. Die Amphibien und Reptilien der Inseln Bali, Lombok, Sumbawa und Flores. Abh. Senckenb. Naturforsch. Ges. (423):117–342.
———. 1934. Die Insel-Reptilien. Zoologica (Stuttgart), 32(6):1–209.
———. 1942. Die Familie der Warane (Varanidae). Abh. Senckenb. Naturforsch. Ges., Parts 1–3, (465):1–39.
———. 1946a. Die Warn- und Drohreaktionen der Reptilien. Abh. Senckenb. Naturforsch. Ges. (471):1–103.
———. 1946b. Über den Komodo-Waran Berliner Aquarium, besonders seinen Schädel. Senckenbergiana 27:153–57.
———. 1959. Liste der Warane Asiens und der Indo-Australischen Inselwelt mit systematischen Bemerkungen. Senckenbergiana 40(5/6):112–47.
Miller, M. R. 1969. The cochlea duct of lizards. Proc. Calif. Acad. Sci. 33(11):255–359.
Milstead, W. W. 1961. Observations of the activities of small animals (Reptilia and Mammalia) on a quadrat in southwest Texas. Am. Midl. Nat. 65:127–38.
———(ed.). 1967. Lizard Ecology. Univ. Missouri Press, Columbia. 302 p.
Minton, S. 1966. A contribution to the herpetology of West Pakistan. Bull. Am. Mus. Nat. Hist. 134(2):31–184.
Mitchell, B. L. 1967. Growth layers in dental cement for determining the age of red deer. J. Anim. Ecol. 36:279–93.
Mohr, C. O. 1943. A comparison of North American small mammal censuses. Am. Midl. Nat. 29(3):545–87.
———. 1947. Table of equivalent densities of North American small mammals. Am. Midl. Nat. 37:223–38.
Morris, D. J. 1956. The function and causation of courtship ceremony. *In* P. Grass (ed.) L'Instinct dans le Compartement des Animaux et de l'Homme. Masson and Cie., Paris p. 261–87.
Morrison, D. W. 1978. On the optimal searching strategy for refuging predators. Am. Nat. 112(987):925–34.
Mudrack, W. 1969. Paarung und Ei-Ablag bie *Varanus storri* Mertens. Aquar. Terr. 6(3):125–38.
Müller, L. 1923. Aquarium und Reptilien haus der Zooligischen Gartens zu Frankfurt. Blat. Aquar. Terr. Kunde 20:289–91, 308–13, 324–29.
Murdoch, W. W. 1969. Switching in general predators: Experiments on predator specificity of prey populations. Ecol. Monogr. 39:335–54.
———. 1971. The developmental response of predators to changes in prey density. Ecology 52(1):132–37.

Murie, A. 1944. The wolves of Mt. McKinley. U.S. Park Serv. Fauna of the Natl. Parks, Fauna Series (5). 238 p.
Murphy, J. B., and L. A. Mitchell. 1974. Ritualized combat behavior of the pygmy mulga monitor lizard, *Varanus gilleni* (Sauria: Varanidae). Herpetologica 30(1):90–97.
Myers, J. H., and C. J. Krebs. 1971. Genetic, behavioral, and reproductive attributes of dispersing field voles, *Microtus pennsylvanicus* and *Microtus ochrogaster*. Ecol. Monogr. 41(1):53–78.
Noble, G. K., and H. T. Bradley. 1933. The mating behavior of lizards, its bearing on the theory of sexual selection. Ann. N.Y. Acad.Sci. 35:25–100.
Norris, K. S., and W. R. Dawson. 1964. Observations on the water economy and electrolyte excretion of Chuckwallas (Lacertilia, *Sauromalus*). Copeia (4):638–46.
Oesman, H. 1967. Breeding behavior of *V. komodoensis* at the Jogjakarta Zoo. Int. Zoo Yearb. 7:10–11.
———. 1970. Collecting little giants. Animals 13(4):158–61.
Ogilby, G. 1936. Cutaneous glands of cervids. Proc. Zool. Soc. (London), 32:131–39.
Oliver, J. A. 1955. The Natural History of North American Amphibians and Reptiles. Van Nostrand Co., New York. 358 p.
Oosting, P. 1958. A Study of Plant Communities. W. H. Freeman and Co., San Francisco. 440 p.
O'Roke, E. C., and F. N. Hamerstrom, Jr. 1948. Productivity and yield of the George Reserve deer herd. J. Wildl. Mgt. 12:78–86.
Ouwens, P. A. 1912. On a large *Varanus* species from the island of Komodo. Bull. Jard. Bot. Buit. 2(6):1–3.
Parrington, F. R. 1955. On the cranial anatomy of some gorgonopsids and the synapsid middle ear. Proc. Zool. Soc. (London), 125 (1):1–40.
Parry, N. E. 1932. Some notes on the water monitor in the Garo Hills, Assam. J. Bombay Natl. Hist.Soc. 35:901–5.
Parsons, T. S. 1970. The origin of Jacobson's organ. Forma Functio 3:105–11.
Pearson, O. P. 1964. Carnivore-mouse predation: an example of its intensity and bioenergetics. J. Mammal. 45(2):177–88.
Peters, W. 1870. Über die africanischen Warneidechsen, monitors, und ihre geographische Verbreitung. Akad. Wiss. Berlin 81:106–10.
Petrides, G. A. 1956. Big game densities and range carrying capacities in East Africa. Trans. 21st No. Amer. Wildl. Conf. p. 525–37.
Pfeffer, P. 1959. Observations sur le Varan de Komodo (*Varanus komodoensis* Ouwens, 1912). Terre Vie 106(2/3):195–243.
———. 1965. Aux Iles du Dragon. Flammarion, Paris. 203 p.
Phillips, J. 1936. Fire in vegetation. J. S. Afr. Botany 12:36–45.
———. 1965. Fire—as master and servant; its influence in the bioclimatic regions of trans-Saharan Africa. Proc. 4th Ann. Tall Timbers Fire Ecol. Conf. (Tallahassee), 6–109.
Pianka, E. R. 1968. Notes on the biology of *Varanus eremius*. West. Aust. Nat. 11(3):39–44.
———. 1969. Notes on the biology of *Varanus gouldi flavirufus*. West. Aust. Nat. 11(6):141–44.
———. 1971. Notes on the biology of *Varanus tristis*. West. Aust. Nat. 11(7):180–83.
Piazzini, G. 1960. The Children of Lilith. E.P. Dutton and Co., New York (English Transl.). 192 p.
Pienaar, U. 1968. Predator-prey relationships amongst larger mammals of the Kruger National Park. Koedoe 18:108–56.
Pimlot, D. M. 1967. Wolf predation and ungulate populations. Am. Zool. 7:267–78.
Pitelka, F. A., H. Green, and R. Smithe. 1955. Ecological relations of jaegers and owls as lemming predators near Barrow, Alaska. Ecol. Monogr. 25:35–117.
Pitman, C. R. S. 1931. A Game Warden among His Charges. Nisbet, London. 336 p.
Pocock, R. I. 1910. On the specialized cutaneous glands of ruminants. Proc. Zool. Soc. (London), 18:840–986.
Popham, E. J. 1942. Further experimental studies of the selective action of predators. Proc. Zool. Soc. (London), (A) 112:105–17.
Prater, S. M. 1965. The Book of Indian Animals. Bombay Nat. Hist. Soc. 316 p.
Procter, J. B. 1929. On a living Komodo dragon (*Varanus komodoensis* Ouwens) exhibited at the scientific meeting, October 23, 1928. Proc. Zool. Soc. (London), 23:1017–19.
Pruitt, W. O. 1960. Behavior of the barren-ground caribou. Biol. Pap. Univ. Alaska (3):1–15.

Quay, W. B. 1959. Microscopic structure and variation in the cutaneous glands of the deer, *Odocoileus virginianus*. J. Mammal. 4(1):114–28.

———. 1966. Histology and cytochemistry of skin gland areas in the caribou, *Rangifer*. J. Mammal. 36(2):187–201.

Ramanathan, K. R., and L. A. Ramdas. 1935. Derivation of Angstrom's formula ... and some general considerations regarding nocturnal cooling of air-layers near the ground. Proc. Indian Acad. Sci. 1:822–29.

Rand, A. S. 1967. The adaptive significance of territoriality in iguanid lizards. *In* W. W. Milstead (ed.) Lizard Ecology: a symposium. Univ. Texas Press, Austin, p. 106–15.

Rapport, D. J. 1971. An optimization model of food selection. Am. Nat. 105(946):575–87.

Regal, P. J. 1966. Thermophilic response following feeding in certain reptiles. Copeia (3):588–90.

Rensch, B., and M. Eisentraut. 1927. Experimentelle Untersuchungen über den Geschmackssin der Reptilien. Z. Vergl. Physiol. 5:607–12.

Rice, S. A. 1932. Rodents and fire. Nevada Lumber J. 14(8):78–91.

Richards, P. W. 1952. The Tropical Rain Forest: An Ecological Study. Cambridge Univ. Press, Cambridge. 450 p.

Robinette, W. L., G. Winaker, and G. Jones. 1957. Notes on tooth development and wear for Rocky Mountain mule deer. J. Wildl. Mgt. 21(2):134–53.

———. 1959. Food habits of the cougar in Utah and Nevada. J. Wildl. Mgt. 23:261–73.

Robinson, P. L. 1967. The evolution of the Lacertilia. *In* Problemes actuels de paleontologie (Evolution des Vertebres). Coll. Internatl. du Centre Natl. Rech. Sci., Paris (163). p. 395–407.

Romanov, A. M. 1956.Automatic tagging of wild animals and prospects for its use. Zool. J. Moscow 35:1902.

Roosevelt, Th. 1910. African Game Trails. Macmillan Co., New York. 562 p.

Rose, W. 1950. The Reptiles and Amphibians of Southern Africa. Robertson, Cape Town. 378 p.

Rosenzweig, M. L. 1966. Community structure in sympatric Carnivora. J. Mammal. 47:602–12.

Rudebeck, G. 1950. The choice of prey and modes of hunting of predatory birds, with special reference to their selective effect. Oikos 2:65–88.

———. 1951. The choice of prey and modes of hunting of predatory birds, with special reference to their selective effect (continued). Oikos 3:200–231.

Ruibal, R. 1961. Thermal regulations of five species of tropical lizards. Evolution 15(1):98–111.

Sachs, W. B. 1927. Neues und Zusammengefasstes vom Riesenwaren der Komodo Inseln, *Varanus komodoensis* Ouwens. Blat. Aquar. Terrar. 38(22):450–56.

Saunders, J. K., Jr. 1963. Food habits of the lynx in Newfoundland. J. Wildl. Mgt. 27:384–90.

Schaffer, J. 1940. Die Haut drüsenorgane der Saügetiere mit besonderer Berucksichtigung ihres histologischer Aufbaues und Bermerkungen über die Praktodäaldrüsen. Urbau and Schwarzenburg, Berlin. 464 p.

Schaller, G. B. 1966. The tiger and its prey. Nat. Hist. 42:31–37.

———. 1967. The Deer and the Tiger. Univ. Chicago Press, Chicago. 370 p.

———. 1972. The Serengeti Lion. Univ. Chicago Press, Chicago. 480 p.

———, and G. R. Lowther. 1969. The relevance of carnivore behavior to the study of early homonids. Southwest. J. Anthropol. 25:307–41.

Schenkel, R. 1967. Submission patterns and functions in the wolf and dog. Am. Zool. 7:319–29.

Scherer, J. 1903. Der Nilwaran (*Varanus niloticus*). Nat. Haus 11:249–51.

———. 1907. Eine Jagd auf den Nilwaran am Senegal. Blat. Aquar. Terrar. Kunde 18:1–4.

Schimper, A. F. W. 1935. Pflanzengeographie auf physiologischer Grundlage. 3d ed. Rev. by F.C. van Faber. Jena:1–401.

Schmidt, K. P. 1927. The reptiles of Hainan. Bull. Am. Mus. Nat. Hist. 54(3):395–465.

———, and R. F. Inger. 1957. Living Reptiles of the World. Hanover House, Garden City. 236 p.

Schmidt-Nielsen, K. 1964. Desert Animals. Oxford Univ. Press, Oxford. 326 p.

Schneider, A. 1900. (*In* Werner, 1900)

Schoener, T. W. 1968. Sizes of territories among birds. Ecology 49:123–41.

———. 1969. Models of optimal size for solitary predators. Am. Nat. 103(931):277–311.

Schoenmakers, J. 1935. Hoe de Komodo-varanen gevangen werden. De Telegraaf (Surabaja), 30 June :3.

Selander, R. K. 1966. Sexual dimorphism and differential niche utilization in birds. Condor 68:113–51.
Self, J. T., and R. E. Kuntz. Pentastomids from African reptiles and mammals and from reptiles of Florida Island, British Solomon Islands. J. Parasitol. 43(2):194–200.
———. 1966. New Pentastomidae. *Sambonia parapodum* n. sp. from *Varanus salvator* and *Armillifer agkistrodontis* n. sp. from *Agkistrodon acutus*. Trans. Am. Microsc. Soc. 85(2):256–60.
Seton, E. T. 1909. Life Histories of Northern Animals, an Account of the Mammals of Manitoba. 2 vols. Macmillan Co., New York. 587 p.
Sharma, R. C., and T. G. Vazirani. 1977. Food and feeding habits of some reptiles of Rajasthan. Rec. Zool. Surv. India 73:77–93.
Smith, H. C. 1931. The monitor lizards of Burma. J. Bombay Nat. Hist. Soc. 34:367–73.
Smith, M. A. 1927. Contributions to the herpetology of the Indo-Australian region. Proc. Zool. Soc. (London), 19:199–225.
———. 1932. Some notes on the monitors. J. Bombay Nat. Hist. Soc. 35:615–19.
———. 1935. The Fauna of British India. Reptiles and Amphibians. Vol. 2, Sauria. Taylor and Francis, London. 305 p.
Sody, K. 1941. Notes on the zoogeography of the Sunda Islands. Treubia 18(3):301–12.
Soule, M. 1963. Aspects of thermoregulation in nine species of lizards from Baja California. Copeia (2):107–15.
Southwood, T. R. E. 1966. Ecological Methods. Methuen and Co., Ltd., London. 391 p.
Sparrowe, R. D. 1972. Prey-catching behaviour in the sparrowhawk. J. Wildl. Mgt. 36:297–308.
Spatz, P. 1931. Wüstenwaran (*Varanus griseus* Daudin). Das Aquar. 10:160–62.
Spieth, P. T. 1974. Theoretical considerations of unequal sex ratios. Am. Nat. 108(964):837–49.
Stadt, P. A. 1939. Batavische planten-en dierentium, 1864–1939. K.v. Aemstel, Batavia. 137 p.
Stamps, J. A., and G. W. Barlow. 1973. Variation and stereotypy in the displays of *Anolis aeneus* (Sauria: Iguanidae). Behavior 47(1/2):67–94.
Staring, M. 1926. Wild und Hund. Berlin, December 1926. (53):13–18.
Stebbins, R. C., and R. E. Barwick. 1968. Radiotelemetric study of thermoregulation in a lace monitor. Copeia (3):541–47.
Sternfield, R. 1924. Beiträge zur Herpelologie Inner-Australiens. Abh. Senckenb. Naturforsch. Ges. 38:221–51.
Stevenson-Hamilton, B. 1949. (see Wynne-Edwards)
Stirling E. C. 1912. Observations on the habits of the large central Australian monitor (*Varanus giganteus*), with a note on the "fat bodies" of this species. Trans. Proc. Soc. So. Australia 36:26–33.
Stokes, R. A., and K. Smiley. 1968. An Introduction to Tree-ring Dating. Univ. Chicago Press, Chicago. 73 p.
Strode, V. 1954. Use of artificial watering tanks for wildlife in arid zones. J. Wildlife Management 21(3):201–33.
Suzuki, H. K. 1963. Studies on the osseous system of the slider turtle. Ann. N.Y. Acad. Sci. 109:351–410.
Swank, W. G. 1958. The mule deer in Arizona chaparral. Arizona Game and Fish Dept. Wildl. Bull. (3). 109 p.
Sweeney, J. R., and J. Marchison. 1971. Responses of radio-monitored white-tailed deer chased by hunting dogs. J. Wildl. Mgt. 35(4):707–16.
Symons, C. T. 1912. Note on the arboreal habits of the Kabaragoya (*Varanus salvator*) and the Talgoya (*V. bengalensis*). Spolia Zeylanica 8:65–66.
Taber, R. D. 1961. The black-tailed deer: A review of ecology and management. Terre Vie 1961:221–45.
———, and R. Cowan. 1963. Wildlife Investigational Techniques. Wildl. Soc., Edwards Bros., Ann Arbor. 419 p.
———., and R. F. Dasmann. 1958. The black-tailed deer of the chaparral. Calif. Dept. Fish Game, Game Bull. (8):1–163.
Talbot, L. M., and M. H. Talbot. 1963. The high biomass of wild ungulates on East African savanna. Trans. N. Am. Wildl. Conf. 28:465–76.

Tanzer, S. L., and J. H. R. W. C. van Heurn. 1938. Observations made by S. L. Tanzer and J. H. R. W. C. van Heurn with reference to the propagation of the *Varanus komodoensis* Ouw. Treubia 16(3):365–68.
Taylor, E. H. 1922. The lizards of the Philippine Islands. Philipp. Bur. Sci. Monogr. (17):13–269.
Taylor, R. J. 1976. Value of clumping to prey and the evolutionary response of predators. Am. Nat. 110(971):13–29.
Templeton, J. R. 1960. Respiration and water loss at the higher temperature in the desert iguana, *Dipsoaurus dorsalis*. Physiol. Zool. 33:136–45.
———. 1964. Nasal salt excretion in terrestrial lizards. Comp. Biochem. Physiol. 11:223–29.
———. 1967. Nasal salt gland excretion and adjustment to sodium loading in *Ctenosaura pectinata*.Copeia (1):136–40.
Theobald, W., Jr. 1868. Catalogue of the reptiles of British Burma. J. Linn. Soc.Zool. 10:4–67.
Thilenius, G. 1897. Der Farbenwecksel von *Varanus griseus* bei *Uromastix acanthurus* und *Agama inermis*. Morph. Jarb. 7:515–45.
Thorarinsson, S. 1956. Hekla on Fire. Hans Reich Verlag, Munich. 53 p.
Tinbergen, N. 1953. Social Behaviour in Animals. Doubleday, New York. 201 p.
———. 1969. The Study of Instinct. Oxford Univ. Press, New York. 228 p. (reprint of the 1951 ed., with new introduction).
Tinkle, D. W. 1961. Geographic variation in reproduction, size, sex ratio and maturity in *Sternotherus odoratus* (Testudinata: Chelydridae). Ecology 42:68–76.
———. 1967. The life and demography of the side-blotched lizard, *Uta stansburiana*. Misc. Publ. Mus. Zool. Univ. Michigan 132:1–182.
———. 1969. The concept of reproductive effort and its relation to the evolution of life histories of lizards. Am. Nat. 103(933):501–16.
———, H. M. Wilbur, and S. G. Tilley. 1970. Evolutionary strategies in lizard reproduction. Evolution 24(1):55–74.
Tofahr, O. 1908. Warane. Blat. Aquar. Terrar. Kunde 19:677–80, 698–701.
Toumanoff, C. 1944. Les tigues (Ixodoidea) de l'Indochine. Inst. Pasteur l'Indochine. S.I.L.I., Saigon. 220 p.
Turner, F. G., R. I. Jennrich, and J. D. Weintraub. 1969. Home ranges and body sizes of lizards. Ecology 50:1076–81.
Umbgrove, J. H. 1949. Structural history of the East Indies. Cambridge Univ. Press, Cambridge. 63 p.
Van Bemmel, A. C. V. 1949. Revision of the rudine deer in the Indo-Australian Archipelago. Treubia 20(2):191–262.
Van Dijk, A. 1935. Ausfuhrgange der Leben und das Pankreas bei *Varanus komodoensis* und einigen anderen Reptilien. Anat. Anz. 6(17/20):347–55.
Van Heurn, W. C. 1932. Vluchtigezoologische waarnemingen gedaan een kort verbliff op Timor en Flores. Natuurk. Tdschr. Ned. Indies (92):18–30.
Van Huut, M. A. 1937. Capturing giant dragon lizards for the World's zoos… . Illus. London News 191:174–75.
Van Kampen, P. N. 1923. The Amphibia of the Indo-Australian Archipelago. E. J. Brill, Leiden. 208 p.
Van Steenis, C. G. G. J. 1933. Report of a botanical trip to the Rama region. Bull. Jardin Bot. (Buitenzorg), Ser. 3. 13:1–56.
———. 1935. Maleische vegetatieschetsen. Tijd. Kon. Ned. Aardrijksk. Genootsch. Ser. 2. 52:46–52.
Verheijen, J. A. J. (S.V.D.). 1967. Kamus Mangarrai I. S-Gravenhage, Martinus Nijhoff. 772 p.
Vogel, P. 1979. Zu biologie des Binderwarens (*Varanus salvator*) in Westjavanischen Naturschutsgebiet Ujung Kulon. Ph.D. dissertation. Univ. Basel. 139 p.
Von Buddenbrock, W. 1958. The Senses. Univ. Michigan Press, Ann Arbor. 167 p.
Von Frisch, K. 1967. The Dance Language and Orientation of Bees. (Transl. by L. E. A. Chadwick) Belnap Press, Cambridge. 566 p.
Von Rathgen, W. 1894.Der Wüstenwaren im Terrarium. Nat. Haus 2:169–71.
Vorstman, A. G. 1933. The septa in the ventricle of the heart of *Varanus komodoensis*. Proc. Kon. Akad. Wet. 36(10):911–13.

Waite, E. R. 1927. The fauna of Kangaroo Island, South Australia. Part 3. The reptiles and amphibians. Trans. Soc. So. Australia 51:326–29.
———. 1929. The reptiles and amphibians of South Australia. Govn. Printer, Adelaide. 183 p.
Warburg, M. R. 1965. The influence of ambient temperatures and humidity on the body temperature and water loss from two Australian lizards. Australian J. Zool. 13:331–50.
Watts, L. F., R. F. Morris, and R. Bellman. 1936. The management of range lands. *In* The Western Range. Senate Document 199. Washington, D.C. 58 p.
Weaver, W. J. 1970. Courtship and combat behavior in *Gopherus berlandieri*. Bull. Fla. State Mus., Biol. Sci. 15(1):1–43.
Weber, J. 1894. Zoological Ergebnesse einer Reise in Nederlandische Ost-Indes. Vol. 3. Buitenzorg. 203 p.
Wendt, F. 1956. Auf Noah's spuren. G. Groetsche Verlagsbucherie. 418 p.
———. 1959. Out of Noah's ark. Houghton-Mifflin Co., Boston. 407 p.
Werner, F. 1893. Bemerkungen über Reptilien und Batrachier aus dem tropischen Asien und vor der Sinae-Halbinsel. Verh. Zool., Bot. Ges. (Wien), 43:349–359.
———. 1900. Reptilien und Batrachier aus Sumatra, gesammelt von Herrn Gustav Schneider jr. im Jahre 1897-98. Zool. Jarhb. Syst. 13:479–508.
———. 1904. Die warane. Blat. Aquar. Terrar. Kunde 15:84–87, 99–101.
Wharton, C. 1966. Man, fire and wild cattle in North Cambodia. 5th Ann. Tall Timbers Fire Ecol. Conf. (Tallahassee), 23–65.
Wheeler, M. S., and W. A. Volk. 1969. Basic Microbiology. J. B. Lippincott Co., Philadelphia. 410 p.
Wight, N. 1938. Field and Laboratory Techniques in Wildlife Management. Univ. Michigan Press, Ann Arbor. 126 p.
Wilbur, H. M., D. W. Tinkle, and J. P. Collins. 1974. Environmental certainty, trophic level, and resource availability in life history evolution. Am. Nat. 108:805.
Wilcox, J. 1959. Effects of vulcanism on plants. J. Vulcan. 13(3):133–41.
Wilson, E. O. 1975. Sociobiology. Harvard Univ. Press, Cambridge. 697 p.
Williams, G. P. 1966. Adaptation and Natural Selection. Princeton Univ. Press, Princeton. 307 p.
Witte, G. F. 1941. Batraciens et reptiles. Exploration du Parc National Albert Miss. G. F. Witte (1933-5). 33:1–261.
Woodland, B. 1938. The genus *Duthiersia*. J. Parasitol. 11(2):110–27.
Worrell, E. 1963. Reptiles of Australia. Angus and Robertson, Sydney. 206 p.
Wright, B. S. 1960. Predation on big game in East Africa. J. Wildl. Mgt. 24:1–15.
Wynne-Edwards, V. 1962. Animal Dispersion in Relation to Social Behavior. Oliver and Boyde, Edinburgh. 571 p.
Yamaguti, S. 1959. Systema Helminthum. Vol. II. The Cestodes of Vertebrates. Interscience, New York. 860 p.
Zakhidov, T. 1938. Biology of reptiles of southern Kyzl-Kumy desert and the Noura-Tan Mountains. Acta Univ. Asiae Zool. 54 (in Russian).
Zander, A. 1895. Einige transkapische Reptilien, *Varanus griseus*. Zool. Gart. 36:298–301.
Zippin, C. 1958. The removal method of population estimation. J. Wildl. Mgt. 22:82–90.
Zollinger, H. 1850. Verslag van eene reis naar Bima en Sumbawa en naar eenige plaatsen op Celebes, Saleyer en Flores gedurende de maanden Mei tot December, 1847. Verhl. Batav. Gennotschap. 23:10–22.

Index